BASIC
ELECTRIC CIRCUIT
ANALYSIS

BASIC
ELECTRIC CIRCUIT
ANALYSIS

D.E. JOHNSON, J.L. HILBURN, and **J.R. JOHNSON**

Department of Electrical Engineering
Louisiana State University

PRENTICE-HALL, INC. *Englewood Cliffs, New Jersey* 07632

Library of Congress Cataloging in Publication Data

Johnson, David E
    Basic electric circuit analysis.

    Includes index.
    1. Electric circuits. I. Hilburn, John L., 1938-
       joint author. II. Johnson, Johnny Ray, joint
author. III. Title.
TK454.J56   621.319'2   77-24210
ISBN 0-13-060137-3

© 1978 by Prentice-Hall, Inc., Englewood Cliffs, N.J. 07632

Printed in the United States of America

10 9 8 7 6

PRENTICE-HALL INTERNATIONAL, INC., *London*
PRENTICE-HALL OF AUSTRALIA PTY. LIMITED, *Sydney*
PRENTICE-HALL OF CANADA, LTD., *Toronto*
PRENTICE-HALL OF INDIA PRIVATE LIMITED, *New Delhi*
PRENTICE-HALL OF JAPAN, INC., *Tokyo*
PRENTICE-HALL OF SOUTHEAST ASIA PTE. LTD., *Singapore*
WHITEHALL BOOKS LIMITED, *Wellington, New Zealand*

To *Frances, Meme,* and *Betty*

CONTENTS

PREFACE

This book was written for a one-year course in linear circuit analysis in the sophomore year. Such a course is basic in electrical engineering and is usually the first encounter of the student with his or her chosen field of specialization. It is imperative, therefore, for the textbook used to cover thoroughly the fundamentals of the subject and at the same time be as easy to understand as it is possible to make it. These have been our objectives throughout the writing of the book.

Most students, when they take this subject, will have studied electricity and magnetism in a physics course. This background is helpful, of course, but is not a prerequisite for reading the book. The material presented here may be easily understood by a student who has had a basic course in differential and integral calculus. The differential equations theory required in circuit analysis is fully developed in the book and integrated with the appropriate circuit theory topics. Even determinants, Gaussian elimination, and complex number theory are presented in appendices.

The operational amplifier is introduced immediately after the discussion of the resistor, and appears, as a matter of course, along with resistors, capacitors, and inductors, as a basic element throughout the book. Likewise, dependent sources and their construction using operational amplifiers are discussed early and are encountered routinely in almost every chapter.

To aid the reader in understanding the textual material, examples are liberally supplied and numerous exercises, with answers, are given at the end of virtually every section. Problems, some more difficult and some less difficult than the exercises, are also given at the end of every chapter. A special effort has been made to include a number of problems and exercises with realistic element values. Of course, network scaling, which is also presented, can be used to make almost all of the remaining problems practical. In particular, in the chapter on amplitude and phase responses, problems are given that relate to electric filters, which, of course, are very useful circuits. Active filters, using operational amplifiers, as well as passive filters, are used as examples. Finally, a select few of the exercises and problems are used to extend the theory discussed in the chapters. In this way, optional material is included without adding to the text of the chapter.

The first nine chapters of the book are devoted to terminology and time-domain analysis and the last nine chapters deal with frequency-domain analysis. Some sections and chapters, identified by asterisks, may be omitted without any loss in continuity. Among these is the chapter on network topology, an interesting subject which could be covered entirely, in part, or not at all. Also those on Fourier methods and Laplace transforms are often reserved for a succeeding course in linear systems, but their essential ingredients are here if there is time to cover them.

There are many people who have provided invaluable assistance and advice concerning this book. We are indebted to our colleagues and our students for the form the book has taken, and to Mrs. Dana Brown and Mrs. Norma Duffy, who provided the expert typing and draftsmanship, respectively. A special note of thanks is due Professors M. E. Van Valkenburg, A. P. Sage, S. R. Laxpati, and S. K. Mitra, who reviewed the manuscript and made many helpful comments and suggestions.

Louisiana State University   DAVID E. JOHNSON
Baton Rouge, La.   JOHN L. HILBURN

                                                              JOHNNY R. JOHNSON

1

INTRODUCTION

Electric circuit analysis, in nearly every electrical engineering curriculum, is the f[...]
course taken in the major area by an electrical engineering student. Virtually [...]
branches of electrical engineering, such as electronics, power systems, communicati[...]
systems, rotating machinery, and control theory, are based on circuit theory. The on[...]
topic in electrical engineering more basic than circuits is electromagnetic field theor[...]
and even there many problems are solved by means of equivalent electric circuits. Thu[...]
it is no exaggeration to say that the basic circuit theory course a student first encounte[...]
in electrical engineering is the most important course in his or her curriculum.

To begin our study of electric circuits we need to know what an electric circuit is[...]
what we mean by its analysis, what quantities are associated with it, in what units[...]
these quantities are measured, and the basic definitions and conventions used in[...]
circuit theory. These are the topics we shall consider in this chapter.

1.1 DEFINITIONS AND UNITS

An electric *circuit*, or electric *network*, is a collection of electrical elements interconnected in some specified way. Later we shall define the electrical elements in a formal
manner, but for the present we shall be content to represent a general *two-terminal*
element as shown in Fig. 1.1. The terminals a and b are accessible for connections with
other elements. Examples with which we are all familiar, and which we shall formally
consider in later sections, are resistors, inductors, capacitors, batteries, generators, etc.

More complicated circuit elements may have more than two terminals. Transistors
and operational amplifiers are common examples. Also a number of simple elements
may be combined by interconnecting their terminals to form a single package having
any number of accessible terminals. We shall consider some multiterminal elements
later, but our main concern will be simple two-terminal devices.

An example of an electric circuit with six elements is shown in Fig. 1.2. Some authors distinguish a circuit from a network by requiring a circuit to contain at least one closed path such as path *abca*. We shall use the terms interchangeably, but we may note that without at least one closed path the circuit is of little or no practical interest.

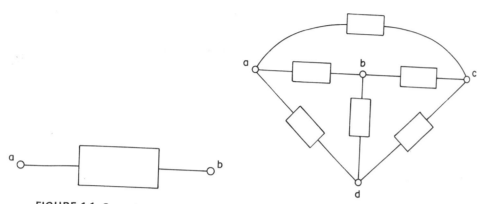

FIGURE 1.1 *General two-terminal electrical element*

FIGURE 1.2 *Electric circuit*

To be more specific in defining a circuit element we shall need to consider certain quantities associated with it, such as *voltage* and *current*. These quantities and others, when they arise, must be carefully defined. This can be done only if we have a standard system of units so that when a quantity is described by measuring it we can all agree on what the measurement means. Fortunately, there is such a standard system of units that is used today by virtually all the professional engineering societies and the authors of most modern engineering textbooks. This system, which we shall use throughout the book, is the *International System of Units* (abbreviated SI), adopted in 1960 by the General Conference on Weights and Measures.

There are six basic units in the SI, and all other units are derived from them. Four of the basic units, the meter, kilogram, second, and coulomb, are important to circuit theorists, and we shall consider them in some detail. The remaining two basic units are the degree Kelvin and the candela, which are important to such people as the electron device physicist and the illumination engineer.

The SI units are very precisely defined in terms of permanent and reproducible quantities. However, the definitions are highly esoteric and in some cases are comprehensible only to atomic scientists.[1] Therefore we shall be content to name the basic units and relate them to the very familiar *British System of Units*, which includes inches, feet, pounds, etc.

The basic unit of length in the SI is the *meter*, abbreviated m, which is related to the British system by the fact that 1 inch is 0.0254 m. The basic unit of mass is the *kilogram*

[1]Complete definitions of the basic units may be found in a number of sources, such as, for example, "IEEE Recommended Practice for Units in Published Scientific and Technical Work," by C. H. Page et al. (*IEEE Spectrum*, vol. 3, no. 3, pp. 169–173, March 1966).

BASIC
ELECTRIC CIRCUIT
ANALYSIS

(kg), and the basic unit of time is the *second* (s). In terms of the British units, 1 pound-mass is exactly 0.45359237 kg, and the second is the same in both systems.

The fourth unit in the SI is the *coulomb* (C), which is the basic unit used to measure electric charge. We shall defer the definition of this unit until the next section when we consider charge and current. The name coulomb was chosen to honor the French scientist, inventor, and army engineer Charles Augustin de Coulomb (1736–1806), who was an early pioneer in the fields of friction, electricity, and magnetism.

We might note at this point that all SI units named for famous people have abbreviations that are capitalized. Otherwise, lowercase abbreviations are most often used. It is also worth mentioning that we could choose units other than the ones we have selected to form the basic units. For example, instead of the coulomb we could take the *ampere* (A), the unit of electric current to be considered later. In this case the coulomb could then be obtained as a derived unit.

There are three derived units in addition to the ampere that we shall find useful in circuit theory. They are the units used to measure force, work or energy, and power. The fundamental unit of force is the *newton* (N), which is the force required to accelerate a 1-kg mass by 1 meter per second per second (1 m/s^2). Thus $1\text{ N} = 1$ kg-m/s^2. The newton is named, of course, for the great English scientist, astronomer, and mathematician Sir Isaac Newton (1642–1727). Newton's accomplishments are too numerous to be listed in a mere chapter.

The fundamental unit of work or energy is the *joule* (J), named for the British physicist James P. Joule (1818–1889), who shared in the discovery of the law of conservation of energy and helped establish the idea that heat is a form of energy. A joule is the work done by a constant 1-N force applied through a 1-m distance. Thus $1\text{ J} = 1$ N-m.

The last derived unit we shall consider is the *watt* (W), which is the fundamental unit of power, the rate at which work is done or energy is expended. The watt is defined to be 1 J/s and is named in honor of James Watt (1736–1819), the Scottish engineer whose engine design first made steam power practicable.

Before we leave the subject of units we should point out that one of the greatest advantages the SI has over the British system is its incorporation of the decimal system to relate larger and smaller units to the basic unit. The various powers of 10 are denoted by standard prefixes, some of which are given, along with their abbreviations, in Table 1.1.

TABLE 1.1 *Prefixes in the SI*

Multiple	*Prefix*	*Symbol*
10^9	Giga	G
10^6	Mega	M
10^3	Kilo	k
10^{-3}	Milli	m
10^{-6}	Micro	μ
10^{-9}	Nano	n
10^{-12}	Pico	p

As an example, at one time a second was thought to be a short time, and fractions such as 0.1 or 0.01 of a second were unimaginably short. Nowadays in some applications, such as digital computers, the second is an impractically large unit. As a result, times such as 1 nanosecond (1 ns or 10^{-9} s) are in common use. Another common example is 1 gram (g) = 10^{-3} kg.

EXERCISES

1.1.1 Find the number of millimeters in 10 km.   *Ans.* 10^7

1.1.2 If a mile equals 5280 ft, how many miles in 10 km?   *Ans.* 6.2137

1.1.3 Find the work in millijoules done by a constant force of 25 μN applied to a mass of 4 g for a distance of 10 m.   *Ans.* 0.25

1.2 CHARGE AND CURRENT

We are all familiar with gravitational forces of attraction between bodies, which are responsible for holding us on the earth and which cause an apple dislodged from a tree to fall to the ground rather than to soar upward into the sky. There are bodies, however, that attract each other by forces far out of proportion to their masses. Also, such forces are observed to be repulsive as well as attractive and are clearly not gravitational forces.

We explain these forces by saying that they are electrical in nature and caused by the presence of *electrical charges*. We explain the existence of forces of both attraction and repulsion by postulating that there are two kinds of charges, positive and negative, and that unlike charges attract and like charges repel.

As we know, according to modern theory, matter is made up of atoms, which are composed of a number of fundamental particles. The most important of these particles are protons (positive charges) and neutrons (neutral, with no charge) found in the nucleus of the atom and electrons (negative charges) moving in orbit about the nucleus. Normally the atom is electrically neutral, the negative charge of the electrons balancing the positive charge of the protons. Particles may become positively charged by losing electrons to other particles and become negatively charged by gaining electrons from other particles.

As an example, we may produce a negative charge on a balloon by rubbing it against our hair. The balloon will then stick to a wall or the ceiling, which are uncharged. Relative to the negatively charged balloon the neutral wall and ceiling are oppositely charged.

We now define the *coulomb* (C), discussed in the previous section, by stating that the charge of an electron is a negative one of 1.6021×10^{-19} coulombs. Putting it another way, a coulomb is the charge of about 6.24×10^{18} electrons. These are, of course, mind-boggling numbers, but their sizes enable us to use more manageable numbers, such as 2 C, in the circuit theory to follow.

The symbol for charge will be taken as Q or q, the capital letter usually denoting constant charges such as $Q = 4\,C$, and the lowercase letter indicating a time-varying charge. In the latter case we may emphasize the time dependency by writing $q(t)$. This practice involving capital and lowercase letters will be carried over to the other electrical quantities as well.

The primary purpose of an electric circuit is to move or transfer charges along specified paths. This motion of charges constitutes an *electric current*, denoted by the letters i or I, taken from the French word "intensité." Formally, current is the time rate of change of charge, given by

$$i = \frac{dq}{dt} \tag{1.1}$$

The basic unit of current is the *ampere* (A), named for André Marie Ampère (1775–1836), a French mathematician and physicist who formulated laws of electromagnetics in the 1820s. An ampere is 1 coulomb per second.

In circuit theory current is generally thought of as the movement of positive charges. This convention stems from Benjamin Franklin (1706–1790), who guessed that electricity traveled from positive to negative. We now know that in metal conductors the current is the movement of electrons that have been pulled loose from the orbits of the atoms of the metal. Thus we should distinguish *conventional* current (the movement of positive charges), which is used in electric network theory, and *electron* current. Unless otherwise stated, our concern will be with conventional current.

As an example, suppose the current in the wire of Fig. 1.3(a) is $I = 3\,A$. That is, 3 C/s pass some specific point in the wire. This is symbolized by the arrow labeled 3 A, whose direction indicates that the motion is from left to right. This situation is equivalent to that depicted by Fig. 1.3(b), which indicates -3 C/s or -3 A in the direction from right to left.

Figure 1.4 represents a general circuit element with a current i flowing from the left toward the right terminal. The total charge entering the element between time t_0 and t is found by integrating (1.1). The result is

$$q_T = q(t) - q(t_0) = \int_{t_0}^{t} i\,dt \tag{1.2}$$

We should note at this point that we are considering the network elements to be *electrically neutral*. That is, no net positive or negative charge can accumulate in the element. A positive charge entering must be accompanied by an equal positive

(a)   (b)

FIGURE 1.3 *Two representations of the same current*

FIGURE 1.4 *Current flowing in a general element*

charge leaving (or, equivalently, an equal negative charge entering). Thus the current shown entering the left terminal in Fig. 1.4 must leave the right terminal.

There are several types of current in common use, some of which are shown in Fig. 1.5. A constant current, as shown in Fig. 1.5(a), will be termed a *direct current*, or dc. An *alternating current*, or ac, is a sinusoidal current, such as that of Fig. 1.5(b). Figures 1.5(c) and (d) illustrate, respectively, an *exponential* current and a *sawtooth* current.

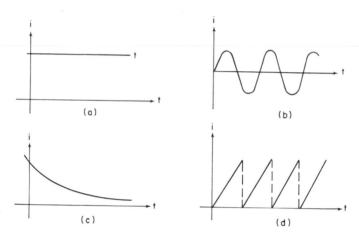

FIGURE 1.5 *(a) dc; (b) ac; (c) Exponential current; (d) Sawtooth current*

There are many commercial uses for dc, such as in flashlights and in power supplies for electronic circuits, and, of course, ac is the common household current found all over the world. Exponential currents appear quite often (whether we want them or not!) when a switch is actuated to close a path in an energized circuit. Sawtooth waves are useful in equipment, such as oscilloscopes, used for displaying electrical characteristics on a screen.

EXERCISES

1.2.1 Find the charge in picocoulombs represented by 10,000 electrons.

Ans. 0.0016021

1.2.2 The total charge entering a terminal of an element is given by

$$q(t) = 10t^2 - 2t \text{ C}$$

Find $i(t)$ at $t = 0$ and at $t = 1$ s.

Ans. -2 A, 18 A

1.2.3 The current entering a terminal is given by

$$i(t) = 30t^2 - 4t \text{ A}$$

Find the total charge entering the terminal between $t = 1$ s and $t = 3$ s.

Ans. 244 C

1.3 VOLTAGE, ENERGY, AND POWER

Charges in a conductor, exemplified by free electrons, may move in a random manner. However, if we want some concerted motion on their part, such as is the case with an electric current, we must apply an external or so-called *electromotive force* (EMF). Thus work is done on the charges. We shall define *voltage* "across" an element as the work done in moving a unit charge ($+1$ C) through the element from one terminal to the other. The unit of voltage, or *potential difference*, as it is sometimes called, is the *volt* (V), named in honor of the Italian physicist Alessandro Guiseppe Antonio Anastasio Volta (1745–1827).

Since voltage is the number of joules of work performed on 1 coulomb, we may say that 1 V is 1 J/C. Thus the volt is a derived SI unit, expressible in terms of other units.

We shall represent a voltage by v or V and use the $+$, $-$ polarity convention shown in Fig. 1.6. That is, terminal A is v volts positive with respect to terminal B. Putting it another way in terms of potential difference, terminal A is at a *potential* of v volts higher than terminal B. In terms of work, it is clear that moving a unit charge from B to A requires v joules of work.

Some authors prefer to describe the voltage across an element in terms of voltage *drops* and *rises*. Referring to Fig. 1.6, a voltage drop of v volts occurs in moving from A to B. In contrast, a voltage rise of v volts occurs in moving from B to A.

As examples, Figs. 1.7(a) and (b) are two versions of exactly the same voltage. In (a), terminal A is $+5$ V above terminal B, and in (b), terminal B is -5 V above A (or $+5$ V below A).

In transferring charge through an element work is being done, as we have said. Or, putting it another way, energy is being supplied. To know whether energy is being

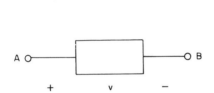

FIGURE 1.6 *Voltage polarity convention*

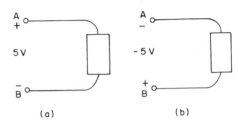

FIGURE 1.7 *Two equivalent voltage representations*

supplied *to* the element or *by* the element to the rest of the circuit, we must know not only the polarity of the voltage across the element but also the direction of the current through the element. If a positive current enters the positive terminal, then an external force must be driving the current and is thus supplying or *delivering* energy to the element. The element is *absorbing* energy in this case. If, on the other hand, a positive current leaves the positive terminal (enters the negative terminal), then the element is delivering energy to the external circuit.

As examples, in Fig. 1.8(a) the element is absorbing energy. A positive current enters the positive terminal. This is exactly the case also in Fig. 1.8(b). In Figs. 1.8(c) and (d) a positive current enters the negative terminal, and therefore the element is delivering energy in both cases.

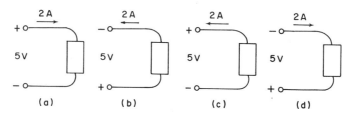

Figure 1.8 *Various v-i relationships*

Let us consider now the *rate* at which energy is being delivered to or by a circuit element. If the voltage across the element is v and a small charge Δq is moved through the element from the positive to the negative terminal, then the energy absorbed by the element, say Δw, is given by

$$\Delta w = v\, \Delta q$$

If the time involved is Δt, then the rate at which the work is being done, or the energy w is being expended, is given by

$$\lim_{\Delta t \to 0} \frac{\Delta w}{\Delta t} = \lim_{\Delta t \to 0} v \frac{\Delta q}{\Delta t}$$

or

$$\frac{dw}{dt} = v \frac{dq}{dt} = vi \tag{1.3}$$

Since by definition the rate at which energy is expended is power, denoted by p, we have

$$p = \frac{dw}{dt} = vi \tag{1.4}$$

We might observe that (1.4) is dimensionally correct since the units of vi are (J/C)(C/s) or J/s, which is watts (W), defined earlier.

Since v and i are generally functions of time, p given by (1.4) is a time-varying quantity. It is sometimes called the *instantaneous* power because its value is the power at the instant of time at which v and i are measured.

Summarizing, the typical element of Fig. 1.9 is absorbing power, given by $p = vi$. If either the polarity of v or i (but not both) is reversed, then the element is delivering power, $p = vi$, to the external circuit. Of course, to say that an element delivers a negative power, say -10 W, is equivalent to saying that it absorbs a positive power, in this case $+10$ W.

FIGURE 1.9 *Typical element with voltage and current*

As examples, in Figs. 1.8(a) and (b) the element is absorbing power of $p = (5)(2) = 10$ W. [In Fig. 1.8(b) the 2 A leave the negative terminal, and thus 2 A enter the positive terminal.] In Figs. 1.8(c) and (d) it is delivering 10 W to the external circuit, since the 2 A leave the positive terminal, or, equivalently, -2 A enter the positive terminal.

Before ending our discussion of power and energy, let us solve (1.4) for the energy w delivered to an element between time t_0 and t. We have, upon integrating both sides between t_0 and t,

$$w(t) - w(t_0) = \int_{t_0}^{t} vi \, dt \tag{1.5}$$

For example, the energy delivered to the element of Fig. 1.8(a) between $t = 0$ and $t = 2$ s is given by

$$w(2) - w(0) = \int_{0}^{2} (5)(2) \, dt = 20 \text{ J}$$

Since the left member of (1.5) represents the energy delivered to the element between t_0 and t, we may interpret $w(t)$ as the energy delivered to the element between the beginning of time and t and $w(t_0)$ as the energy between the beginning of time and t_0. In the beginning of time, which let us say is $t = -\infty$, the energy delivered to the element was zero; that is,

$$w(-\infty) = 0$$

If $t_0 = -\infty$ in (1.5), then we shall have the energy delivered to the element from the beginning up to t, given by

$$w(t) = \int_{-\infty}^{t} vi \, dt \tag{1.6}$$

This is consistent with (1.5) since

$$w(t) = \int_{-\infty}^{t} vi \, dt$$

$$= \int_{-\infty}^{t_0} vi \, dt + \int_{t_0}^{t} vi \, dt$$

By (1.6) this may be written

$$w(t) = w(t_0) + \int_{t_0}^{t} vi \, dt$$

which is (1.5).

EXERCISES

1.3.1 The element shown is absorbing power of $p = 20$ W. Find the current entering terminal b.

Ans. -4 A

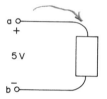

a ○
+

5 V

b ○
−

EXERCISE 1.3.1

1.3.2 Find the energy delivered to the element of Ex. 1.3.1 between 0 and 2 s.

Ans. 40 J

1.4 PASSIVE AND ACTIVE ELEMENTS

We may classify circuit elements into two broad categories, *passive* elements and *active* elements, by considering the energy delivered to or by them.

A circuit element is said to be passive if the total energy delivered to it from the rest of the circuit is always nonnegative. That is, referring to (1.6), for all t we have

$$w(t) = \int_{-\infty}^{t} p(t) \, dt = \int_{-\infty}^{t} vi \, dt \geq 0 \tag{1.7}$$

The polarities of v and i are as shown in Fig. 1.9. As we shall see later, examples of passive elements are resistors, capacitors, and inductors.

An active element is one that is not passive, of course. That is, (1.7) does not hold for all time. Examples of active elements are generators, batteries, and electronic devices that require power supplies.

We are not ready at this stage to begin a formal discussion of the various passive elements. This will be done in later chapters. In this section we shall give a brief discussion of two very important active elements, the independent voltage source and the independent current source.

An *independent voltage source* is a two-terminal element, such as a battery or a generator, that maintains a specified voltage between its terminals. The voltage is completely independent of the current through the element. The symbol for a voltage source having v volts across its terminals is shown in Fig. 1.10. The polarity is as shown, indicating that terminal a is v volts above terminal b. Thus if $v > 0$, then terminal a is at a higher potential than terminal b. The opposite is true, of course, if $v < 0$.

In Fig. 1.10, the voltage v may be time varying, or it may be constant, in which case we would probably label it V. Another symbol that is often used for a constant voltage source, such as a battery with V volts across its terminals, is shown in Fig. 1.11. In the case of constant sources we shall use Figs. 1.10 and 1.11 interchangeably.

We might observe at this point that the polarity marks on Fig. 1.11 are redundant since the polarity could be defined by the positions of the longer and shorter lines. We shall leave the polarity marks off in most cases in the future. There are times, however, in analyzing circuits when it is convenient to use the polarity marks.

FIGURE 1.10 *Independent voltage*   FIGURE1.11 *Constant voltage source*
source

An *independent current source* is a two-terminal element through which a specified current flows. The current is completely independent of the voltage across the element. The symbol for an independent current source is shown in Fig. 1.12, where i is the specified current. The direction of the current is indicated by the arrow.

Independent sources are usually meant to deliver power to the external circuit and not to absorb it. Thus if v is the voltage across the source and its current i is directed out of the positive terminal, then the source is delivering power, given by $p = vi$, to the external circuit. Otherwise it is absorbing power. For example, in Fig. 1.13(a) the

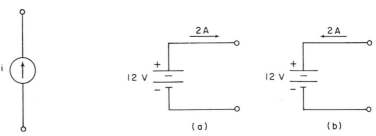

FIGURE1.12 *Independent current*   FIGURE 1.13 *(a) A Source delivering and (b) Absorb-*
source   *ing power*

battery is delivering 24 W to the external circuit. In Fig. 1.13(b) the battery is absorbing 24 W, as would be the case when it is being charged.

The sources that we have discussed here, as well as the circuit elements to be considered later, are *ideal elements*. That is, they are *mathematical models* that approximate the actual or physical elements only under certain conditions. For example, an ideal automobile battery supplies a constant 12 V, no matter what external circuit is connected to it. Since its current is completely arbitrary, it could theoretically deliver an infinite amount of power. This, of course, is not possible in the case of an actual device. A real 12-V automobile battery supplies approximately constant voltage only as long as the current it delivers is low. When the current exceeds a few hundred amperes, the voltage drops appreciably from 12 V.

We shall consider practical sources in a later chapter and see under what conditions they may be approximated by ideal sources. Also later, we shall consider *dependent* sources whose voltage (or current) is controlled by another voltage or current somewhere else in the circuit.

EXERCISE

1.4.1 Find the power being supplied by the sources shown.

Ans. (a) 36 W, (b) −40 W, (c) −2 W

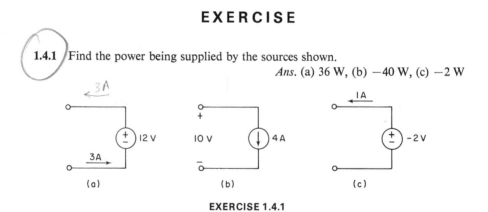

EXERCISE 1.4.1

1.5 CIRCUIT ANALYSIS

Let us now look at the words *circuit analysis*, which are contained in the title of the book, and see what they mean. Generally, if an electric circuit is subjected to an *input* or *excitation* in the form of, say, a voltage or a current provided by an independent source, then an *output* or *response* is produced. The output or response may also be a voltage or a current associated with some element in the circuit. There may be, of course, more than one input and more than one output.

There are two main branches of circuit theory, and they are derived from the following three key words: input, output, and circuit. The first branch is *circuit analysis,* which, given the circuit and the input, is concerned with finding the output. The other branch is *circuit synthesis,* which, given the input and output, is concerned with finding the circuit itself.

Network synthesis is much more complex in general than analysis and will probably be encountered by the student in a later course. Circuit analysis will be our concern in this book. We may be interested only in finding one or more outputs, such as a voltage or current existing somewhere in the circuit, or in determining the energy or power delivered to one element or another. Or we may wish to perform a complete analysis, finding every unknown current and voltage in the circuit. In any case, in the chapters to come we shall develop systematic methods of analysis that can be applied generally to any circuit of the type we consider.

PROBLEMS

1.1   Let $f(t)$ in the graph shown be the charge $q(t)$ in millicoulombs that has entered the positive terminal of an element as a function of time. (a) Find the total charge that has entered the element between 3 and 7 s, (b) the current entering the element at 5 s, and (c) the energy absorbed by the element between 1 and 7 s if the voltage across the element is 10 V.

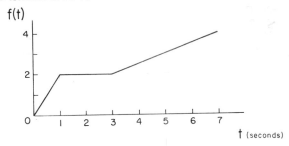

PROBLEM 1.1

1.2   Let $f(t)$ in Prob. 1.1 be the current $i(t)$ in milliamperes entering an element terminal as a function of time. (a) Find the charge that enters the terminal between 0 and 7 s and (b) the rate at which the charge is entering at $t = 5$ s.

1.3   The terminal voltage of a voltage source is given by $v = 10 \cos 100t$ V. If the charge leaving the positive terminal is $q = 6 \sin 100t$ mC, find the power being supplied by the source at $t = \pi/100$ s.

1.4   For the element shown, the charge entering the positive terminal is $q = -2e^{-0.5t}$ C. Find the power delivered to the element as a function of time and the total energy delivered to the element between 0 and 2 s if (a) $v = 6i$, (b) $v = 3 \, di/dt$, and (c) $v = 2 \int_0^t i \, dt + 4$. (Note: v is in volts if i is in amperes.)

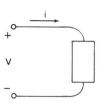

PROBLEM 1.4

1.5 If the element in Prob. 1.4 is a battery with a constant terminal voltage of $v = 12$ V
and the current i is 2 A, then the battery is in the process of being charged. (It is
absorbing rather than delivering power.) (a) Find the energy supplied to the battery in 2 h (hours). (b) Find the charge delivered to the battery in this time. Note
the consistency of the units, 1 V = 1 J/C.

1.6 Suppose the voltage v in Prob. 1.5 varies linearly from 4 to 12 V during the 2 h
that the battery is being charged. If $i = 2$ A is constant during this period, find
(a) the total energy supplied and (b) the total charge delivered to the battery.

1.7 In the element of Prob. 1.4, let $i(t) = 0$ for $t < 0$ and $i(t) = 6 \sin 2t$ A for $t > 0$.
If $v = 4\, di/dt$ V, show that the energy delivered to the element is nonnegative for
all $t > 0$.

1.8 In Prob. 1.7, find the total charge delivered to the element and the power absorbed
at $t = \pi/4$ s.

2

RESISTIVE CIRCUITS

The simplest and most commonly used circuit element is the resistor. All electrical conductors exhibit properties which are characteristic of a resistor. When currents flow in conductors, electrons which make up the current collide with the lattice of atoms in the conductor. This, of course, on the average, impedes the motion of the electrons. The larger the number of collisions, the greater the *resistance* of the conductor. We shall consider a *resistor* to be any device which exhibits solely a resistance. Materials which are commonly used in fabricating resistors include metallic alloys and carbon compounds.

In this chapter, we shall first introduce the terminal relations for a resistor based on Ohm's law. Two laws necessary for systematic solutions of networks, known as Kirchhoff's laws, are then examined. With these laws, we shall begin our study of circuit analysis by finding solutions for "single-loop" and "single-node-pair" resistive networks having independent sources as inputs. We shall conclude the chapter with a discussion of simple measuring instruments followed by a discussion of practical resistors.

2.1 OHM'S LAW

Georg Simon Ohm (1787–1854), a German physicist, is credited with formulating the current-voltage relationship for a resistor based on experiments performed in 1826. In 1827 he published the results in a paper titled "The Galvanic Chain, Mathematically Treated." As a result of this work, the unit of resistance is called the *ohm*. It is ironic, however, that Henry Cavendish (1731–1810), a British chemist, discovered the same results 46 years earlier. Had he not failed to publish his findings, the unit of resistance might well be known as the *caven*.

Ohm's law states that the voltage across a resistor is directly proportional to the current flowing through the resistor. The constant of proportionality is the resistance value of the resistor in ohms. The circuit symbol for the resistor is shown in Fig. 2.1. For the current and voltage shown, Ohm's law is

$$v(t) = Ri(t) \tag{2.1}$$

where $R \geq 0$ is the resistance in ohms.

Rearranging (2.1) into the form $R = v(t)/i(t)$, we see that

$$1 \text{ ohm} = 1 \text{ V/A}$$

The symbol used to represent the ohm is the capital Greek omega (Ω).

Since R is constant, (2.1) is the equation of a straight line. For this reason, the resistor is called a *linear resistor*. A graph of $v(t)$ versus $i(t)$ is shown in Fig. 2.2, which is a line passing through the origin with a slope of R. Obviously, a straight line is the only graph possible for which the ratio of $v(t)$ to $i(t)$ is constant for all $i(t)$.

Resistors whose resistances do not remain constant for different terminal currents are known as *nonlinear resistors*. For such a resistor, the resistance is a function of the current flowing in the device. A simple example of a nonlinear resistor is an incandescent lamp. A typical voltage-current characteristic for this device is shown in Fig. 2.3, where we see that the graph is no longer a straight line. Since R is not a constant, the analysis of a circuit containing nonlinear resistors is more difficult.

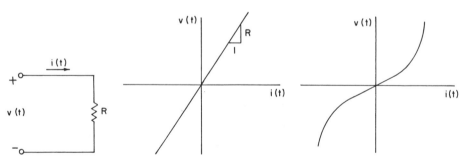

FIGURE 2.1 *Circuit symbol for the resistor*   FIGURE 2.2 *Voltage-current characteristic for a linear resistor*   FIGURE 2.3 *Typical voltage-current characteristic for a nonlinear resistor*

In reality, all practical resistors are nonlinear because the electrical characteristics of all conductors are affected by environmental factors such as temperature. Many materials, however, closely approximate an ideal linear resistor over a desired operating region. We shall concentrate on these types of elements and simply refer to them as resistors.

An examination of (2.1), in conjunction with Fig. 2.1, shows if $i(t) > 0$ (current entering the upper terminal), then $v(t) > 0$. Thus the current enters the terminal of higher potential and exits from that of the lower potential. Next, suppose that $i(t) < 0$

(current entering the lower terminal). Then $v(t) < 0$, and the lower terminal is higher in potential than the upper one. Once again, the current enters the terminal of higher potential. Since charges are transported from a higher to a lower potential in passing through the resistor, the energy lost by a charge q [energy $= qv(t)$] is absorbed by the resistor in the form of heat. The rate at which energy is dissipated is, by definition, the instantaneous power

$$p(t) = v(t)\ i(t) = Ri^2(t) = \frac{v^2(t)}{R} \qquad (2.2)$$

A graph of (2.2), shown in Fig. 2.4, reveals that $p(t)$ is a parabolic (and thus nonlinear) function of $i(t)$ or $v(t)$ which is always positive. (The horizontal scales are, of course, different in the two cases.) Thus, for a linear resistor, the instantaneous power is nonlinear even though the voltage-current relationship is linear.

The condition for passivity, given in (1.7), is

$$w(t) = \int_{-\infty}^{t} p(t)\ dt \geq 0$$

Therefore, since $p(t)$ is always positive, we see that the above integral is positive and that the resistor is, indeed, a passive element.

The beginning student often encounters difficulty in determining the proper algebraic sign in applying (2.1) when the voltage assignment differs from that of Fig. 2.1. Consider, for example, the assignment of Fig. 2.5. Comparing these figures, we see that $v_1(t) = -v(t) = -Ri(t)$. Therefore, when the voltage assignment is such that the assumed current enters the terminal of lower potential $(-)$, a minus sign must be employed in using (2.1). It should be noted that

$$p(t) = \frac{v^2(t)}{R} = \frac{[-v_1(t)]^2}{R} = \frac{v_1^2(t)}{R}$$

Hence the voltage assignment has no effect on the sign of $p(t)$.

Another important quantity which is very useful in circuit analysis is known as *conductance*, defined by

$$G = \frac{1}{R} \qquad (2.3)$$

The unit of conductance is the *mho* (A/V), which is ohm spelled backwards. The symbol representing the mho is an inverted omega ($\mho$). Combining (2.1)–(2.3), we see that alternative expressions for Ohm's law and instantaneous power are

$$i(t) = Gv(t) \qquad (2.4)$$

and

$$p(t) = \frac{i^2(t)}{G} = Gv^2(t) \qquad (2.5)$$

As an example of the application of Ohm's law, consider finding the current for the circuit of Fig. 2.6 in which a 12-V storage battery is connected to a 1-kΩ resistor. From (2.3) and (2.4), $G = \frac{1}{1000} = 10^{-3}$ ℧ and $I = 10^{-3} \times 12$ A $= 12$ mA. Also, (2.5) yields $p(t) = 10^{-3} \times 12^2 = 144$ mW, which is the minimum power rating or *wattage* required for the resistor to ensure that it will not be damaged as a result of over-heating.[1]

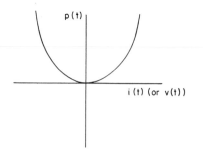

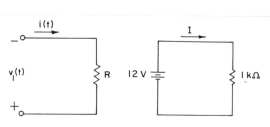

FIGURE 2.4 *Graph of the instan-taneous power for a resistor*

FIGURE 2.5 *Resistor with a reversed voltage assignment*

FIGURE 2.6 *Simple resistive circuit*

The current in this example is a direct current since its value is invariant in time. Suppose we now replace the 12-V battery by the time-varying voltage $v(t) = 10 \cos t$ V and repeat the above procedure. The current is

$$i(t) = \frac{10 \cos t}{10^3} \text{ A} = 10 \cos t \text{ mA}$$

and the instantaneous power is

$$p(t) = 0.1 \cos^2 t \text{ W}$$

which is always positive. The current, in this case, is an alternating current.

EXERCISES

2.1.1 The terminal current of a 20-kΩ resistor is 10 mA. Find (a) the conductance, (b) the terminal voltage, and (c) the minimum wattage of the resistor.

Ans. (a) 5×10^{-5} ℧, (b) 200 V, (c) 2 W

2.1.2 The instantaneous power absorbed by a 1-kΩ resistor is 10 sin² 377t W. Find $v(t)$ and $i(t)$.   *Ans.* 100 sin 377t V, 0.1 sin 377t A

[1]The wattage of a resistor is based on the *average power* (to be discussed in Chapter 12). For direct currents, the instantaneous power equals the average power.

2.1.3 Find $v(t)$ and $p(t)$.   *Ans.* -50 V, 250 W

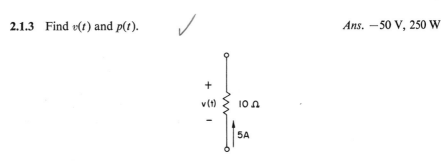

EXERCISE 2.1.3

2.2 KIRCHHOFF'S LAWS

The solution of single-resistor circuits, such as those in the previous examples, may be found using Ohm's law; however, more complicated circuits require the use of two laws first stated by the German physicist Gustav Kirchhoff (1824–1887) in 1847.[2] The two laws are formally known as Kirchhoff's current law and Kirchhoff's voltage law. These laws, together with the terminal characteristics for the various circuit elements, permit systematic methods of solutions for any electrical network. We shall not attempt to prove Kirchhoff's laws here since the concepts necessary for the proof are developed in a later, interesting study of electromagnetic field theory.

A circuit consists of two or more circuit elements connected by means of perfect conductors. Perfect conductors are zero-resistance wires which allow current to flow freely but accumulate no charge and no energy. In this case, the energy can be considered to reside, or be lumped, entirely within each circuit element, and thus the network is called a *lumped-parameter circuit*.

A point of connection of two or more circuit elements is called a *node*. An example of a circuit with three nodes is shown in Fig. 2.7(a). Node 1 consists of the entire con-

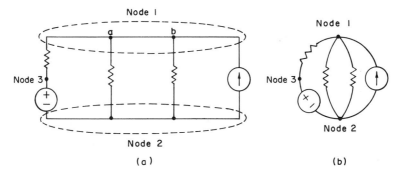

FIGURE 2.7 *(a) Three-node circuit; (b) Three-node circuit redrawn*

[2]Actually, even single-resistor circuits tacitly require Kirchhoff's laws.

nection at the top of the circuit. The beginner quite often mistakes points a and b for nodes. It should be noticed, however, that a and b are connected by a perfect conductor and can be considered electrically as being identical points. This is readily demonstrated by redrawing the circuit in the form of Fig. 2.7(b), where at node 1 all connections are shown at a single point. Similar comments apply for node 2. Node 3 is required for the interconnection of the independent voltage source and the resistor. With these concepts, we are now ready to discuss the all-important laws of Kirchhoff.

Kirchhoff's current law (KCL) states that

The algebraic sum of the currents entering any node is zero.

To demonstrate the use of this law, consider currents flowing into a node, as shown in Fig. 2.8. KCL states that

$$i_1 + i_2 + (-i_3) + i_4 = 0$$

where we recall that i_3 flowing out of the node is equivalent to $-i_3$ entering the node.

For the sake of argument, let us suppose that the sum is not zero. In such a case, we would have

$$i_1 + i_2 - i_3 + i_4 = \Psi \neq 0$$

where Ψ has units of C/s and hence must be the rate at which charges are accumulated in the node. However, a node consists of perfect conductors and cannot accumulate charges. In addition, a basic principle of physics states that charges can neither be created nor destroyed (conservation of charge). Therefore, our assumption is not valid, and Ψ must be zero, demonstrating the plausibility of KCL.

Suppose in our example we now consider the sum of the currents leaving the node. From Fig. 2.8, the sum is

$$-i_1 - i_2 + i_3 - i_4 = 0$$

or, multiplying both sides by -1, we see that

$$i_1 + i_2 - i_3 + i_4 = 0$$

which is identical to our previous result. This demonstrates an equivalent statement for KCL, which states that

The algebraic sum of the currents leaving any node is zero.

Let us now rearrange the above equation in the form

$$i_1 + i_2 + i_4 = i_3$$

where i_1, i_2, and i_4 are entering the node and i_3 is leaving. This form of the equation illustrates another statement for KCL, stated as

The sum of the currents entering any node equals the sum of the currents leaving the node.

In general, a mathematical expression of KCL is

$$\sum_{n=1}^{N} i_n = 0 \qquad\qquad (2.6)$$

where i_n is the nth current entering (or leaving) the node and N is the number of node currents.

As an example of KCL, let us find the current i in Fig. 2.9. Summing the currents entering the node, we have

$$5 + i - (-3) - 2 = 0$$

or

$$i = -6 \text{ A}$$

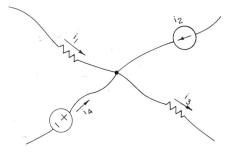

FIGURE 2.8 *Currents flowing into a node*

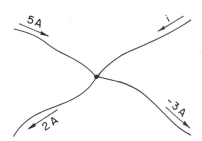

FIGURE 2.9 *Example of KCL*

We note that $i = -6$ A entering the node is equivalent to 6 A leaving the node. Therefore it is not necessary to guess the correct current direction prior to solving the problem. We still arrive at the correct answer in the end.

If we sum the currents leaving the node, we have

$$-5 - i + (-3) + 2 = 0$$

or

$$i = -6 \text{ A}$$

which completes a demonstration of two of the three forms of KCL.

We now move on to Kirchhoff's voltage law (KVL), which states that

The algebraic sum of the voltages around any closed path is zero.

As an illustration, application of this statement to the closed path *abcda* of Fig. 2.10 gives

$$-v_1 + v_2 - v_3 = 0 \qquad\qquad (2.7)$$

where the algebraic sign for each voltage has been taken positive when going from +
to − (higher to lower potential) and negative when going from − to + (lower to
higher potential) in traversing the element.

As in the case of KCL, we shall not attempt a proof of KVL. However, to illustrate
the plausibility of (2.7), let us assume that its right member is not zero. That is,

$$-v_1 + v_2 - v_3 = \Phi \neq 0$$

The left member of this equation is by definition the work required to move a unit
charge around the path *adcba*. A lumped-parameter circuit is a conservative system,
which means that the work required to move a charge around any closed path is zero.
(Proof of this will come in a later course involving an interesting study of electromagnetic theory.) Thus, our assumption is not valid, and Φ is indeed zero.

We should point out, however, that all electrical systems are not conservative. In
fact, electrical power generation, radio waves, and sunlight, to mention only a few, are
consequences of nonconservative systems.

The application of KVL is independent of the direction in which the path is
traversed. Consider, for example, the path *adcba* in Fig. 2.10. Summing the voltages, we
find

$$v_3 - v_2 + v_1 = 0$$

which is equivalent to (2.7).

In general, a mathematical representation for KVL is

$$\sum_{n=1}^{N} v_n = 0 \tag{2.8}$$

where v_n is the nth voltage in a loop of N voltages. The sign of each voltage is chosen
as described earlier for (2.7).

As an example of the use of KVL, consider finding v in Fig. 2.11. Traversing the
circuit in a clockwise direction, we find

$$-5 + v + 10 - 2 = 0$$

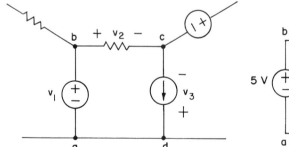

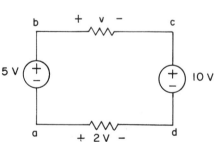

FIGURE 2.10 *Voltages around a closed path*   FIGURE 2.11 *Circuit to illustrate KVL*

or $v = -3$ V. Suppose we now perform a counterclockwise traversal. In such a case,

$$5 + 2 - 10 - v = 0$$

or $v = -3$ V, which, of course, is the same result obtained for the clockwise traversal.

Still another version of KVL for Fig. 2.11 yields

$$10 + v = 2 + 5$$

where the sum of the voltages with one polarity is equated to the sum of the voltages with the opposite polarity. (Stated another way, the voltage rises equal the voltage drops.)

In each of the previous examples, KVL has been applied around conducting paths, such as *abcda* above. The law, however, is valid for *any* closed path. Consider, for instance, the path *acda* of Fig. 2.11. We note that movement directly from *a* to *c* is not along a conducting path. Applying KVL to this closed path yields $v_{ac} + 10 - 2 = 0$, where v_{ac} is the potential of point *a* with respect to *c*. Thus, $v_{ac} = -8$ V. We could also have chosen the path *abca*, for which

$$-5 + v - v_{ac} = -5 - 3 - v_{ac} = 0$$

Therefore, $v_{ac} = -8$ V, which demonstrates the use of different closed paths to obtain the same result.

As another example of the application of KCL and KVL, consider finding i_x and v_x in the network of Fig. 2.12. Summing the currents entering node *a* gives $-4 + 1 + i_1 = 0$, or $i_1 = 3$ A. At node *b*, $-i_1 + 2 - i_2 = 0$, or $i_2 = -1$ A. At node *c*, $i_2 + i_3 - 3 = 0$, or $i_3 = 4$ A. Therefore, at node *d*, $-i_x - 1 - i_3 = 0$, or $i_x = -5$ A. Next, KVL about the path *abcda* gives $-10 + v_2 - v_x = 0$. From Ohm's law, $v_2 = 5i_2 = -5$ V. Therefore, $v_x = -15$ V.

Before concluding our discussion of Kirchhoff's laws, consider the network of Fig. 2.13, in which several elements are shown within a closed surface *S*. We recall that the current entering each element equals that leaving the device so that each element stores zero net charge. Therefore, the total net charge stored within the surface is zero, requiring that

$$i_1 + i_2 + i_3 + i_4 = 0$$

This result illustrates a generalization of KCL, which states

The algebraic sum of the currents entering any closed surface is zero.[3]

To illustrate the plausibility of the generalized KCL, let us write KCL equations at nodes *a*, *b*, *c*, and *d* of Fig. 2.13. The results are

---

[3]The surface cannot pass through an element which is considered to be concentrated at a point in lumped-parameter circuits.

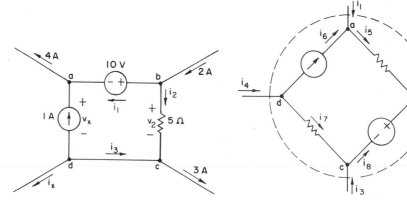

FIGURE 2.12 *Network for example of KCL and KVL*

FIGURE 2.13 *Network for generalized KCL*

$$i_1 = i_5 - i_6$$
$$i_2 = -i_5 - i_8$$
$$i_3 = -i_7 + i_8$$
$$i_4 = i_6 + i_7$$

Adding these equations yields

$$i_1 + i_2 + i_3 + i_4 = 0$$

as previously noted.

From the generalized KCL, we see immediately in Fig. 2.12 for a surface enclosing points a, b, c, and d that $-i_x - 4 + 2 - 3 = 0$, or $i_x = -5$ A.

EXERCISES

2.2.1 Identify the nodes in the figure.

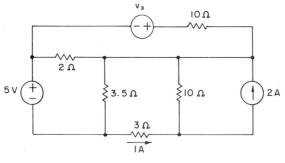

EXERCISE 2.2.1

2.2.2   Determine (a) i_x, (b) i_y, and (c) the potential of a with respect to b.

Ans. (a) -3 A, (b) 4 A, (c) -8 V

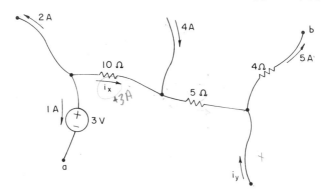

EXERCISE 2.2.2

2.2.3   In Ex. 2.2.1, find v_x.   *Ans.* 22 V

2.3 SERIES RESISTANCE AND VOLTAGE DIVISION

Now that the laws of Ohm and Kirchhoff have been introduced, we are prepared to begin analyzing simple resistive circuits. We shall consider a *simple circuit* to be one that can be completely described by a single equation. One type, which we shall consider in this section, is a circuit consisting of a single closed path, or loop, of elements. By KCL each element has a common current, say i. Then Ohm's law and KVL applied around the loop yield a single equation in i that completely describes the circuit.

Elements are said to be connected in *series* when they all carry the same current. Clearly, the networks of this section consist entirely of elements connected in series. An important circuit of this type, consisting of two resistors and an independent voltage source, provides an excellent starting point. We shall first analyze this special case and then develop the more general case.

A single-loop circuit having two resistors and an independent voltage source is shown in Fig. 2.14(a). The first step in the analysis procedure is the assignment of currents and voltages to all elements in the network. In this circuit, it is obvious from KCL that all elements carry the same current. We may *arbitrarily* call this current i in the direction shown (clockwise). The novice often attempts to guess the true current direction in making assignments. Such an assignment is not necessary, and is not usually possible, even for the expert. We next make the voltage assignments for R_1 and R_2 as v_1 and v_2, respectively. These assignments are also arbitrary but in the figure have been chosen to satisfy Ohm's law for a positive algebraic sign.

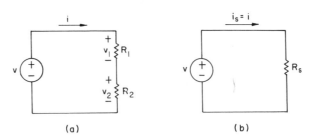

FIGURE 2.14 *(a) Single-loop circuit; (b) Equivalent circuit*

The second step in the analysis is the application of KVL, which yields

$$v = v_1 + v_2$$

where, from Ohm's law,

$$v_1 = R_1 i$$
$$v_2 = R_2 i$$

(2.9)

Combining these equations, we find

$$v = R_1 i + R_2 i$$

Solving for i yields

$$i = \frac{v}{R_1 + R_2}$$

(2.10)

Let us now consider a simple circuit consisting of the voltage source v connected to a resistance R_s, as shown in Fig. 2.14(b). If R_s is selected such that

$$i_s = \frac{v}{R_s} = i$$

(2.11)

then the network is called an *equivalent circuit* of Fig. 2.14(a) because an identical current (response) is produced for a voltage v (excitation). In general, two circuits are said to be equivalent when they exhibit identical voltage-current relationships at their terminals.

Comparing (2.10) and (2.11), it is obvious that

$$R_s = R_1 + R_2$$

(2.12)

Interpreting (2.12), we see that if the series combination of R_1 and R_2 is replaced by a resistor R_s, the same current flows from the source v. Therefore R_s is the *equivalent resistance* of the series connection.

Combining (2.9) and (2.10), we see that

$$v_1 = \frac{R_1}{R_1 + R_2} v$$
$$v_2 = \frac{R_2}{R_1 + R_2} v$$

(2.13)

The potential of source v divides between resistances R_1 and R_2 in direct proportion to their resistances, demonstrating the *principle of voltage division* for two series resistors. We see in (2.13) that the greater voltage appears across the larger resistor.

The instantaneous powers absorbed by R_1 and R_2 are

$$p_1 = \frac{v_1^2}{R_1} = \frac{R_1}{(R_1 + R_2)^2} v^2$$

and

$$p_2 = \frac{v_2^2}{R_2} = \frac{R_2}{(R_1 + R_2)^2} v^2$$

respectively. The total power absorbed is

$$p_1 + p_2 = \frac{v^2}{R_1 + R_2} = v\left(\frac{v}{R_1 + R_2}\right) = vi$$

The power delivered by the source also equals vi, indicating that the power delivered by the source equals that absorbed by R_1 and R_2. This result is known as *conservation of power* (sometimes also referred to as Tellegen's theorem), a property which is often useful in circuit analysis.

Let us now digress briefly and repeat the analysis for the counterclockwise current assignment of Fig. 2.15. Application of KVL yields

$$v = v_1 + v_2$$

in which

$$v_1 = -R_1 i_1$$
$$v_2 = -R_2 i_1$$

Combining these equations, we have

$$v = -R_1 i_1 - R_2 i_1$$

from which

$$i_1 = -\frac{v}{R_1 + R_2}$$

Comparing our result with (2.10), we see that $i_1 = -i$, which is also evident from an inspection of Figs. 2.14(a) and 2.15. Hence, if the proper direction for positive current happens not to be chosen, a negative sign will result. This, of course, by definition means the positive current flows in the opposite direction. Similar statements apply to voltage assignments. If the polarity for positive voltage happens not to be chosen, a negative sign will occur in its solution, leading, of course, to the correct answer.

As an example of the utility of the preceding analysis, suppose $v = 120 \sin t$ V, $R_1 = 90\ \Omega$, and $v_1 = 72 \sin t$ V in Fig. 2.14(a). Let us now determine R_2, i, and the instantaneous power associated with each element. From (2.13)

$$72 \sin t = \frac{90}{90 + R_2} \, 120 \sin t$$

which yields $R_2 = 60 \, \Omega$. Hence, $R_s = R_1 + R_2 = 150 \, \Omega$, and, from (2.10), $i = 120 \sin t/150 = 0.8 \sin t$ A. The instantaneous power to R_1 and R_2 is $p_1 = R_1 i^2 = 57.6 \sin^2 t$ W and $p_2 = R_2 i^2 = 38.4 \sin^2 t$ W. Thus, the power delivered by the source is $96 \sin^2 t$ W. We also observe that the power absorbed by R_1 and R_2 should equal that which is absorbed by R_s in Fig. 2.14(b). The power to R_s is $R_s i^2 = 96 \sin^2 t$ W.

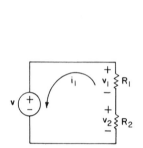

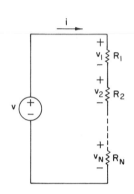

FIGURE 2.15 *Single-loop circuit with counterclockwise current*

FIGURE 2.16 *Single-loop circuit with N series resistors*

Let us now extend our analysis to include the series connection of N resistors and an independent voltage source, as shown in Fig. 2.16. KVL gives

$$v = v_1 + v_2 + \ldots + v_N$$

in which

$$v_1 = R_1 i$$
$$v_2 = R_2 i$$
$$\vdots \qquad\qquad (2.14)$$
$$v_N = R_N i$$

Therefore

$$v = R_1 i + R_2 i + \ldots + R_N i$$

Solving this equation for i yields

$$i = \frac{v}{R_1 + R_2 + \cdots + R_N} \qquad\qquad (2.15)$$

Let us now select R_s in the circuit of Fig. 2.14(b) so that (2.11) is satisfied. Equivalence of (2.11) and (2.15) requires that

$$R_s = R_1 + R_2 + \ldots + R_N = \sum_{n=1}^{N} R_n \qquad\qquad (2.16)$$

Therefore the equivalent resistance of N series resistors is simply the sum of the individual resistances.

Substituting (2.15) and (2.16) into (2.14), we find

$$v_1 = \frac{R_1}{R_s} v$$

$$v_2 = \frac{R_2}{R_s} v$$

$$\cdot$$
$$\cdot$$
$$\cdot$$

$$v_N = \frac{R_N}{R_s} v$$

(2.17)

which are the equations describing the voltage division property for N series resistors. Again, we see that the voltage divides in direct proportion to the resistance.

The instantaneous power delivered to the series combination, from (2.2) and (2.17), is

$$
\begin{aligned}
p &= \frac{v_1^2}{R_1} + \frac{v_2^2}{R_2} + \cdots + \frac{v_N^2}{R_N} \\
&= \frac{R_1}{R_s^2} v^2 + \frac{R_2}{R_s^2} v^2 + \cdots + \frac{R_N}{R_s^2} v^2 \\
&= \frac{v^2}{R_s} = vi
\end{aligned}
$$

This power is equal to that delivered by the source, verifying conservation of power for the series connection of N resistors.

EXERCISES

2.3.1 For the network shown, find (a) an equivalent circuit, (b) the current i, (c) the power delivered by the source, (d) v_1, and (e) the minimum wattage required for R_1.   *Ans.* (b) 0.5 A, (c) 5 W, (d) -4 V, (e) 2 W

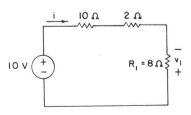

EXERCISE 2.3.1

2.3.2 In Fig. 2.14 (a), $v = 10e^{-t}$, $v_2 = 2e^{-t}$, and $R_1 = 16 \, \Omega$. Find (a) R_2, (b) the instantaneous power delivered to R_2, and (c) the current i.

Ans. (a) $4 \, \Omega$, (b) e^{-2t} W, (c) $0.5e^{-t}$ A

2.3.3 A resistive load[4] requires 8 V and dissipates 2 W. A 12-V storage battery is available to operate the load. Referring to Fig. 2.14 (a), if R_2 represents the load and v the 12-V battery, find (a) the current i, (b) the necessary resistance R_1, and (c) the minimum wattage of R_1.   *Ans.* (a) 0.25 A, (b) 16 Ω, (c) 1 W

2.4 PARALLEL RESISTANCE AND CURRENT DIVISION

Another important simple circuit is the single-node-pair resistive circuit. In analyzing these networks, we shall first examine a special case and then develop the more general case, as was performed for the single-loop network.

Elements are connected in *parallel* when the same voltage is common to each of them. A single-node-pair circuit consisting of the parallel connection of two resistors and an independent current source is shown in Fig. 2.17(a). To begin the analysis, we first assign voltages and currents to each circuit element. As in the case of the single-loop circuit, the assignments are completely arbitrary. We have chosen the assignments given by i_1, i_2, and v.

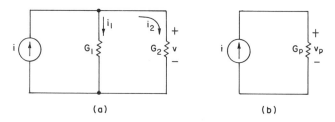

(a)   (b)

FIGURE 2.17 *(a) Single node-pair circuit; (b) Equivalent circuit*

Next, we need to apply either KCL or KVL. Inspection of the circuit shows that a single node (either the upper or the lower) is common to all elements. This suggests that for single-node-pair circuits KCL is the most efficient choice. Applying KCL at the upper node yields

$$i = i_1 + i_2$$

where, from Ohm's law,

$$i_1 = G_1 v$$
$$i_2 = G_2 v$$
(2.18)

Combining these equations gives

$$i = G_1 v + G_2 v$$

[4]A *load* is an element or collection of elements connected between the output terminals. In this case, the load is a resistor.

and solving for v, we find

$$v = \frac{i}{G_1 + G_2} \qquad (2.19)$$

In the circuit of Fig. 2.17(b), if G_p is selected such that

$$v_p = \frac{i}{G_p} = v \qquad (2.20)$$

then the network is an equivalent circuit to that of Fig. 2.17(a). Comparing (2.19) and (2.20), we see that

$$G_p = G_1 + G_2 \qquad (2.21)$$

Clearly, G_p is the equivalent conductance of the two parallel conductances. In terms of resistances, (2.21) becomes

$$G_p = \frac{1}{R_p} = \frac{1}{R_1} + \frac{1}{R_2}$$

or

$$R_p = \frac{R_1 R_2}{R_1 + R_2} \qquad (2.22)$$

Therefore, the equivalent resistance of two resistors connected in parallel is equal to the product of their resistances divided by their sum. It is interesting to note that G_p is greater than either G_1 or G_2; therefore, R_p is less than either R_1 or R_2. From this result, we see that connecting resistors in parallel reduces the overall resistance.

In the special case $R_2 = R_1$, we see from (2.22) that $R_p = R_1/2$.

Substituting (2.19) into (2.18) gives

$$i_1 = \frac{G_1}{G_1 + G_2} i$$
$$\qquad (2.23)$$
$$i_2 = \frac{G_2}{G_1 + G_2} i$$

The current of the source i divides between conductances G_1 and G_2 in direct proportion to their conductances, demonstrating the *principle of current division*. It is common practice to give the resistor values in circuit diagrams in ohms (resistance) and not mhos (conductance). In terms of resistance values, (2.23) becomes

$$i_1 = \frac{R_2}{R_1 + R_2} i$$
$$\qquad (2.24)$$
$$i_2 = \frac{R_1}{R_1 + R_2} i$$

Therefore the current divides in inverse proportion to the resistances. We see that the larger current flows through the smaller resistor. The power absorbed by the parallel combination is

$$p_1 + p_2 = R_1 i_1^2 + R_2 i_2^2$$

$$= \frac{R_2^2 i^2}{(R_1 + R_2)^2} R_1 + \frac{R_1^2 i^2}{(R_1 + R_2)^2} R_2$$

$$= \frac{R_1 R_2}{R_1 + R_2} i^2 = vi$$

which equals that delivered to the network by the current source.

Suppose in Fig. 2.17(a) that $R_1 = 3\,\Omega$, $R_2 = 6\,\Omega$, and $i = 3$ A. Then, from (2.22), $R_p = (3)(6)/(3 + 6) = 2\,\Omega$. From (2.24), $i_1 = \frac{6}{9}(3) = 2$ A, and $i_2 = \frac{3}{9}(3) = 1$ A. The voltage $v = R_1 i_1 = R_2 i_2 = (3)(2) = 6$ V. The current of the source flows through an equivalent resistance of R_p. Hence the voltage is also given by $v = R_p i = (2)(3) = 6$ V.

Let us now consider the more general case of N parallel conductances and an independent current source, as shown in Fig. 2.18. KCL gives

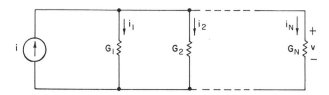

FIGURE 2.18 *Single-node-pair circuit with N parallel conductances*

$$i = i_1 + i_2 + \ldots + i_N$$

for which

$$i_1 = G_1 v$$

$$i_2 = G_2 v$$

$$\vdots \tag{2.25}$$

$$i_N = G_N v$$

Therefore we have

$$i = G_1 v + G_2 v + \ldots + G_N v$$

from which

$$v = \frac{i}{G_1 + G_2 + \ldots + G_N} \tag{2.26}$$

If we now select G_p in Fig. 2.17(b) such that (2.20) is satisfied, then (2.26) requires that

$$G_p = G_1 + G_2 + \ldots + G_N = \sum_{i=1}^{N} G_i \tag{2.27}$$

In terms of resistances, this equation becomes

$$\frac{1}{R_p} = \frac{1}{R_1} + \frac{1}{R_2} + \ldots + \frac{1}{R_N} = \sum_{i=1}^{N} \frac{1}{R_i} \tag{2.28}$$

Hence the reciprocal of the equivalent resistance is simply the sum of the reciprocals of the resistances.

Combining (2.25)–(2.28), we find

$$i_1 = \frac{G_1}{G_p}i = \frac{R_p}{R_1}i$$

$$i_2 = \frac{G_2}{G_p}i = \frac{R_p}{R_2}i$$

$$\cdot$$

$$\cdot \qquad\qquad\qquad\qquad (2.29)$$

$$\cdot$$

$$i_N = \frac{G_N}{G_p}i = \frac{R_p}{R_N}i$$

Again the currents divide in inverse proportion to the resistances.

We observe in (2.28) that for $N > 2$ an expression for R_p is more complicated than (2.22). Formulas could be obtained, of course, for $N = 3, 4$, etc., but it is usually easier to apply (2.28) directly. As an example, suppose for $N = 3$ that $R_1 = 4\,\Omega$, $R_2 = 12\,\Omega$, and $R_3 = 6\,\Omega$. Then

$$\frac{1}{R_p} = \frac{1}{4} + \frac{1}{12} + \frac{1}{6} = \frac{1}{2}\,\mho$$

and $R_p = 2\,\Omega$.

Let us now consider finding the equivalent resistance R_{eq} of the network of Fig. 2.19(a), as viewed from terminals x-y. Such reductions are very helpful in analyzing many types of circuits, as we shall see in the next section. The process is carried out by

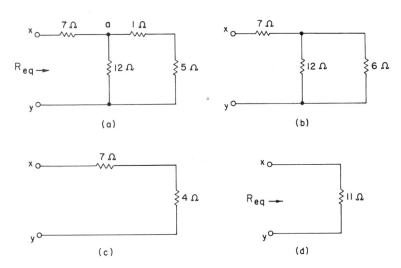

FIGURE 2.19 *Steps in determining the equivalent resistance of a network*

successive combinations of parallel and series connected resistors. In Fig. 2.19(a), the student often errs in taking combinations such as the 7- and 12-Ω resistors to be in series. We see, however, that at node a, a current in the 7-Ω resistor would divide between the 1- and 12-Ω resistor; hence they cannot be in series. The 1- and 5-Ω resistors, however, would carry the same current. Therefore they are in series, having a series resistance of 6 Ω as shown in Fig. 2.19(b). We now observe that the same voltage would occur across the 6- and 12-Ω resistors, indicating a parallel connection having an equivalent resistance of $(6)(12)/(6 + 12) = 4\,\Omega$, as shown in Fig. 2.19(c). It is apparent that the 7- and 4-Ω resistors of this network are in series, yielding an equivalent resistance for the entire network of 11 Ω [Fig. 2.19(d)]. Therefore, from terminals x-y, the network could be replaced by a single resistor of 11 Ω. This is useful in determining, for instance, the power delivered by a source connected to terminals x-y. Suppose a 22-V source is applied. Then the current flowing from the source is $i = \frac{22}{11} = 2$ A, which gives an instantaneous power $p(t) = (22)(2) = 44$ W delivered to the resistor network.

EXERCISES

2.4.1 In the circuit shown, find (a) the equivalent resistance seen from the source terminals, (b) v_1, (c) i_2, and (d) i.

$Ans.$ (a) 20 Ω, (b) -120 V, (c) -2 A, (d) -6 A

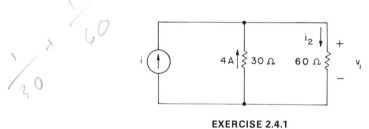

EXERCISE 2.4.1

2.4.2 In Fig. 2.18, if $N = 3$, $R_1 = 120\,\Omega$, $R_2 = 60\,\Omega$, and $v = 10 \sin t$ V and the instantaneous power delivered by the current source is $5 \sin^2 t$ W, find (a) R_3, (b) i, and (c) i_3. $Ans.$ (a) 40 Ω, (b) 0.5 sin t A, (c) 0.25 sin t A

2.4.3 A load requires 4 A and absorbs 16 W. A current source of 10 A is available. For the circuit of Fig. 2.17 (a), if G_2 represents the load and i the current source, find (a) the voltage v and (b) the required conductance G_1.

$Ans.$ (a) 4 V, (b) 1.5 ℧

2.4.4 Find the equivalent resistance, R_{eq}. $Ans.$ 9.4 Ω

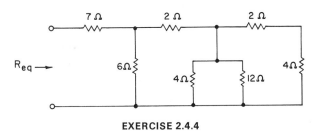

EXERCISE 2.4.4

2.5 ANALYSIS EXAMPLES

For our first example, let us consider a single-loop circuit containing three resistors and two independent voltage sources, as shown in Fig. 2.20(a). KVL and Ohm's law give

$$-10 + 20i + 30i + 20 + 50i = 0$$

Thus the reduced equation is

$$10 + 100i = 0$$

for which $i = -0.1$ A and $v_1 = 30i = -3$ V.

We see that the reduced equation is satisfied by the equivalent circuit of Fig. 2.20(b), where the voltage sources are replaced by their algebraic sum and the resistances by their series equivalent. From this circuit, the power absorbed by all resistors is $p_{100\,\Omega} = 100i^2 = 1$ W. Therefore the net power delivered by the two voltage sources must be 1 W. We recall that the power absorbed by an element is the product of the voltage and the current, where the current enters the positive voltage terminal. For the 10-V source, we see that $p_{10\,V} = (10)(-i) = (10)(0.1) = 1$ W absorbed. Physically, this means that current is flowing into the positive terminal and that the source is receiving power, or is being charged. The absorbed power in the 20-V source is $p_{20\,V} = 20i = (20)(-0.1) = -2$ W. The minus sign indicates that the source is delivering 2 W to the circuit. Hence the net power from the two sources is 1 W, which is the power delivered to the resistors. Finally, let us find the potential of point *a* with

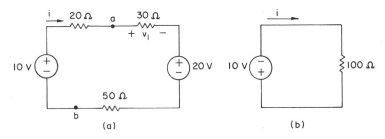

FIGURE 2.20 *(a) Single-loop circuit; (b) Equivalent circuit*

respect to point b, which we denote as v_{ab}. From KVL, $-10 + 20i + v_{ab} = 0$, or $v_{ab} = 10 - 20i = 10 - (20)(-0.1) = 12$ V.

For a second example, let us find i_1 and $v(t)$ for the circuit of Fig. 2.21. In the circuit, three conductances and two independent current sources are connected in parallel. Applying KCL to the upper node yields

$$10 \sin \pi t - 0.01v(t) - 0.02v(t) - 5 - 0.07v(t) = 0$$

or

$$(10 \sin \pi t - 5) - 0.1v(t) = 0$$

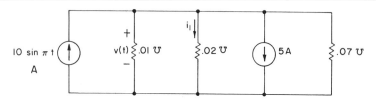

FIGURE 2.21 *Single-node-pair circuit*

It is apparent that a current source of $(10 \sin \pi t - 5)$ A connected to a conductance of 0.1 ℧ (a 10-Ω resistor) would be an equivalent circuit for the network [see Fig. 2.17(b)]. The total instantaneous power absorbed by the conductances is therefore

$$p_{abs} = \frac{(10 \sin \pi t - 5)^2}{0.1}$$

$$= 1000 \sin \pi t (\sin \pi t - 1) + 250 \text{ W}$$

Solving for $v(t)$ above and subsequently for i_1, we have

$$v(t) = 100 \sin \pi t - 50 \text{ V}$$

$$i_1 = 0.02(100 \sin \pi t - 50) = 2 \sin \pi t - 1 \text{ A}$$

We observe $v(0) = -50$ V, $v(0.5) = 50$ V, and $v(1.5) = -150$ V.

The power delivered by the leftmost source is

$$p_1 = 10 \sin \pi t (100 \sin \pi t - 50) \text{ W}$$

Similarly, the 5-A source delivers

$$p_2 = -5(100 \sin \pi t - 50) \text{ W}$$

Thus the total power delivered by these sources is

$$p_{del} = p_1 + p_2 = 1000 \sin \pi t (\sin \pi t - 1) + 250 \text{ W}$$

which equals that absorbed by the conductances.

As a third example, consider finding i, v, the power delivered by the source, and the power absorbed by the 8-Ω load of Fig. 2.22(a). We begin by obtaining successive combinations of parallel and series resistor connections. The 4- and 8-Ω resistances (in series) add to give 12 Ω. These 12 Ω are in parallel with the 6-Ω resistor, giving an equivalent value of $(12)(6)/(12 + 6) = 4\,\Omega$ [Fig. 2.22(b)]. We now add the 12- and 4-Ω resistances, which are in parallel with the 16 Ω, giving $(16)(16)/(16 + 16) = 8\,\Omega$, as shown in Fig. 2.22(c). This is the equivalent resistance as seen looking into the circuit at x-x. Applying KVL to the simplified circuit, we have

$$-30e^{-2t} + 2i + 8i = 0$$

from which

$$i = 3e^{-2t} \text{ A}$$

Therefore the power delivered by the source is

$$p_s = (30e^{-2t})(3e^{-2t}) = 90e^{-4t} \text{ W}$$

By voltage division we see that

$$v_1 = \left(\frac{8}{2 + 8}\right) 30e^{-2t} = 24e^{-2t} \text{ V}$$

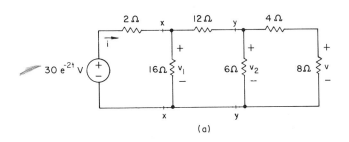

(a)

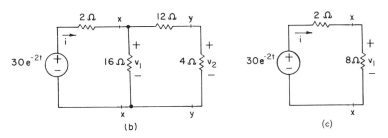

(b)   (c)

FIGURE 2.22 *Circuit for analysis example using voltage division*

which is the voltage across points x-x in the circuit. Proceeding to Fig. 2.22(b), we see that v_1 is the voltage across the series combination of the 12- and 4-Ω resistors;

hence, again using voltage division, we find

$$v_2 = \left(\frac{4}{12 + 4}\right)v_1 = 6e^{-2t} \text{ V}$$

which is the voltage across points y-y in the circuit. In Fig. 2.22(a), it is obvious that v_2 is the voltage across the series connection of the 4- and 8-Ω resistors. Therefore voltage division requires that

$$v = \left(\frac{8}{8 + 4}\right)v_2 = 4e^{-2t} \text{ V}$$

The power to the 8-Ω load is

$$p_2 = \frac{v^2}{8} = 2e^{-4t} \text{ W}$$

As a final example, let us find the current i in Fig. 2.23(a). We first combine the two 6-Ω resistors on the right of Fig. 2.23(a) and obtain the equivalent resistance which is in parallel with the 4-Ω resistor. The result, $(4)(12)/(4 + 12) = 3$ Ω, is shown in Fig. 2.23(b). If we now replace the two series resistors to the right of points x-x and the parallel 3- and 6-Ω resistors to the left of x-x by their respective series and parallel equivalents, we obtain the circuit of Fig. 2.23(c). Using current division, we find

$$i_1 = \left(\frac{2}{2 + 6}\right) \times 12 = 3 \text{ A}$$

A second application of current division to Fig. 2.23(a) reveals immediately that

$$i = \left(\frac{4}{4 + 6 + 6}\right)i_1 = \left(\frac{1}{4}\right) \times 3 = \frac{3}{4} \text{ A}$$

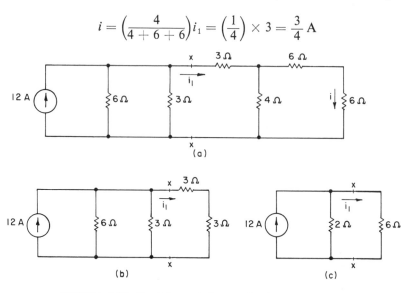

FIGURE. 2.23 *Circuit for analysis example using current division*

EXERCISES

2.5.1   Find v, v_{ab}, and the power delivered by the 5-V source.

Ans. -10 V, -8.33 V, -0.167 W

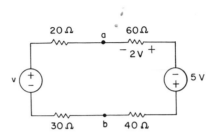

EXERCISE 2.5.1

2.5.2   Find G and construct an equivalent circuit having one current source and a single conductance.   *Ans.* $G = 0.03$ ℧, $i = 3 \sin t$ A directed upward

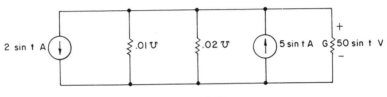

EXERCISE 2.5.2

2.5.3   Find the power absorbed by the 90-Ω resistor.   *Ans.* 3.6 W

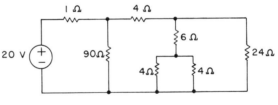

EXERCISE 2.5.3

2.5.4   Find v and i.   *Ans.* 6 V, $-\frac{1}{6}$ A

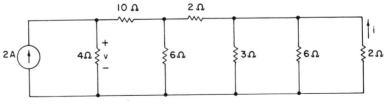

EXERCISE 2.5.4

2.6   AMMETERS, VOLTMETERS, AND OHMMETERS*

A good example of the usefulness of current and voltage division is demonstrated in the design of simple two-terminal measuring instruments, such as ammeters, voltmeters, and ohmmeters. An *ideal ammeter* measures the current flowing through its terminals and has zero voltage across its terminals. In contrast, an *ideal voltmeter* measures the voltage across its terminals and has a terminal current of zero. An *ideal ohmmeter* measures the resistance connected between its terminals and delivers zero power to the resistance.

The practical measuring instruments that we shall consider only approximate the ideal devices. The ammeters, for instance, will not have zero terminal voltages. Likewise the voltmeters will not have zero terminal currents, and the ohmmeters will not have zero power delivered from their terminals.

A popular type of ammeter consists of a mechanical movement known as a D'Arsonval meter. This device is constructed by suspending an electrical coil between the poles of a permanent magnet. A dc current passing through the coil causes a rotation of the coil, as a result of magnetic forces, that is proportional to the current. A pointer is attached to the coil so that the rotation, or meter deflection, can be visually observed. D'Arsonval meters are characterized by their *full-scale current*, which is the current that will cause the meter to read its greatest value. Meter movements are common having full-scale currents from 10 μA to 10 mA.

An equivalent circuit for the D'Arsonval meter consists of an ideal ammeter in series with a resistance R_M, as shown in Fig. 2.24. In this circuit, R_M represents the resistance of the electrical coil. Clearly, a voltage appears across the ammeter terminals as a result of the current i flowing through R_M. R_M is usually a few ohms, and the terminal voltage for a full-scale current is nominally from 20 to 200 mV.

The D'Arsonval meter of Fig. 2.24 is an ammeter which is suitable for measuring dc currents not greater than the full-scale current I_{FS}. Suppose we wish, however, to measure a current which exceeds I_{FS}. It is apparent that we must not allow a current greater than I_{FS} to flow through the device. A circuit to accomplish this is shown in Fig. 2.25, where R_p is a parallel resistance which reduces the current flowing through the meter coil.

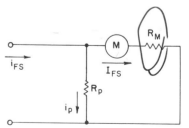

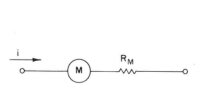

FIGURE 2.24 *Equivalent circuit for a D'Arsonval meter*

FIGURE 2.25 *Ammeter circuit*

From current division, we see that

$$I_{FS} = \frac{R_p}{R_M + R_p} i_{FS}$$

where i_{FS} is the current which produces I_{FS} in the D'Arsonval meter. (Clearly, this is the maximum current the ammeter can measure.) Solving for R_p, we have

$$R_p = \frac{R_M I_{FS}}{i_{FS} - I_{FS}} \tag{2.30}$$

A dc voltmeter can be constructed using the basic D'Arsonval meter by placing a resistance R_s in series with the device, as shown in Fig. 2.26. It is obvious that the full-scale voltage, $v = v_{FS}$, occurs when the meter current is I_{FS}. Therefore, from KVL,

$$-v_{FS} + R_s I_{FS} + R_M I_{FS} = 0$$

from which

$$R_s = \frac{v_{FS}}{I_{FS}} - R_M \tag{2.31}$$

The *current sensitivity* of a voltmeter, expressed in ohms per volt, is the value obtained by dividing the resistance of the voltmeter by its full-scale voltage. Therefore

$$\Omega/V \text{ rating} = \frac{R_s + R_M}{v_{FS}} \approx \frac{R_s}{v_{FS}} \tag{2.32}$$

(Note: "$\approx$" means approximately equal to.)

A simple ohmmeter circuit employing a D'Arsonval meter for measuring an unknown resistance R_x is shown in Fig. 2.27. In this circuit, the battery E causes a current i to flow when R_x is connected into the circuit. Applying KVL, we have

$$-E + (R_s + R_M + R_x)i = 0$$

from which

$$R_x = \frac{E}{i} - (R_s + R_M)$$

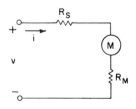

FIGURE 2.26 *Voltmeter circuit*

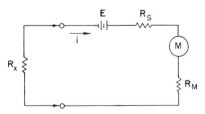

FIGURE 2.27 *Ohmmeter circuit*

We select E and R_s such that for $R_x = 0$, $i = I_{FS}$. Therefore

$$I_{FS} = \frac{E}{R_s + R_M}$$

Combining the last two equations, we find

$$R_x = \left(\frac{I_{FS}}{i} - 1\right)(R_s + R_M) \tag{2.33}$$

A very popular general-purpose meter which combines the three previously described circuits is the *VOM* (voltmeter-ohmmeter-milliammeter). In the VOM, provisions are made for changing R_p and R_s so that a wide dynamic range of operation is provided.

EXERCISES

2.6.1 A D'Arsonval meter has $I_{FS} = 1$ mA and $R_M = 50\ \Omega$. Determine R_p in Fig. 2.25 so that i_{FS} is (a) 1.0 mA, (b) 10 mA, and (c) 100 mA.

Ans. (a) infinite, (b) 5.556 Ω, (c) 0.505 Ω

2.6.2 In Fig. 2.26, determine R_s and the Ω/V rating for a voltmeter to have a full-scale voltage of 100 V using a D'Arsonval meter with (a) $R_M = 100\ \Omega$ and $I_{FS} = 50\ \mu$A and (b) $R_M = 50\ \Omega$ and $I_{FS} = 1$ mA.

Ans. (a) 2 MΩ, 20 kΩ/V, (b) 100 kΩ, 1 kΩ/V

2.6.3 What voltage would each meter design of Ex. 2.6.2 measure in the circuit below? Why are the two measurements different?

Ans. 99.5 V, 90.9 V

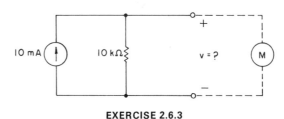

EXERCISE 2.6.3

2.6.4 The meter movement of Ex. 2.6.1 is used to form the ohmmeter circuit of Fig. 2.27. Determine R_s and E so that $i = I_{FS}/2$ mA when $R_x = 10$ kΩ.

Ans. 9.95 kΩ, 10 V

2.7 PHYSICAL RESISTORS*

Resistors are manufactured from a variety of materials and are available in many sizes and values. Their characteristics include a nominal resistance value, an accuracy with which the actual resistance approximates the nominal value (known as *tolerance*), a power dissipation, and a stability as a function of temperature, humidity, and other environmental factors.

The most common type of resistor found in electrical circuits is the carbon composition or carbon film resistor. The composition type is made of hot-pressed carbon granules. The carbon film device consists of carbon powder which is deposited on an insulating substrate. A typical resistor of this type is shown in Fig. 2.28. Multicolored

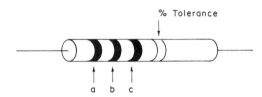

FIGURE 2.28 *Carbon resistor*

bands, shown as *a*, *b*, *c*, and % tolerance, are painted on the resistor body to indicate the nominal value of the resistance. The color code for the bands is given in Table 2.1.

Bands *a*, *b*, and *c* give the nominal resistance of the resistor, and the tolerance band gives the percent that the resistor may deviate from the nominal value. Referring to Fig. 2.28, the resistance is

$$R = (10a + b)10^c \pm \% \text{ tolerance} \tag{2.34}$$

TABLE 2.1 *Color code for carbon resistors*

Color	Value	Color	Value
Bands a, b, and c			
Gold*	−2	Yellow	4
Silver*	−1	Green	5
Black	0	Blue	6
Brown	1	Violet	7
Red	2	Gray	8
Orange	3	White	9
% Tolerance Band			
Gold	±5%		
Silver	±10%		

* These colors apply to band *c* only.

by which we mean that the % tolerance of the nominal resistance is to be added or subtracted to give the range in which the resistance lies.

As an example, suppose we have a resistor with band colors of yellow, violet, red, and silver. The resistor will have a value given by

$$R = (4 \times 10 + 7) \times 10^2 \pm 10\%$$
$$= 4700 \pm 470 \; \Omega$$

Therefore the resistance value lies between 4230 and 5170 Ω.

Values of carbon resistors range from 2.7 to $2.2 \times 10^7 \; \Omega$, with wattages from $\frac{1}{8}$ to 2 W. For resistance values less than 10 Ω, we see from (2.34) that the third band must be gold or silver. Carbon resistors are inexpensive but have the disadvantage of a relatively high variation of resistance with temperature.

Another resistor type which is commonly used in applications requiring a high power dissipation is the wire-wound resistor. These devices consist of a metallic wire, usually a nickel-chromium alloy, wound on a ceramic core. Low-temperature-coefficient wire permits the fabrication of resistors that are very precise and stable, having accuracy and stability of the order of $\pm 1\%$ to $\pm 0.001\%$.

The metal film resistor is another valuable and useful resistor type. These resistors are made by vacuum-depositing a thin metal layer on a low-thermal-expansion substrate. The resistance is then adjusted by etching or grinding a pattern through the film. Accuracy and stability for these resistors approaches that of wire-wound types, and high resistance values are much easier to attain.

In the future, the reader may be more likely to encounter resistors in *integrated circuits*, which were developed in the late 1950s and came into their own in the 1960s. An integrated circuit is a single monolithic chip of *semiconductor* (material with conducting properties between those of a conductor and an insulator) in which active and passive elements are fabricated by diffusion and deposition processes. Integrated circuits containing hundreds of elements are available on chips about $\frac{1}{8}$ in. square. An integrated-circuit resistor has the typical structure shown in Fig. 2.29.

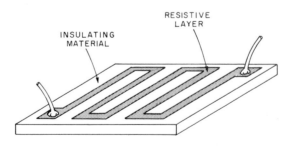

FIGURE 2.29 *Integrated-circuit resistor*

EXERCISE

2.7.1 Find the resistance range of carbon resistors having color bands of (a) brown, black, red, silver; (b) red, violet, yellow, silver; and (c) blue, gray, silver, gold.
Ans. (a) 900–1100 Ω, (b) 243–297 kΩ, (c) 6.46–7.14 Ω

PROBLEMS

2.1 A 5-V battery is connected to the ends of a 1000-ft length of conducting wire and a 10-A current flows. What is the resistance per foot of the wire?

2.2 A 2-kΩ resistor is connected to a battery and 10-mA flow. What current will flow if the battery is connected to a 500-Ω resistor?

2.3 A 200-Ω resistor is connected to a 12-V source. Find the current and the minimum wattage of the resistor. If a 100-Ω resistor is inserted in series in the circuit, what is the voltage across it?

2.4 Find v and i.

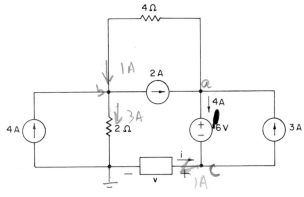

PROBLEM 2.4

2.5 Several series resistors draw 100 mA from a 10-V source. If a 1-kΩ resistor is inserted in series in the circuit, find the new current.

2.6 Two resistors connected in series draw 25 mA from a 100-V source. The voltage across one of the resistors is 25 V. Find the value of each resistor.

2.7 Design a voltage divider to provide 1, 5, 10 and 20 V from a 25-V source. The source is to deliver 25 mW of power to the divider.

2.8 A 1-A current source is connected to the parallel combination of 20- and 30-Ω resistors. Find the voltage, current, and minimum wattage of each resistor.

2.9 A 6-Ω resistor is added in parallel with the combination of Prob. 2.8. Find the voltage and current for this resistor.

2.10 Find the equivalent resistance of 10 1-kΩ resistors connected in parallel.

2.11 A 4-Ω resistor and a 12-Ω resistor are in parallel. The parallel combination is in series with a 7-Ω resistor and a voltage source of 10 sin 2*t* V. Find the current in the 4-Ω resistor and the instantaneous power from the source.

2.12 Two 1-kΩ resistors are in series. When a resistor *R* is connected in parallel with one of them, the resistance of the combination is 1750 Ω. Find *R*.

2.13 Find R_{eq}.

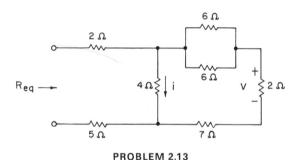

PROBLEM 2.13

2.14 A voltage of $20e^{-t}$ V is connected to the network of Prob. 2.13. Find *v*.

2.15 A current source is connected to the network of Prob. 2.13 and $i = 3 \cos t$ A. What is the value of the current source?

2.16 Find *i*.

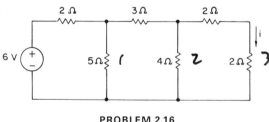

PROBLEM 2.16

2.17 Repeat Prob. 2.16 with the 6-V source replaced by a current source of 6 A.

2.18 Find *R*.

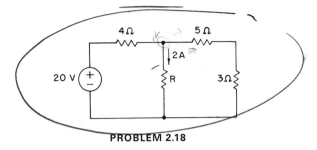

PROBLEM 2.18

2.19 Find the power delivered to the 1-Ω resistor.

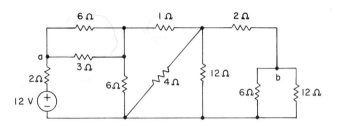

PROBLEM 2.19

2.20 Find the potential of *a* with respect to *b* in Prob. 2.19.

2.21 Find *R*.

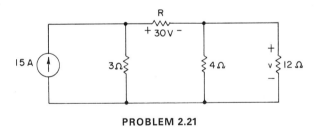

PROBLEM 2.21

2.22 Find *v* in Prob. 2.21.

2.23 Find i_1 and i_2.

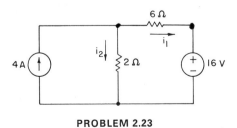

PROBLEM 2.23

2.24 Find $v(t)$ and the instantaneous power to the 3-Ω resistor.

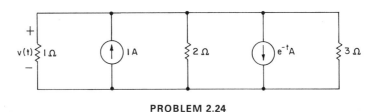

PROBLEM 2.24

2.25 Find $v(t)$.

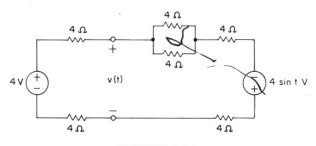

PROBLEM 2.25

2.26 A D'Arsonval meter has a full-scale current of 1 mA and a resistance of 4.9 Ω. If a 0.1-Ω parallel resistor is used in Fig. 2.25, what is i_{FS}? What voltage occurs across the meter?

2.27 A 20,000-Ω/V voltmeter has a full-scale voltage of 120 V. What current flows in the meter when measuring 90 V?

2.28 Two 10-kΩ resistors are connected in series across a 100-V source. What voltage will the voltmeter of Prob. 2.27 measure across one of the 10-kΩ resistors? Repeat for two 1-MΩ resistors in series.

2.29 The D'Arsonval meter of Prob. 2.26 is used for the ohmmeter of Fig. 2.27. What value of series resistance is required if $E = 1.5$ V? What value of unknown resistance will cause a one-quarter full-scale deflection?

2.30 Determine the color codes for resistors having the following resistance ranges: (a) 4.23–5.17 Ω, (b) 6460–7140 Ω, and (c) 3.135 –3.465 MΩ.

3

DEPENDENT SOURCES

The voltage and current sources of Chapters 1 and 2 are independent sources, as defined earlier in Sec. 1.4. We may also have *dependent* sources, which are very important in circuit theory, particularly in electronic circuits. In this chapter we shall define dependent sources and consider an additional circuit element, the operational amplifier, which may be used to obtain dependent sources.

We shall also analyze a few simple circuits containing resistors and sources, both independent and dependent. As we shall see, the analysis is very similar to that performed in Chapter 2, and the results may be used to construct a number of important circuits, such as amplifiers and inverters, which will be defined in the chapter.

3.1 DEFINITIONS

A *dependent* or *controlled voltage source* is one whose terminal voltage depends on, or is controlled by, a voltage or a current existing at some other place in the circuit. A *voltage-controlled voltage source* (VCVS) is a voltage source controlled by a voltage, and a *current-controlled voltage source* (CCVS) is one controlled by a current. The symbol for a dependent voltage source with terminal voltage v is shown in Fig. 3.1(a).

A *dependent*, or *controlled current source*, symbolized by Fig. 3.1(b), is one whose current is dependent on a voltage or a current existing elsewhere in the circuit. A *voltage-controlled current source* (VCCS) is controlled by a voltage, and a *current-controlled current source* (CCCS) is controlled by a current.

Figure 3.2 illustrates the four types of controlled sources and shows the voltage or current on which they are dependent. The quantities μ and β are dimensionless constants, commonly referred to as the voltage and current *gain*, respectively. The constants r and g have units of ohms and mhos, respectively.

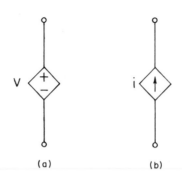

FIGURE 3.1 *(a) Dependent voltage source; (b) Dependent current source*

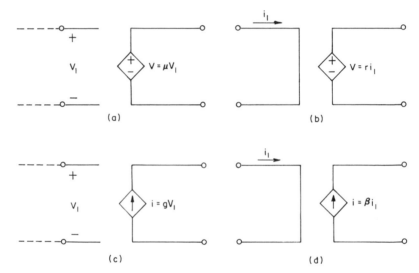

FIGURE 3.2 *(a) VCVS; (b) CCVS; (c) VCCS; (d) CCCS*

As an example, in the circuit of Fig. 3.3 we have an independent source, a dependent source, and two resistors. The dependent source is a voltage source controlled by the current i_1. The value of r for the dependent source is 0.5 V/A.

3.2 CIRCUITS WITH DEPENDENT SOURCES

Circuits containing dependent sources are analyzed in the same manner as those without dependent sources. That is, Ohm's law for resistors and Kirchhoff's voltage and current laws apply, as well as the concepts of equivalent resistance and voltage and current division.

As an illustration of the procedure, let us find the current i in the circuit of Fig. 3.4.

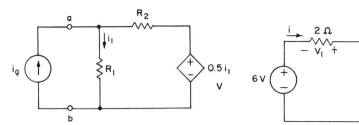

FIGURE 3.3 *Circuit containing a dependent source*   **FIGURE 3.4** *Dependent source example*

The dependent source is a voltage source controlled by the voltage v_1 as shown. Applying Kirchhoff's voltage law around the circuit, we have

$$-v_1 + 3v_1 + v_2 = 6 \tag{3.1}$$

and by Ohm's law we have

$$v_1 = -2i, \qquad v_2 = 6i \tag{3.2}$$

Using (3.2), we may eliminate v_1 and v_2 in (3.1), which results in

$$2(-2i) + 6i = 6$$

or $i = 3$ A. Thus the dependent source has complicated matters only to the extent of requiring an extra equation, the first of (3.2).

As another example, let us find the voltage v in Fig. 3.5. Applying Kirchhoff's current law to the currents leaving the top node, we have

$$-4 + i_1 - 2i_1 + \frac{v}{2} = 0 \tag{3.3}$$

Also, by Ohm's law we have

$$i_1 = \frac{v}{6} \tag{3.4}$$

Substituting (3.4) into (3.3) yields

$$-4 - \frac{v}{6} + \frac{v}{2} = 0$$

or $v = 12$ V.

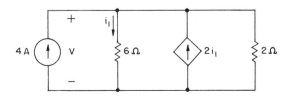

FIGURE 3.5 *Another dependent source example*

EXERCISES

3.2.1   In Fig. 3.3, let $R_1 = 3\,\Omega$, $R_2 = 1\,\Omega$, and $i_1 = 2$ A. Find i_g.   *Ans.* 7 A

3.2.2   In Ex. 3.2.1, find the resistance seen by the source looking in terminals *a-b*.
Ans. $\frac{6}{7}\,\Omega$

3.2.3   In Fig. 3.3, let $R_1 = R_2 = \frac{1}{10}\,\Omega$. Find the resistance seen by the source.
Ans. $-\frac{1}{30}\,\Omega$

3.3   OPERATIONAL AMPLIFIERS

A logical question at this point might be, How do we obtain dependent sources? One answer is that they arise as parts of equivalent circuits of electronic devices operating under certain conditions. Another answer is that they can be deliberately constructed by means of certain electronic devices in conjunction with passive elements.

We shall not be interested here in undertaking a study of electronic devices. However, there is one such device that is extremely useful in the construction of dependent sources and whose ideal mathematical model is both simple and elegant. This is the *operational amplifier*, or op amp, the ideal model of which we shall consider in this section.

The symbol that we shall use for an operational amplifier is shown in Fig. 3.6. The op amp is a multiterminal device, but for simplicity we shall show only the three terminals indicated. Terminal 1 (marked $-$) is the *inverting input terminal*, terminal 2 (marked $+$) is the *noninverting input terminal*, and terminal 3 is the *output terminal*. The purposes of the terminals that are not shown include, in general, dc power supply connections, frequency compensation terminals, and offset null terminals. We shall not be interested in discussing these other terminals here, but the interested student may find their purposes and how they are used in any op amp user's manual.

Operational amplifiers are commonly available in integrated circuit form and are normally fabricated in packages having 8 to 14 terminals and containing 1 to 4 op amps.

The operational amplifier has many characteristics that are important to designers, but the *ideal* model of the op amp has only two properties that the circuit analyst needs to know: that the currents into both input terminals are zero and that the voltage between the input terminals is zero.

As an example of a circuit with an operational amplifier, let us consider Fig. 3.7. It is desired to find the current i and the voltage v_3, considering v_g to be a known generator voltage. Let us write KVL around the loop *abca* through the source. Since the voltage across terminals *a* and *b* is zero, we have

$$v_1 - v_g = 0 \tag{3.5}$$

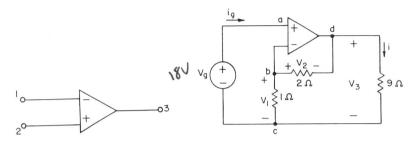

FIGURE 3.6 *Operational amplifier*   **FIGURE 3.7** *Circuit containing an op amp*

or $v_1 = v_g$. Applying KCL at node b and noting that the current into the negative terminal of the op amp is zero, we have

$$\frac{v_1}{1} + \frac{v_2}{2} = 0 \tag{3.6}$$

or $v_2 = -2v_1 = -2v_g$. Next, KVL around loop $cbdc$ through the 9-Ω resistor yields

$$-v_1 + v_2 + v_3 = 0$$

or

$$v_3 = v_1 - v_2 = 3v_g \tag{3.7}$$

Finally, Ohm's law yields

$$i = \frac{v_3}{9} = \frac{3v_g}{9} = \frac{v_g}{3}$$

Thus if $v_g = 12 \cos 10t$, for example, we have $i = 4 \cos 10t$. Also, we note that KCL is not valid at the bottom node c. This is because node c (referred to as *ground,* to be discussed in Chapter 4) is connected to the unshown power supply terminals.

As a by-product of this example, let us note that $i_g = 0$ (the current into an input lead of the op amp) and $v_3 = 3v_g$ from (3.7). Thus we may draw an equivalent circuit, insofar as v_g, i_g, v_3, and i are concerned, as shown in Fig. 3.8. Analysis of the equivalent circuit yields exactly the same v_3 and i for the same v_g and i_g as in the case of Fig. 3.7. Thus the op amp has been used to obtain a controlled source with a gain of 3. (In this case the controlled source is a VCVS, in which v_g controls v_3.)

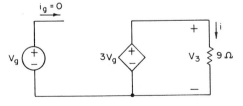

FIGURE 3.8 *Circuit equivalent to that of Fig. 3.7*

Before we leave this section, let us observe that the op amp in Fig. 3.7 is being operated in a *feedback* mode. That is, the output v_3 at node d is fed back to the inverting input terminal through the 2-Ω resistor. A practical op amp is a very high-gain device and is generally never used without feedback. In cases when the feedback is to one input terminal rather than to both, it is always to the inverting terminal, for the simple reason that otherwise the op amp will not work. We are not interested here in the reason for this, which is a consequence of the op amp's design. The interested student, in all probability, will have occasion to thoroughly study op amps and their construction in a later course in electronic devices.

3.4 AMPLIFIER CIRCUITS

Equations (3.6) and (3.7), which were the basis for the VCVS circuit of Fig. 3.8, are independent of the current i in the 9-Ω load of Fig. 3.7. This is true in general of VCVS circuits of this type. To see this, let us consider the circuit of Fig. 3.9. The voltage v_2 is the output voltage of the op amp and, as we shall see, is a function only of the input voltage v_1 and the two resistors.

Since there is no voltage across the input terminals of the op amp, we have $v_{ba} = v_1$. Also, KVL around the loop *abca* containing v_2 yields

$$-v_{ba} + v_{bc} + v_2 = 0$$

or

$$v_{bc} = v_{ba} - v_2 = v_1 - v_2$$

Therefore applying KCL at node b results in

$$\frac{v_1}{R_1} + \frac{v_1 - v_2}{R_2} = 0$$

Solving for v_2, we have

$$v_2 = \mu v_1 \tag{3.8}$$

where

$$\mu = 1 + \frac{R_2}{R_1} \tag{3.9}$$

Figure 3.9 is therefore a VCVS with gain μ. Since it is a three-terminal network (the output and input share a common terminal) and the currents are zero into the input leads of the op amp, an equivalent circuit may be drawn as in Fig. 3.10. We note that since R_1 and R_2 are nonnegative, we have $\mu \geq 1$.

A special case of Fig. 3.9 is the case $R_2 = 0$ (short circuit) and $R_1 = \infty$ (open circuit), shown in Fig. 3.11. This circuit has $\mu = 1$, or $v_2 = v_1$, and is called a *voltage follower*; that is, v_2 *follows* v_1. It is also called a *buffer amplifier* because it may be used

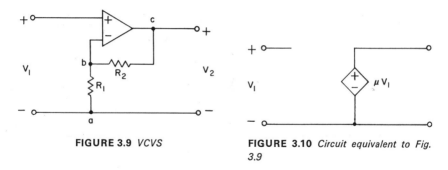

FIGURE 3.9 *VCVS*   **FIGURE 3.10** *Circuit equivalent to Fig. 3.9*

to isolate, or *buffer*, one circuit from another. (The voltages at the two pairs of terminals are the same, but no current can flow from one pair to the other.)

Let us consider next the circuit of Fig. 3.12. Applying KCL at the inverting input terminal of the op amp, we have

$$-\frac{v_1}{R_1} - \frac{v_2}{R_2} = 0$$

or

$$v_2 = -\frac{R_2}{R_1} v_1 \tag{3.10}$$

(Recall that the voltage across and the currents into the input terminals of the op amp are zero.)

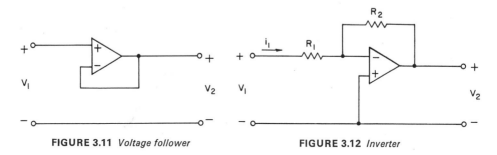

FIGURE 3.11 *Voltage follower*   **FIGURE 3.12** *Inverter*

This circuit is called an *inverter* because the polarity of v_2 is opposite that of v_1, as seen in (3.10). It is also a VCVS, but in this case the input current i_1 is not zero, being given by

$$i_1 = \frac{v_1}{R_1} \tag{3.11}$$

An equivalent circuit is shown in Fig. 3.13.

By (3.11) $v_1 = R_1 i_1$, and thus we may eliminate v_1 in Fig. 3.13, resulting in another equivalent circuit, shown in Fig. 3.14. This circuit is evidently a CCVS, since the voltage v_2 is controlled by the current i_1.

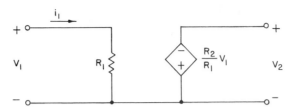

FIGURE 3.13 *Equivalent circuit of the inverter*

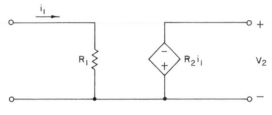

FIGURE 3.14 *CCVS*

We may also obtain dependent current sources from Fig. 3.12, which we redraw as shown in Fig. 3.15. Since there is no current into the op amp terminals, we have

$$i_2 = i_1 = \frac{v_1}{R_1} \tag{3.12}$$

Insofar as terminals 1, 2, 3, and 4 are concerned, an equivalent circuit is then that of Fig. 3.16, which is a CCCS with a gain of 1.

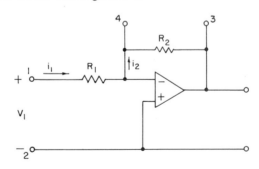

FIGURE 3.15 *Inverter circuit redrawn*

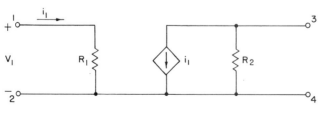

FIGURE 3.16 *CCCS*

Finally, substituting for i_1 in (3.12), we may redraw Fig. 3.16 as a VCCS, shown in Fig. 3.17. In this case, $g = 1/R_1$.

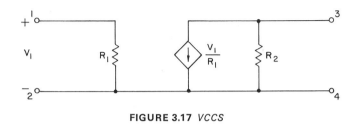

FIGURE 3.17 *VCCS*

EXERCISES

3.4.1 If in Fig. 3.7 $v_g = 18$ V, find v_1, v_2, and i.   *Ans.* 18 V, −36 V, 6 A

3.4.2 If in Fig. 3.12 $i_1 = 2$ A, $v_2 = -10$ V, and $R_1 = 2$ Ω, find R_2 and v_1.
          Ans. 5 Ω, 4 V

3.4.3 If in Fig. 3.15 the terminals 3-4 are open, find R_1 and v_{43} if $v_1 = 12$ V, $i_1 = 3$ A, and $R_2 = 2$ Ω.   *Ans.* 4 Ω, 6 V

PROBLEMS

3.1 Find the power delivered to the 8-Ω resistor in the circuit shown.

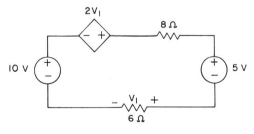

PROBLEM 3.1

3.2 Find v in the circuit shown if $R = 4$ Ω.

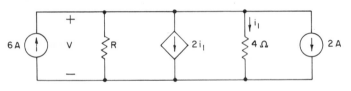

PROBLEM 3.2

3.3 Find R in the figure for Prob. 3.2 so that $v = 2$ V.

3.4 Find R in the figure for Prob. 3.2 so that $i_1 = 1$ A.

3.5 Find v_1 and i in the circuit shown if $R = 4\,\Omega$.

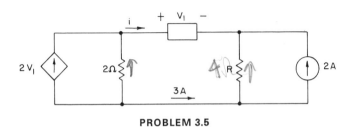

PROBLEM 3.5

3.6 Find R in the figure for Prob. 3.5 so that $v_1 = -4$ V.

3.7 Find i in the circuit shown if $i_g = 5e^{-t}$ A and $R = 4\,\Omega$.

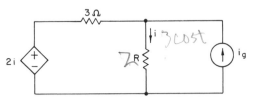

PROBLEM 3.7

3.8 Find i_g in the figure for Prob. 3.7 if $R = 2\,\Omega$ and $i = 3 \cos t$ A.

3.9 In the circuit given, show that

$$v_3 = -R_0\left(\frac{v_1}{R_1} + \frac{v_2}{R_2}\right)$$

(This circuit is called a *summer*, since the output voltage is the negative of a weighted sum of the input voltages. Note that the result is independent of the output connections at terminals *a-b*.)

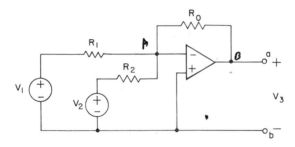

PROBLEM 3.9

3.10 In the figure for Prob. 3.9, connect a resistance of $R = 5\ k\Omega$ across terminals *a-b* and find the resulting current i_{ab} (from *a* to *b*) if $v_1 = 10\ V$, $v_2 = 5\ V$, $R_0 = 2\ k\Omega$, $R_1 = 1\ k\Omega$, and $R_2 = 2\ k\Omega$.

3.11 In the circuit given, show that regardless of the load at terminals *c-d*, we have

$$v_1 = v_2$$

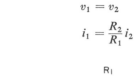

$$i_1 = \frac{R_2}{R_1} i_2$$

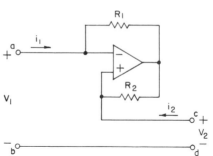

PROBLEM 3.11

3.12 In the figure for Prob. 3.11, let $R_1 = R_2$ and connect a resistor R between terminals *c-d*. Show that the resistance seen at input terminals *a-b* is $R_{ab} = -R$. (The figure for Prob. 3.11 thus converts positive resistance to negative resistance.)

3.13 Find i in the circuit shown if $R_2 = 2R_1 = 1\ k\Omega$.

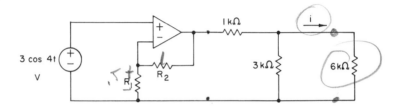

PROBLEM 3.13

3.14 Find R_2/R_1 in the figure for Prob. 3.13 so that $i = 2 \cos 4t$ mA.

3.15 Find i in the circuit shown if $R = 6\,\Omega$ and $v_g = 5 \cos 10t$ V.

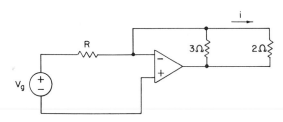

PROBLEM 3.15

3.16 In the figure for Prob. 3.15, find v_g so that $R = 12\,\Omega$ and $i = 100$ mA.

4

ANALYSIS METHODS

In Chapter 2 we considered methods of analyzing simple circuits, which we recall are those that may be described completely by a single equation. The analysis of more general circuits entails the solution of a set of simultaneous equations, as we shall see in this chapter. As an example, the reader may have noticed that Prob. 2.23 required two equations in its solution. Also, most of the circuits of Chapter 3 generally involved more than one equation, but the equations were of a type that were easily solved.

In this chapter we shall consider systematic ways of formulating and solving the equations that arise in the analysis of more complicated circuits. We shall consider two general methods, one based primarily on Kirchhoff's current law and one on Kirchhoff's voltage law. As we shall see, KCL generally leads to equations in which the unknowns are voltages, whereas KVL leads to equations in which the unknowns are currents.

It should be evident from our work in previous chapters that a complete analysis of a circuit can be performed by finding a relatively few key voltages and/or currents. For example, in a simple circuit consisting of a single loop, a key variable is the current, for if we know the current, we may find every voltage around the loop, and, of course, the current around the loop is the current in every element.

In Sec. 4.1 we shall discuss the case where the selected unknowns are voltages. Quite naturally, our choice of voltages should lead to a set of independent equations. This technique will be referred to as *nodal* analysis. In Sec. 4.5 we shall consider *mesh* analysis, in which the unknowns are currents.

In this chapter we shall discuss the techniques for selecting the voltages or currents to be found and the formulation of the circuit equations. The justification of the methods in the general case will be left until Chapter 6.

4.1 NODAL ANALYSIS

In this section we shall consider methods of circuit analysis in which voltages are the unknowns to be found. A convenient choice of voltages for many networks is the set of *node voltages*. Since a voltage is defined as existing between two nodes, it is convenient to select one node in the network to be a *reference node* or *datum node* and then associate a voltage or a potential with each of the other nodes. The voltage of each of the nonreference nodes with respect to the reference node is defined to be a *node voltage*. It is common practice to select polarities so that the node voltages are positive relative to the reference node. For a circuit containing N nodes, there will be $N - 1$ node voltages, some of which may be known, of course, if voltage sources are present.

Frequently the reference node is chosen to be the node to which the largest number of branches are connected. Many practical circuits are built on a metallic base or chassis, and usually there are a number of elements connected to the chassis, which is often then connected to the earth. The chassis may then be called *ground*, and it becomes the logical choice for the reference node. For this reason, the reference node is frequently referred to as ground. The reference node is thus at ground potential or zero potential, and the other nodes may be considered to be at some potential above zero.

Since the circuit unknowns are to be voltages, the describing equations are obtained by applying KCL at the nodes. The currents in the elements are proportional to the element voltages, which are themselves either a node voltage (if one element node is ground) or the difference of two node voltages. For example, in Fig. 4.1 the reference node is node 3 with zero or ground potential. The symbol shown attached to node 3 is the standard symbol for ground. The nonreference nodes 1 and 2 have node voltages v_1 and v_2. Thus the element voltage v_{12} with the polarity shown is

$$v_{12} = v_1 - v_2$$

The other element voltages shown are

$$v_{13} = v_1 - 0 = v_1$$

and

$$v_{23} = v_2 - 0 = v_2$$

These equations may be established by applying KVL around the loops (real or imagined).

The application of KCL results in an equation relating node voltages. Clearly, simplification in writing the resulting equations is possible when the reference node is chosen to be a node with a large number of elements connected to it. As we shall see, however, this is not the only criterion for selecting the reference node, but it is frequently the overriding one.

Since we are going to apply KCL, it seems that the simplest networks to consider

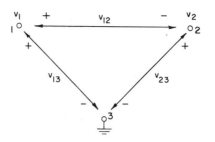

FIGURE 4.1 *Reference and nonreference nodes*

are those whose only sources are independent current sources. This is not always true, as we shall see, but we shall begin with an example of this type. In the network shown in Fig. 4.2(a), there are three nodes, dashed and numbered as shown. [This is easier to see in the redrawn version of Fig. 4.2(b).] Since there are four elements connected to node 3, we select it as the reference node, identifying it by the ground connection shown.

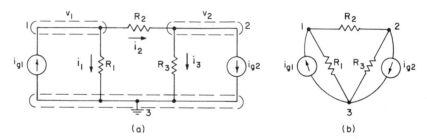

FIGURE 4.2 *Circuit containing independent current sources*

Before writing the node equations, consider the element shown in Fig. 4.3, where v_1 and v_2 are node voltages. The element voltage v is given by

$$v = v_1 - v_2$$

and thus by Ohm's law we have

$$i = \frac{v}{R} = \frac{v_1 - v_2}{R}$$

or

$$i = G(v_1 - v_2)$$

where $G = 1/R$ is the conductance. That is, the current from node 1 to node 2 is the difference of the *node voltage at node 1* and the *node voltage at node 2* divided by the resistance R, or multiplied by the conductance G. This relation will allow us to rapidly write the node equations by inspection directly in terms of the node voltages.

Now returning to the circuit of Fig. 4.2, the sum of the currents leaving node 1 must be zero, and this results in the equation

$$i_1 + i_2 - i_{g1} = 0$$

In terms of the node voltages, this equation becomes

$$G_1 v_1 + G_2(v_1 - v_2) - i_{g1} = 0$$

We could have obtained this equation directly using the procedure of the previous paragraph. Applying KCL at node 2 in a similar manner, we obtain

$$-i_2 + i_3 + i_{g2} = 0$$

or

$$G_2(v_2 - v_1) + G_3 v_2 + i_{g2} = 0$$

Again, the G's are the conductances (reciprocals of the R's).

We could have equated the sum of currents leaving the node to the sum of currents entering the node. Had we done so, the terms i_{g1} and i_{g2} would have appeared on the right-hand side:

$$G_1 v_1 + G_2(v_1 - v_2) = i_{g1}$$
$$G_2(v_2 - v_1) + G_3 v_2 = -i_{g2}$$

Rearranging these two equations results in

$$(G_1 + G_2)v_1 - G_2 v_2 = i_{g1} \tag{4.1}$$
$$-G_2 v_1 + (G_2 + G_3)v_2 = -i_{g2} \tag{4.2}$$

These equations exhibit a symmetry that may be used to write the equations in the rearranged form by inspection. In (4.1) the coefficient of v_1 is the sum of conductances of the elements connected to node 1, while the coefficient of v_2 is the negative of the conductance of the element connecting node 1 to node 2. The same statement holds for (4.2) if the numbers 1 and 2 are interchanged. Thus node 2 plays the role in (4.2) of node 1 in (4.1). That is, it is the node at which KCL is applied. In each equation the right-hand side is the current from a current source which enters the corresponding node.

In general, in networks containing only conductances and current sources, KCL applied at the kth node, with node voltage v_k, may be written as follows. In the left member the coefficient of v_k is the sum of the conductances connected to node k, and the coefficients of the other node voltages are the negatives of the conductances between those nodes and node k. The right member of the equation consists of the net current flowing into node k due to current sources.

To illustrate the process, consider Fig. 4.4, which is a portion of a circuit. The dashed lines indicate connections to nodes other than node 2 (labeled v_2). At node 2 we

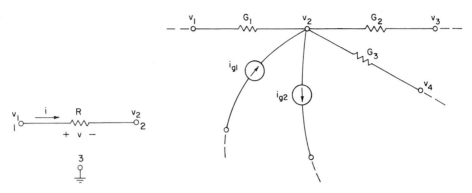

FIGURE 4.3 *Single element*   FIGURE 4.4 *Part of a circuit*

have, using the shortcut procedure,

$$-G_1v_1 + (G_1 + G_2 + G_3)v_2 - G_2v_3 - G_3v_4 = i_{g1} - i_{g2} \qquad (4.3)$$

This may be checked by applying KCL in the usual way, equating currents leaving node 2 to those entering node 2. The result is

$$G_1(v_2 - v_1) + G_2(v_2 - v_3) + G_3(v_2 - v_4) + i_{g2} = i_{g1}$$

Rearranging this result leads to (4.3).

4.2 AN EXAMPLE

To illustrate the method of nodal analysis, let us consider the circuit of Fig. 4.5. We have taken the reference node as shown and labeled the nonreference nodes as v_1, v_2, and v_3. We note that the conductances are specified for the resistors.

Since there are three nonreference nodes, there will be three nodal equations. At node v_1, using the shortcut method, we have

$$4v_1 - v_2 = 2 \qquad (4.4)$$

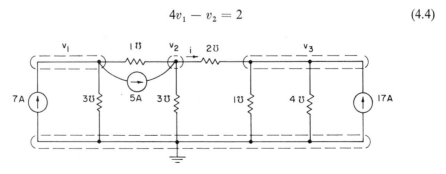

FIGURE 4.5 *Example of a circuit*

The sum of the conductances at node v_1 is $3 + 1 = 4$, the conductance between nodes 1 and 2 is 1, the conductance between nodes 1 and 3 is 0, and the net current into the node through the sources is $7 - 5 = 2$. If we had used the conventional KCL method at v_1, we would have obtained

$$1(v_1 - v_2) + 3v_1 + 5 = 7$$

which is equivalent to (4.4).

At nodes v_2 and v_3, we have

$$-v_1 + 6v_2 - 2v_3 = 5$$
$$-2v_2 + 7v_3 = 17 \tag{4.5}$$

We may solve (4.4) and (4.5) for the node voltages using any one of a variety of methods for solving simultaneous equations. Two such methods are Cramer's rule, which employs determinants, and Gaussian elimination. For the reader who is not familiar with these two methods, a complete discussion is given in Appendices A and B.

Using Cramer's rule, we first find the coefficient determinant, given by

$$\Delta = \begin{vmatrix} 4 & -1 & 0 \\ -1 & 6 & -2 \\ 0 & -2 & 7 \end{vmatrix} = 145$$

Then we have

$$v_1 = \frac{\begin{vmatrix} 2 & -1 & 0 \\ 5 & 6 & -2 \\ 17 & -2 & 7 \end{vmatrix}}{145} = \frac{145}{145} = 1 \text{ V}$$

$$v_2 = \frac{\begin{vmatrix} 4 & 2 & 0 \\ -1 & 5 & -2 \\ 0 & 17 & 7 \end{vmatrix}}{145} = \frac{290}{145} = 2 \text{ V}$$

and

$$v_3 = \frac{\begin{vmatrix} 4 & -1 & 2 \\ -1 & 6 & 5 \\ 0 & -2 & 17 \end{vmatrix}}{145} = \frac{435}{145} = 3 \text{ V}$$

Now that we have the node voltages we may completely analyze the circuit. For example, if we want the current i in the 2-℧ element, it is given by

$$i = 2(v_2 - v_3) = 2(2 - 3) = -2 \text{ A}$$

Equations (4.4) and (4.5) are symmetrical in a way that facilitates the writing of the equations. This symmetry also is present in the general case. For example, the coefficient of v_2 in the first equation is the same as that of v_1 in the second equation. Also, the coefficient of v_3 in the first equation is that of v_1 in the third equation. Finally, the coefficient of v_3 in the second equation is that of v_2 in the third equation. These results follow from the fact that the conductance between nodes 1 and 2 is that between 2 and 1, the conductance between nodes 1 and 3 is that between 3 and 1, etc.

This symmetry shows up also in the coefficient determinant Δ. Its diagonal elements, 4, 6, and 7, are the sums of the conductances connected to the three non-reference nodes, and the off-diagonal elements are symmetrical about the diagonal. The latter elements are, of course, the negatives of the conductances between nodes.

EXERCISES

4.2.1 Using nodal analysis, find v_1 and v_2 in Fig. 4.2 (a) if $R_1 = 2\,\Omega$, $R_2 = 2\,\Omega$, $R_3 = 4\,\Omega$, $i_{g1} = 2$ A, and $i_{g2} = -4$ A.   *Ans. 7 V, 10 V*

4.2.2 Repeat Ex. 4.2.1 with $i_{g2} = 0$. Check by using current division.

Ans. 3 V, 2 V

4.2.3 Using nodal analysis, find v_1, v_2, and v_3.   *Ans. 8 V, -3 V, 5 V*

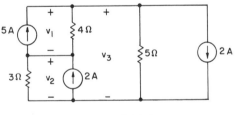

EXERCISE 4.2.3

4.3 CIRCUITS CONTAINING VOLTAGE
SOURCES

At first glance it may seem that the presence of voltage sources in a circuit complicates the nodal analysis. We can no longer write the equations using the shortcut method because we do not know the currents through the voltage sources. However, nodal analysis is no more complicated and in many cases is even easier to apply when voltage sources are present, as we shall see.

To illustrate the procedure, let us consider the circuit of Fig. 4.6. We have labeled the nonreference nodes as v_1, v_2, v_3, v_4, and v_5 and have taken the sixth node as reference, as indicated. Again, the resistors are labeled by their conductances.

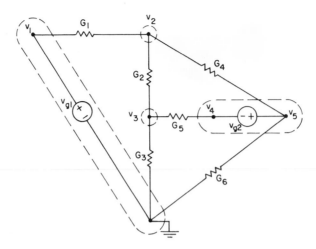

FIGURE 4.6 *Circuit containing voltage sources*

Since there are five nonreference nodes, we need five equations. Without writing any KCL equations we may note that we have, by inspection,

$$v_1 = v_{g1}$$
$$v_5 - v_4 = v_{g2} \tag{4.6}$$

We thus need only three KCL equations. To be systematic and at the same time eliminate the need to know the currents in the voltage sources, let us enclose the voltage sources by dashed lines as shown in Fig. 4.6. We may think of these surfaces as *generalized* nodes or, as some authors prefer, *super* nodes. We recall that KCL holds for such a generalized node as well as for an ordinary node. We have, therefore, two generalized nodes and two regular nodes, labeled v_2 and v_3, a total of four nodes. Thus we need only three KCL equations, which together with (4.6) constitute the required set of five equations in the node voltages.

To complete the formulation of the nodal equations, let us apply KCL at nodes v_2 and v_3 and the generalized node containing v_{g2}. The first two are obtained as before, resulting in

$$(G_1 + G_2 + G_4)v_2 - G_1 v_1 - G_2 v_3 - G_4 v_5 = 0$$
$$(G_2 + G_3 + G_5)v_3 - G_2 v_2 - G_5 v_4 = 0 \tag{4.7}$$

Finally, equating to zero the currents leaving the generalized node, we have

$$G_4(v_5 - v_2) + G_5(v_4 - v_3) + G_6 v_5 = 0 \tag{4.8}$$

The circuit is analyzed by simultaneously solving (4.6), (4.7), and (4.8).

As another example, let us consider the circuit of Fig. 4.7, which contains an independent voltage source and a dependent current source. Anticipating that the

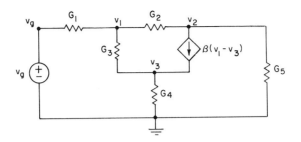

FIGURE 4.7 *Circuit with independent and dependent sources*

presence of the voltage source reduces the number of unknowns by one, we have labeled the node at the top left v_g, the reference node being chosen as indicated. The unknown node voltages are v_1, v_2, and v_3. Applying KCL at these nodes, we have

$$(G_1 + G_2 + G_3)v_1 - G_2v_2 - G_3v_3 - G_1v_g = 0$$
$$-G_2v_1 + (G_2 + G_5)v_2 + \beta(v_1 - v_3) = 0$$
$$-G_3v_1 + (G_3 + G_4)v_3 = \beta(v_1 - v_3)$$

These equations may now be solved for the unknown node voltages (v_g is considered to be known). We note that the presence of the dependent source has destroyed the symmetry that was present in the equations of the previous examples. This is true in the general case involving dependent sources.

As a final example, let us consider Fig. 4.8, which contains independent and dependent current and voltage sources. To simplify matters we have chosen the reference node as shown and labeled the nonreference nodes, which in this case may all be expressed in terms of only the two unknowns v_x and i_y. (We have used Ohm's

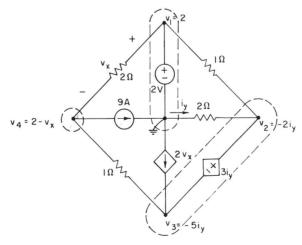

FIGURE 4.8 *More complex circuit*

law to obtain v_2, and v_3 was given by $v_2 - 3i_y = -5i_y$.) We therefore need only two equations. Since we have three nodes, the regular node and the two generalized nodes, there are precisely two independent nodal equations.

At node v_4 we have

$$-\frac{v_x}{2} + 9 + \frac{2 - v_x - (-5i_y)}{1} = 0$$

and at the generalized node containing the dependent voltage source we have

$$\frac{-2i_y - 2}{1} - i_y - 2v_x + \frac{-5i_y - (2 - v_x)}{1} = 0$$

These equations simplify to

$$-3v_x + 10i_y = -22$$
$$-v_x - 8i_y = 4$$

which have solution

$$v_x = 4 \text{ V}, \quad i_y = -1 \text{ A}$$

We may now find all the currents and voltages in the circuit.

EXERCISES

4.3.1   Using nodal analysis, find v_1 if the element x is a 12-V independent voltage source with the positive terminal at the top.   *Ans.* -4 V

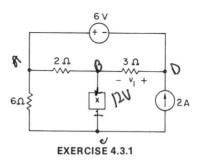

EXERCISE 4.3.1

4.3.2   Repeat Ex. 4.3.1 if element x is a 0.5-A independent current source directed downward.   *Ans.* -3 V

4.3.3   Repeat Ex. 4.3.1 if element x is a dependent voltage source of $6\,v_1$ V with the positive terminal at the top.   *Ans.* -1 V

4.4 CIRCUITS CONTAINING OP AMPS

Nodal analysis very often is the best method of analysis when a circuit contains op amps, because in electronic circuits the reference node is usually shown as grounded and all other elements connected to the reference node are often shown individually grounded. Thus the nodes are easily identified for the nodal method, but the loops are not so easily visualized. Therefore a method based on loops, such as we shall consider in the next section, is not so easy to apply. Also, quite often, only a relatively few nodal equations are required.

As an example, let us consider the VCVS of Fig. 3.9, redrawn as shown in Fig. 4.9. The reference node is shown as grounded, so that the voltages v_1 and v_2 of Fig. 3.9 are now node voltages, as shown. Since the voltage is zero between the input terminals of the op amp and the currents are zero into the input terminals, we see that $v_3 = v_1$ and that the node equation written at node v_3 is

$$\frac{v_1}{R_1} + \frac{v_1 - v_2}{R_2} = 0 \tag{4.9}$$

From this we find

$$v_2 = \left(1 + \frac{R_2}{R_1}\right)v_1 = \mu v_1 \tag{4.10}$$

which is the result given earlier in Sec. 3.4.

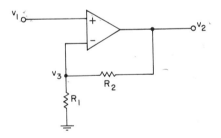

FIGURE 4.9 *VCVS*

As another example, let us consider the circuit of Fig. 4.10. We shall take the ground to be the zero voltage reference point so that the inverting input terminal of the op amp is also at zero potential, as labeled. We shall consider the voltage v_1 to be a known input voltage and solve for the output voltage v_2. At node v_3, KCL yields

$$(G_1 + G_2 + G_3 + G_4)v_3 - G_1v_1 - G_4v_2 = 0$$

using the shortcut procedure of Sec. 4.1. The sum of the currents entering the inverting input node of the op amp is given by

$$G_3v_3 + G_5v_2 = 0$$

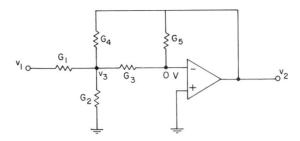

FIGURE 4.10 *Circuit containing an op amp*

Eliminating v_3 from these two equations results in

$$(G_1 + G_2 + G_3 + G_4)\left(-\frac{G_5 v_2}{G_3}\right) - G_1 v_1 - G_4 v_2 = 0$$

from which we have

$$v_2 = \frac{-G_1 G_3 v_1}{G_5(G_1 + G_2 + G_3 + G_4) + G_3 G_4}$$

It is always fruitful to write nodal equations at the inverting input nodes of the op amps, as we have done in this example. However, one generally avoids writing nodal equations at output nodes of op amps because it is difficult to find the current out of an op amp. There is no current into the input terminals, but because there are other terminals not shown, the output terminal carries a current.

EXERCISES

4.4.1 Solve Ex. 3.4.1 using nodal analysis.

4.4.2 Solve Prob. 3.9 using nodal analysis.

4.4.3 Find i in the circuit shown.   *Ans.* $6 \cos 4t$ A

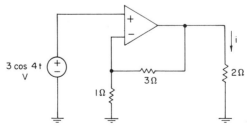

EXERCISE 4.4.3

4.5 MESH ANALYSIS

In the nodal analysis of the previous sections we applied KCL at the nonreference nodes of the circuit. We shall now consider a method, known as *mesh analysis*, or *loop analysis*, in which KVL is applied around certain closed paths in the circuit. As we shall see, in this case the unknowns generally will be currents.

We shall restrict ourselves in this chapter to *planar* circuits, by which we mean circuits that can be drawn on a plane surface in a way such that no element crosses any other element. In this case, the plane is divided by the elements into distinct areas, in the same way that the wooden or metal partitions in a window distinguish the window panes. The closed boundary of each area is called a *mesh* of the circuit. Thus a mesh is a special case of a *loop*, which we consider to be a closed path of elements in the circuit passing through no node or element more than once. In other words, a mesh is a loop that contains no elements within it.

As an example, the circuit of Fig. 4.11 is planar and contains three meshes, identified by the arrows. Mesh 1 contains the elements R_1, R_2, R_3, and v_{g1}; mesh 2 contains R_2, R_4, v_{g2}, and R_5; and mesh 3 contains R_5, v_{g2}, R_6, and R_3.

In the case of nonplanar circuits (that is, those that are not planar), we cannot define meshes. Thus in the analysis using KVL, the closed paths are loops. The procedure is the same, of course, but the equations are not as easily formulated. We shall consider this more general case in detail in Chapter 6.

As an example in which KVL is applied, let us consider the two-mesh circuit of Fig. 4.12. The element currents are I_1, I_2, and I_3. Applying KVL around the first mesh (containing v_{g1}), we have

$$R_1 I_1 + R_3 I_3 = v_{g1} \tag{4.11}$$

Similarly, around the other mesh we have

$$R_2 I_2 - R_3 I_3 = -v_{g2} \tag{4.12}$$

We define a *mesh current* as the current which flows around a mesh. The mesh

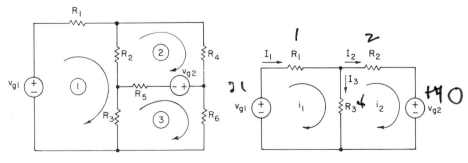

FIGURE 4.11 *Planar circuit with three meshes*   **FIGURE 4.12** *Circuit with two meshes*

current may constitute the entire current in an element of the mesh, or it may be only a portion of the element current. For example, in Fig. 4.12 the currents i_1 and i_2 are mesh currents, with the directions as shown. The element current is the mesh current in R_1 and R_2, but the element current in R_3 is the composite of two mesh currents.

In general, element currents are algebraic sums of mesh currents. This is illustrated in Fig. 4.12 since the element current in R_1 is

$$I_1 = i_1$$

that in R_2 is

$$I_2 = i_2$$

and that in R_3, by KCL, is

$$I_3 = I_1 - I_2 = i_1 - i_2$$

Using these results, we may rewrite (4.11) and (4.12) as

$$R_1 i_1 + R_3(i_1 - i_2) = v_{g1}$$
$$R_2 i_2 - R_3(i_1 - i_2) = -v_{g2}$$

(4.13)

These are the *mesh equations* of the circuit.

There is also a shortcut method of writing mesh equations which is similar to the shortcut nodal method of Sec. 4.1. Rearranging (4.13) in the form

$$(R_1 + R_3)i_1 - R_3 i_2 = v_{g1}$$
$$-R_3 i_1 + (R_2 + R_3)i_2 = -v_{g2}$$

we note that in the first equation, corresponding to the first mesh, the coefficient of the first current is the sum of the resistances in the first mesh, and the coefficients of any other mesh current is the negative of the resistance common to that mesh and the first mesh. The right member of the first equation is the algebraic sum of the voltage sources driving the first mesh current in its assumed direction. Replacing the word *first* by the word *second* everywhere it appears in these last two sentences will describe the second equation, and so forth. This shortcut procedure is a consequence of selecting all the mesh currents in the same direction (clockwise in Fig. 4.12) and writing KVL as the meshes are traversed in the directions of the currents. Of course, the method applies only when no sources are present except independent voltage sources.

As another example, let us return to Fig. 4.11 and define i_1, i_2, and i_3 as the mesh currents shown in meshes 1, 2, and 3, respectively. Then applying the shortcut method to mesh 1, we have

$$(R_1 + R_2 + R_3)i_1 - R_2 i_2 - R_3 i_3 = v_{g1}$$

This result may be checked by applying KVL to mesh 1, resulting in

$$R_1 i_1 + R_2(i_1 - i_2) + R_3(i_1 - i_3) = v_{g1}$$

The two results are evidently the same.

Applying KVL to meshes 2 and 3 yields, in the same manner,

$$-R_2 i_1 + (R_2 + R_4 + R_5)i_2 - R_5 i_3 = -v_{g2}$$
$$-R_3 i_1 - R_5 i_2 + (R_3 + R_5 + R_6)i_3 = v_{g2}$$

The analysis is completed by solving the three mesh equations for the mesh currents.

The same symmetry is present in the mesh equations as was noted in the nodal equations. If Cramer's rule is to be used for solving for the currents, then the coefficient determinant to be calculated is

$$\Delta = \begin{vmatrix} R_1 + R_2 + R_3 & -R_2 & -R_3 \\ -R_2 & R_2 + R_4 + R_5 & -R_5 \\ -R_3 & -R_5 & R_3 + R_5 + R_6 \end{vmatrix}$$

The diagonal elements are the sum of the resistances in the meshes, and the off-diagonal elements are the negatives of the resistances common to the meshes corresponding to the row and column of the determinant. That is, $-R_2$ in row 1, column 2 or in row 2, column 1 is the negative of the resistance common to meshes 1 and 2, etc. Thus the determinant is symmetric about its diagonal.

EXERCISES

4.5.1 Using mesh analysis, find i_1 and i_2 in Fig. 4.12 if $R_1 = 1\ \Omega$, $R_2 = 2\ \Omega$, $R_3 = 4\ \Omega$, $v_{g1} = 21$ V, and $v_{g2} = 14$ V.   *Ans.* 5 A, 1 A

4.5.2 Repeat Ex. 4.5.1 with $v_{g2} = 0$. Check by using voltage division.

                                          Ans. 9 A, 6 A

4.5.3 Using mesh analysis, find i_1 and i_2.   *Ans.* 8 A, 5 A

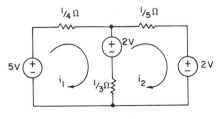

EXERCISE 4.5.3

4.6 CIRCUITS CONTAINING CURRENT SOURCES

As in the case of nodal analysis of circuits with voltage sources, mesh analysis is easier if there are current sources present. To illustrate this point, let us consider the circuit of Fig. 4.13, which has two current sources and a voltage source.

With the mesh currents i_1, i_2, and i_3 chosen as shown, it is clear that we need three independent equations. Not all of these, however, have to be mesh equations. The presence of the two current sources provides us with two constraints which we may obtain by inspection:

$$i_2 = -i_{g1}$$
$$i_3 - i_1 = i_{g2} \qquad (4.14)$$

We need, therefore, only one more equation. Since it will have to come from KVL, we need to select a closed path in which all the voltages are easily obtained. That is, we need to avoid the current sources since their voltages are not readily obtained.

If we imagine for a moment that the two current sources are removed, that is, opened, then we shall have two less meshes. But we already have two equations so there will still be enough meshes left for the required number of equations. Moreover, the loops left (they may not be meshes) will have only resistors and voltage sources in them, and therefore KVL is easily applied. We must stress that we are not taking the current sources out. We are only imagining them out for a moment in order to *locate* the loops around which KVL is to be applied.

Returning to Fig. 4.13 and imagining the current sources open for a moment, we see that we have only one loop left, namely the loop containing v_{g3}, R_1, R_2, and R_3. Applying KVL to this loop, we have our third equation,

$$R_1(i_1 - i_2) + R_2(i_3 - i_2) + R_3 i_3 = v_{g3} \qquad (4.15)$$

The analysis of the circuit can now be completed by solving (4.14) and (4.15).

As a final example, let us apply loop analysis to the circuit of Fig. 4.8, which was analyzed previously by the nodal method. The circuit is redrawn in Fig. 4.14. There are four meshes and two constraints to be satisfied by the controlling variables v_x and i_y. We shall need, therefore, six equations.

We may choose mesh currents and obtain their relations to the two current sources as we did earlier in Fig. 4.13. However, to illustrate the use of loop currents and also to simplify the resulting equations, we have chosen as the unknowns currents i_1, i_2, i_3, and i_4, as shown in the circuit. The selection we have made results in a single-loop current through the independent current source and through the element whose current controls a dependent source. Thus the constraints are simple equations. Before

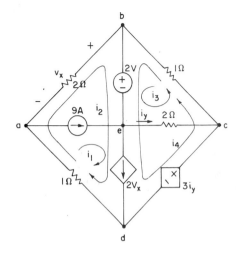

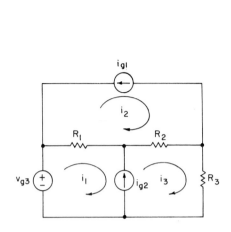

FIGURE 4.13 *Circuit with two current sources and one voltage source*

FIGURE 4.14 *More complex circuit*

applying KVL, we may write, by inspection of the circuit,

$$i_1 = 9$$

$$i_1 + i_2 + i_4 = 2v_x$$

$$i_2 = -\frac{v_x}{2}$$

$$i_3 = i_y$$

From these results we may express all the loop currents in terms of v_x and i_y. The result is

$$i_1 = 9$$

$$i_2 = -\frac{v_x}{2}$$

$$i_3 = i_y \qquad (4.16)$$

$$i_4 = \frac{5v_x}{2} - 9$$

As yet we have not written a single loop equation. We need two more equations since there are basically two unknowns, v_x and i_y. By imagining for the moment the current sources as open, we see the two loops to which we shall apply KVL. They are *abecda* and *bceb*. The respective loop equations are

$$2i_2 + 2 + 2i_3 + 3i_y + 1(i_1 + i_2) = 0$$

and

$$1(i_3 + i_4) + 2 + 2i_3 = 0$$

Substituting (4.16) into these equations, we have

$$-\frac{3v_x}{2} + 5i_y = -11$$

$$\frac{5v_x}{2} + 3i_y = 7$$

The solution is given by

$$v_x = 4 \text{ V}, \qquad i_y = -1 \text{ A}$$

which checks with the result obtained earlier in Sec. 4.3. Clearly the nodal analysis performed earlier was a simpler method.

The loop currents may now be found from (4.16), completing the analysis. Any other current or voltage in the circuit may be readily obtained also.

EXERCISES

4.6.1 Using the methods of this section, find i_1, i_2, and i_3 in Fig. 4.13 if $R_1 = 1 \Omega$, $R_2 = 3 \Omega$, $R_3 = 4 \Omega$, $i_{g1} = 2$ A, $i_{g2} = 8$ A, and $v_{g3} = 24$ V.

Ans. -5 A, -2 A, 3 A

4.6.2 Repeat Ex. 4.6.1 if $v_{g3} = 0$. *Ans.* -8 A, -2 A, 0

4.6.3 Repeat Ex. 4.6.1 if $i_{g1} = 0$. *Ans.* -4 A, 0, 4 A

4.7 DUALITY*

The reader may have noticed a similarity in certain pairs of network equations which we have considered so far. For example, Ohm's law may be stated as

$$v = Ri \tag{4.17}$$

or

$$i = Gv \tag{4.18}$$

In the second case we have solved for i, of course, and used the definition $G = 1/R$. Another way of looking at these equations is to note that the second may be obtained from the first by replacing v by i, i by v, and R by G. In like manner, the first may be obtained from the second by replacing i by v, v by i, and G by R.

Similarly, in the case of *series* resistances $R_1, R_2, \ldots, R_n$, the equivalent resistance was shown in Sec. 2.3 to be

$$R_s = R_1 + R_2 + \ldots + R_n \tag{4.19}$$

and for *parallel* conductances $G_1, G_2, \ldots, G_n$, we saw in Sec. 2.4 that the equivalent conductance was

$$G_p = G_1 + G_2 + \ldots + G_n \qquad (4.20)$$

It is clear that one of these equations may be obtained from the other by interchanging resistances and conductances and the subscripts s and p, i.e., series and parallel.

There is thus a definite *duality* between resistance and conductance, current and voltage, and series and parallel. We acknowledge this by defining these quantities as *duals* of each other. That is, R is the dual of G, i is the dual of v, series is the dual of parallel, and vice versa in each case.

Another simple case of dual equations is

$$v = 0 \qquad (4.21)$$

and its dual

$$i = 0 \qquad (4.22)$$

In the general case, an element described by (4.21) is a *short circuit*, and one described by (4.22) is an *open circuit*. Thus short circuits and open circuits are duals. There are also other dual quantities, as we shall see later.

Every equation in circuit theory has a dual, obtained by replacing each quantity in the equation by its dual quantity. If one equation describes a planar circuit, then the other equation describes the dual of the circuit, or the *dual* circuit. (As the reader may see in a later course, nonplanar circuits do not have duals.) For example, consider the circuit of Fig. 4.15. The mesh equations are given by

$$\begin{aligned}
(R_1 + R_2)i_1 - R_2i_2 &= v_g \\
-R_2i_1 + (R_2 + R_3)i_2 &= 0
\end{aligned} \qquad (4.23)$$

To obtain the dual of (4.23), we simply replace the R's by G's, the i's by v's, and v by i. The result is

$$\begin{aligned}
(G_1 + G_2)v_1 - G_2v_2 &= i_g \\
-G_2v_1 + (G_2 + G_3)v_2 &= 0
\end{aligned} \qquad (4.24)$$

These are the nodal equations of a circuit having two nonreference node voltages, v_1 and v_2, three conductances, and an independent current source i_g. From our shortcut procedure for nodal equations we see that G_1 and G_2 are connected to the first node, G_2 is common to the two nodes, G_2 and G_3 are connected to the second node, and i_g enters the first node. Since G_1 and i_g are not connected to the second node and G_3 is not connected to the first node, these elements are connected to the reference node. Such a circuit is shown in Fig. 4.16 and is a dual of Fig. 4.15.

Figure 4.15 may be described as v_g in series with R_1 and the parallel combination of R_2 and R_3. Replacing the quantities in this statement by their duals, we see correctly

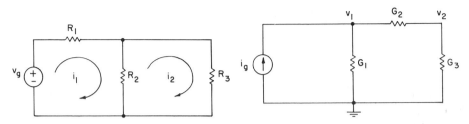

FIGURE 4.15 *Two-mesh circuit*   **FIGURE 4.16** *Dual of Fig. 4.15*

that Fig. 4.16, a dual circuit, may be described as i_g in parallel with G_1 and the series combination of G_2 and G_3.

In this example we note that nodes are duals of meshes and vice versa. This is true in general because of the duality between the mesh currents and the nonreference node voltages. The reference node is the dual of the boundary of the region *outside* the circuit, or what we might call the *outer* mesh.

This last duality (between nodes and meshes) gives us a procedure for finding a dual of a given circuit. We may place one node of the dual network to be obtained inside each mesh, thereby assuring us of the correspondence between each nonreference node and a mesh. Then we may place the reference node outside the circuit (corresponding to the outer mesh). Connecting these nodes together by elements drawn through the elements of the original circuit assures us of correspondence between elements. If these last elements drawn are the duals of those they cross, then the circuit obtained is a dual of the original circuit.

As an example, the circuit of Fig. 4.15 is redrawn using solid lines in Fig. 4.17. The dashed circuit, superimposed in accordance with the dual circuit procedure, is its dual. It should be clear that the latter circuit is the same as that of Fig. 4.16.

In the case of independent or dependent sources, the polarity of the dual source is specified when the dual circuit is obtained from the dual equations, as was the case in Fig. 4.16. However, in the geometric method of Fig. 4.17, the polarity of i_g cannot be determined since the mesh current directions are not shown and thus are completely arbitrary.

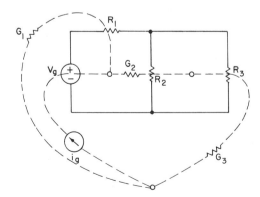

FIGURE 4.17 *Two dual circuits*

If the mesh currents or node voltages are shown in the original circuit, then the polarity of a source obtained in the dual circuit may be determined from the concept of *driving* a node or a mesh. We shall say that a mesh is *driven* by a source in it if the polarity of the source is such as to drive the current in the direction of the mesh current. A node is driven by a source connected to it if the polarity of the source is such as to apply voltage to the node, i.e., send current toward the node. Thus the dual source will drive or not drive a mesh or a node in accordance with whether the source it crosses in the original circuit drives or does not drive the corresponding node or mesh. This is consistent, as it must be, with the two sets of dual equations of the circuits.

In general, when we have analyzed one circuit, the numerical values of its voltages and currents are the same as those of their duals in the dual circuit. When we solve one circuit, therefore, we have really solved two. As an illustration, the numerical values of i_1 and i_2 in Fig. 4.15 are, respectively, the same as those of v_1 and v_2 in Fig. 4.16.

As a final note in this chapter, we shall not attempt to give a dual of an op amp. We may, of course, obtain duals of circuits containing dependent sources which are equivalent to circuits with op amps.

EXERCISE

4.7.1 Find the dual of the circuit in the figure for Ex. 4.5.3.

PROBLEMS

4.1 Using nodal analysis, find the current i in the circuit shown.

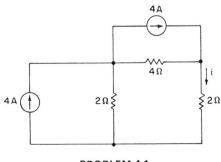

PROBLEM 4.1

4.2 Using nodal analysis, find the voltage v in the circuit shown.

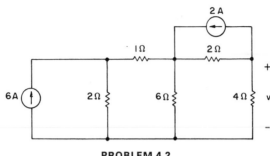

PROBLEM 4.2

4.3 Repeat Prob. 4.2 if the 6-A source is replaced by an independent voltage source having a terminal voltage of 10 V with the positive terminal at the top.

4.4 Using nodal analysis, find v in the circuit shown.

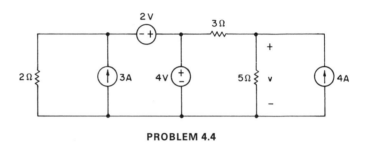

PROBLEM 4.4

4.5 Using nodal analysis, find v in the circuit shown.

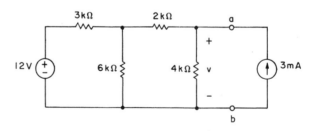

PROBLEM 4.5

4.6   Using nodal analysis, find v in the circuit shown.

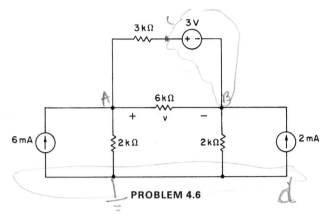

PROBLEM 4.6

4.7   Using nodal analysis, find v in the circuit shown.

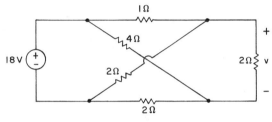

PROBLEM 4.7

4.8   Using nodal analysis, find the power delivered to the 4-Ω resistor in the circuit shown.

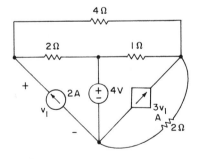

PROBLEM 4.8

4.9   Repeat Prob. 4.8 for the circuit shown.

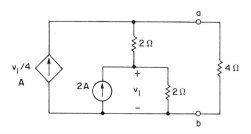

PROBLEM 4.9

4.10   Using nodal analysis, find v_1 in the circuit shown.

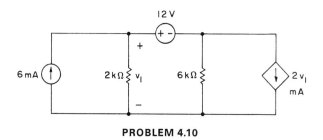

PROBLEM 4.10

4.11   Solve Prob. 3.13 using nodal analysis.

4.12   Solve Prob. 4.1 using mesh analysis.

4.13   Solve Prob. 4.2 using mesh analysis.

4.14   Solve Prob. 4.6 using mesh analysis.

4.15   Solve Prob. 4.9 using mesh or loop analysis.

4.16   Using mesh or loop analysis, find v in the circuit shown. The sources x, y, and z are independent current sources with arrows pointing upward and currents of 10, 15, and 5 A, respectively.

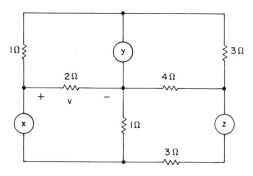

PROBLEM 4.16

4.17 Solve Prob. 4.16 if source *z* is changed to a 12-V voltage source with the positive terminal at the top.

4.18 Solve Prob. 4.16 using any method if *x*, *y*, and *z* are independent voltage sources with the positive terminals at the top and terminal voltages of 39, 10, and 20 V, respectively.

4.19 Find *i* in the circuit shown.

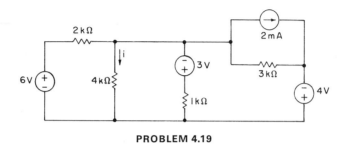

PROBLEM 4.19

4.20 Find *v* in the circuit shown.

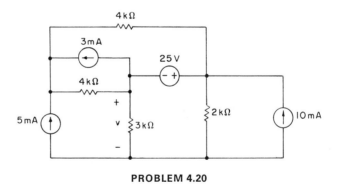

PROBLEM 4.20

4.21 Find *i* in the circuit shown.

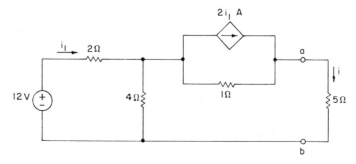

PROBLEM 4.21

4.22   Find v_1 in the circuit shown.

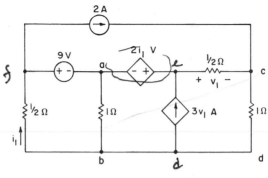

PROBLEM 4.22

4.23   In the circuit shown, find v_2 in terms of R and I_g, and show that if $R = 10\ \text{k}\Omega$, then

$$v_2 \approx -5 \times 10^4 I_g$$

(Note: "$\approx$" means approximately equal to.)

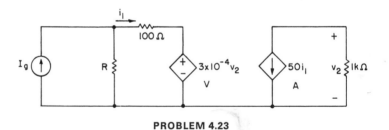

PROBLEM 4.23

4.24   Find v_0 in terms of the node voltages v_1, v_2, and v_3 and the resistances in the circuit shown. (This circuit is a summer, like that of Prob. 3.9, with an additional voltage source.)

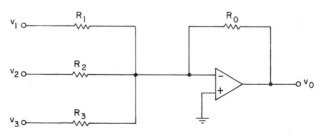

PROBLEM 4.24

4.25 Find v_2 in terms of v_1 in the circuit shown.

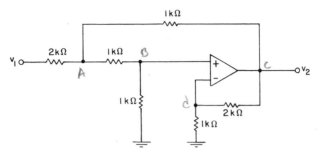

PROBLEM 4.25

4.26 Find v_2 in terms of v_1 and the conductances in the circuit shown.

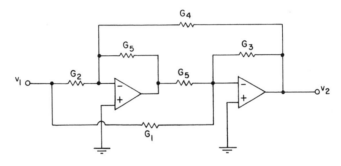

PROBLEM 4.26

4.27 Find the input resistance R_{in}, in terms of the other resistances, in the circuit shown.

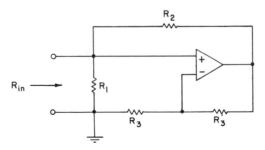

PROBLEM 4.27

*4.28 Draw the dual of the circuit of Prob. 4.1, and find v in the dual circuit corresponding to the unknown i in the original circuit.

*4.29 Draw a circuit and its dual if the mesh equations of the circuit are given by

$$10i_1 - 2i_2 = 4$$
$$-2i_1 + 8i_2 - i_3 = 0$$
$$-i_2 + 11i_3 = -6$$

*4.30 Draw the dual of the circuit of Prob. 4.16. Identify i in the dual circuit corresponding to v in the original circuit and show that they have the same numerical values.

5

NETWORK THEOREMS

In the previous chapters we have considered fairly straightforward methods of analyzing circuits. However, in many cases the analysis can be shortened considerably by the use of certain network theorems. For example, if we are interested only in what happens to one particular element in a circuit, it may be possible by means of a network theorem to replace the rest of the circuit by an equivalent and simpler circuit.

In this chapter we shall introduce a number of network theorems and illustrate their use in simplifying the analysis of certain circuits. The theorems are applicable, in general, to circuits that are *linear*, a term which we also shall discuss.

Finally, we shall use the network theorems as a motivation for introducing *practical* sources, as distinguished from the ideal sources considered thus far. As we would expect, practical sources are capable of delivering only a finite amount of power, and we shall devote a section to determining the maximum power that can be delivered by a given source.

5.1 LINEAR CIRCUITS

In Chapter 2 we defined a linear resistor as one that satisfied Ohm's law

$$v = Ri$$

and we considered circuits that were made up of linear resistors and independent sources. We defined dependent sources in Chapter 3 and analyzed circuits containing both independent and dependent sources in Chapter 4. The dependent sources that we considered all had describing relations of the form

$$y = kx \tag{5.1}$$

where k is a constant and the variables x and y were either voltages or currents. Clearly, Ohm's law is a special case of (5.1).

In (5.1) the variable y is proportional to the variable x, and the graph of y versus x is a straight line passing through the origin. For this reason, some authors refer to elements which are characterized by (5.1) as *linear elements*.

For our purposes we shall define a linear element in a more general way, which includes (5.1) as a special case. If x and y are variables, such as voltages and currents, associated with a two-terminal element, then we shall say that the element is *linear* if multiplying x by a constant K results in the multiplication of y by the same constant K. This feature is called the *proportionality property* and evidently holds for (5.1) since

$$Ky = k(Kx)$$

Thus not only is an element described by (5.1) linear, but, in addition, elements described by relations of the forms

$$\frac{dy}{dt} = ax, \qquad y = b\frac{dx}{dt} \tag{5.2}$$

are also linear if a and b are nonzero constants.

The ideal op amp is a multiterminal element and is described by more than one equation. However, we shall use the op amp only in a feedback mode, as stated in Chapter 3, and in this case the equivalent circuits which result consist of linear elements with describing relations like (5.1). Therefore we may add the ideal op amp to our list of linear elements.

We shall define a *linear circuit* as one containing only independent sources and/or linear elements. As examples, all the circuits we have considered thus far are linear circuits.

The describing equations of a linear circuit are obtained by applying KVL and KCL, and therefore they contain sums of multiples of voltages or currents. For example, a loop equation is of the form

$$a_1 v_1 + a_2 v_2 + \ldots + a_n v_n = f \tag{5.3}$$

where f is the algebraic sum of the voltages of the independent sources in the loop, the v's are the voltages of the remaining loop elements, and the a's are 0 or ± 1. To illustrate, KVL around the loop shown in the circuit of Fig. 5.1 yields

$$v_1 + v_2 - v_3 = v_{g1} - v_{g2}$$

In this case, $a_1 = a_2 = 1$; $a_3 = -1$; the other a's, if any, are all zero; and

$f = v_{g1} - v_{g2}$. We also have

$$v_1 = 2i_1, \qquad v_2 = 5i_2, \qquad v_3 = 3i_6$$

where i_6 is a current elsewhere in the circuit.

The proportionality property of a single linear element also holds for a linear circuit in the sense that if all the independent sources of the circuit are multiplied by a constant K, then all the currents and voltages of the remaining elements are multiplied by this same constant K. This is easily seen in (5.3), which multiplied through by K becomes

$$a_1 K v_1 + a_2 K v_2 + \ldots + a_n K v_n = Kf$$

The right member is the consequence of multiplying the independent sources by K, and for equality to still hold all the v's are multiplied by K. Since the elements are linear, multiplying their voltages by K multiplies their currents by K.

To illustrate the proportionality property, let us find the current i in Fig. 5.2. Since by KCL the current to the right in the 2-Ω resistor is $i - i_{g2}$, KVL around the left mesh yields

$$2(i - i_{g2}) + 4i = v_{g1} \tag{5.4}$$

from which

$$i = \frac{v_{g1}}{6} + \frac{i_{g2}}{3} \tag{5.5}$$

If $v_{g1} = 18$ V and $i_{g2} = 3$ A, then $i = 3 + 1 = 4$ A. If we double v_{g1} to 36 V and i_{g2} to 6 A, then i is doubled to 8 A.

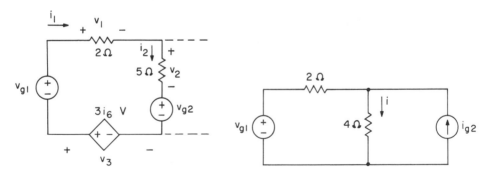

FIGURE 5.1 *Loop of a linear circuit*   **FIGURE 5.2** *Linear circuit with two sources*

As a final example, let us illustrate another use of the proportionality relation by finding v_1 in the circuit of Fig. 5.3. Such a circuit, because it resembles a ladder, is sometimes called a *ladder* network.

We could write mesh or nodal equations, but to illustrate proportionality we shall introduce an alternative method. Let us simply *assume* a solution,

$$v_1 = 1 \text{ V}$$

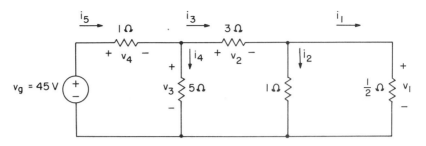

FIGURE 5.3 *Ladder network*

and see where it leads us. Referring to the figure, this assumption on v_1 gives $i_1 = 2$ A and $i_2 = 1$ A, and therefore

$$i_3 = i_1 + i_2 = 3 \text{ A}$$

Proceeding down the ladder toward the source, we have, by KVL and KCL,

$$v_2 = 3i_3 = 9 \text{ V}$$
$$v_3 = v_1 + v_2 = 10 \text{ V}$$
$$i_4 = \frac{v_3}{5} = 2 \text{ A}$$
$$i_5 = i_3 + i_4 = 5 \text{ A}$$
$$v_4 = 1(i_5) = 5 \text{ V}$$

and, finally, if the guess that $v_1 = 1$ V is correct,

$$v_g = v_3 + v_4 = 15 \text{ V}$$

Our guess was not correct, since actually $v_g = 45$ V, but in view of the laws of probability this should not surprise us. However, by the proportionality relation, if a 15-V source gives an output $v_1 = 1$ V, then our 45-V source will give three times as much, so that the correct answer is

$$v_1 = 3 \text{ V}$$

This method of assuming an answer for the output, working backwards to obtain the corresponding input, and adjusting the assumed output to be consistent with the actual input, by means of the proportionality relation, is particularly easy to apply to the ladder network. There are other circuits to which it also applies, but the ladder network is one of the most-often-encountered circuits, and the method is worthwhile for it alone.

A *nonlinear* circuit is, of course, one that is not linear. That is, it has at least one element whose terminal relation is not of the form of (5.1) or (5.2). An example is given in Ex. 5.1.4, for which it is seen that the proportionality property does not apply.

EXERCISES

5.1.1 Show that in Ex. 4.5.3 if all the source voltages are doubled, then the loop currents are doubled.

5.1.2 Show that in Ex. 4.4.3 if the source voltage is doubled, then the response i is doubled.

5.1.3 Find i_1, v_2, i_3, v_3, v_4, and i_5 in Fig. 5.3.

Ans. 6 A, 27 V, 9 A, 30 V, 15 V, 15 A

5.1.4 A circuit is made up of a voltage source v_g, a 2-Ω linear resistor, and a non-linear resistor in series. The nonlinear resistor is described by

$$v = i^2$$

where v is the voltage across the resistor and i, which is constrained to be non-negative, is the current flowing into the positive terminal. Find the current flowing out of the positive terminal of the source if (a) $v_g = 8$ V and (b) $v_g = 16$ V. Note that the proportionality property does not apply.

Ans. (a) 2 A, (b) 3.123 A

5.2 SUPERPOSITION

In this section we shall consider linear circuits with more than one input. The linearity property makes it possible, as we shall see, to obtain the responses in these circuits by analyzing only single-input circuits.

To see how this may be accomplished, let us first consider the circuit of Fig. 5.2, which was analyzed in the previous section. The output i satisfied the circuit equation (5.4), which we repeat as

$$2(i - i_{g2}) + 4i = v_{g1} \tag{5.6}$$

From the solution,

$$i = \frac{v_{g1}}{6} + \frac{i_{g2}}{3} \tag{5.7}$$

also given earlier, we see that i is made up of two components, one due to each input.

If i_1 is the component of i due to v_{g1} alone, i.e., with $i_{g2} = 0$, then by (5.6) we have

$$2(i_1 - 0) + 4i_1 = v_{g1} \tag{5.8}$$

Similarly, if i_2 is the component of i due to i_{g2} alone, i.e., with $v_{g1} = 0$, then by (5.6) we have

$$2(i_2 - i_{g2}) + 4i_2 = 0 \tag{5.9}$$

Adding (5.8) and (5.9), we have

$$2(i_1 - 0) + 4i_1 + 2(i_2 - i_{g2}) + 4i_2 = v_{g1} + 0$$

or

$$2[(i_1 + i_2) - i_{g2}] + 4(i_1 + i_2) = v_{g1} \tag{5.10}$$

Comparing this result with (5.6), we see that

$$i = i_1 + i_2$$

Also solving (5.8) and (5.9), we have

$$i_1 = \frac{v_{g1}}{6}$$

$$i_2 = \frac{i_{g2}}{3} \tag{5.11}$$

which check with the two components given in (5.7).

Alternatively, we may obtain the components of i directly from the circuit of Fig. 5.2. To find i_1 we need to make the current source i_{g2} zero. Since $i_{g2} = 0$ is the equation of an open circuit, this is accomplished by replacing the current source in the circuit by an open circuit. This operation of making a source zero is sometimes referred to, rather grimly, as "killing" the source, or making the source "dead." The resulting circuit, in this case, is shown in Fig. 5.4(a), from which it is easily seen, using Ohm's law, that i_1, the component of i due to v_{g1} alone, is given by the first equation of (5.11).

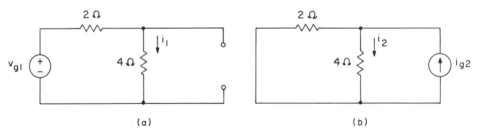

(a)   (b)

FIGURE 5.4 *Circuit of Fig. 5.2 with (a) the current source dead and (b) the voltage source dead*

To find i_2, the component of i due to i_{g2} alone, we must have $v_{g1} = 0$, the equation of a short circuit. Thus to kill a voltage source such as v_{g1}, we replace it by a short circuit, as shown in Fig. 5.4(b). Using current division it is easy to see that i_2 is given by the second equation of (5.11).

The method we have illustrated with this example is called *superposition*, because we have *superposed*, or algebraically added, the components due to each independent source acting alone to obtain the total response. The principle of superposition applies to any linear circuit with two or more sources, because the circuit equations are *linear* equations (first-degree equations in the unknowns). This may be seen from (5.3) and the nature of the *v-i* relations for the linear elements. In particular, in the case of resistive circuits, the response may be found using Cramer's rule and is clearly a sum of components, one due to each independent source alone. (Expanding the numerator determinant by cofactors of the column containing the sources clearly reveals this.)

The linearity of the circuit equations enabled us to add equations (5.8) and (5.9) and to see in (5.10) that the response was the sum of the individual responses. For example, we are able to say that

$$2i_1 + 2i_2 = 2(i_1 + i_2)$$

because these expressions are linear. However, we *cannot* say that

$$2(i_1 + i_2)^2 = 2i_1^2 + 2i_2^2$$

because these are *quadratic* and not linear expressions.

Superposition allows us to analyze linear circuits with more than one independent source by analyzing separately only single-input circuits. This is quite often advantageous since we can use network reduction properties, such as equivalent resistance and voltage division, in analyzing single-input circuits.

Formally, we state the principle of superposition as follows:

> In any linear resistive circuit containing two or more independent sources, any circuit voltage (or current) may be calculated as the algebraic sum of all the individual voltages (or currents) caused by each independent source acting alone, i.e., with all other independent sources dead.

As a second example, let us find the voltage *v* in the circuit with three independent sources, shown in Fig. 5.5. To illustrate the earlier statement that the response is a sum

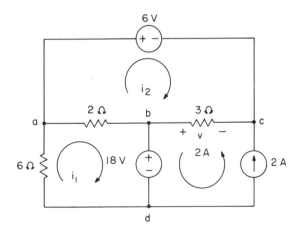

FIGURE 5.5 *Circuit with three sources*

of components, each due solely to an independent source, we shall first solve the circuit by conventional means. Then to contrast the methods we shall use superposition.

To illustrate the role each source plays in the response we shall label the sources as

$$v_{g1} = 6 \text{ V}$$
$$i_{g2} = 2 \text{ A} \tag{5.12}$$
$$v_{g3} = 18 \text{ V}$$

Assigning the mesh currents as indicated, the mesh equations are then given by

$$8i_1 - 2i_2 = -v_{g3}$$
$$-2i_1 + 5i_2 = -v_{g1} - 3i_{g2} \tag{5.13}$$

and the response v is given by

$$v = -3(i_{g2} + i_2) \tag{5.14}$$

Solving (5.13) for i_2 by Cramer's rule we have

$$i_2 = \frac{\begin{vmatrix} 8 & -v_{g3} \\ -2 & -v_{g1} - 3i_{g2} \end{vmatrix}}{\begin{vmatrix} 8 & -2 \\ -2 & 5 \end{vmatrix}} = -\frac{2}{9}v_{g1} - \frac{2}{3}i_{g2} - \frac{1}{18}v_{g3}$$

so that

$$v = \tfrac{2}{3}v_{g1} - i_{g2} + \tfrac{1}{6}v_{g3}$$

Thus we see that v, as well as i_2, is a sum of components due to the individual sources. Finally, substituting (5.12) into the expression for v, we have

$$v = 4 - 2 + 3 = 5 \text{ V} \tag{5.15}$$

Solving for v now by the method of superposition, we may write

$$v = v_1 + v_2 + v_3 \tag{5.16}$$

where v_1 is the component due to the 6-V source alone (the 2-A and 18-V sources dead), v_2 is the component due to the 2-A source alone (the two voltage sources dead), and v_3 is the component due to the 18-V source alone (the 6-V and 2-A sources dead). The circuits showing v_1, v_2, and v_3 are given in Figs. 5.6(a), (b), and (c), respectively. In Fig. 5.6(a), killing the 18-V source ties nodes b and d together; in Fig. 5.6(b), killing the two voltages ties nodes a and c together and b and d together; etc.

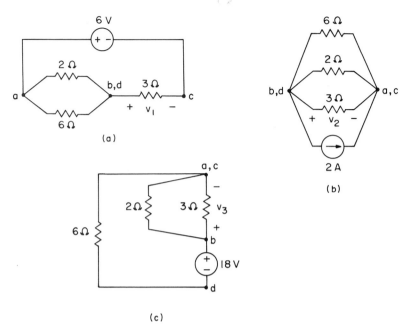

(a)

(b)

(c)

FIGURE 5.6 *Circuit of Fig 5.5 with various sources killed*

From Figs. 5.6(a), (b), and (c), it is almost trivial to obtain

$$v_1 = 4 \text{ V}$$
$$v_2 = -2 \text{ V}$$
$$v_3 = 3 \text{ V}$$

which are the values given in (5.15).

To illustrate superposition when there is a dependent source present, let us consider the circuit of Fig. 5.7, where it is required to find the power delivered to the 3-Ω resistor. At the outset we should make it clear that power is *not a linear combina-*

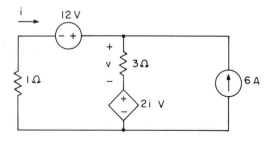

FIGURE 5.7 *Circuit with a dependent source*

tion of voltages or currents. In the case we are considering, we have

$$p = \frac{v^2}{3}$$

which is a quadratic, and not a *linear*, expression. Therefore superposition will *not* apply to obtaining power directly. That is, we *cannot* find the power due to each independent source acting alone and add the results to obtain the total power. However, we *can* find v by superposition and subsequently obtain p.

Letting v_1 be the component of v when the 12-V source is acting alone [Fig. 5.8(a)] and v_2 be the component of v due to the 6-A source alone [Fig. 5.8(b)], we have, as before,

$$v = v_1 + v_2$$

We note that the current i driving the dependent source also has components, which we have labeled i_1 and i_2, due to each source alone. Solving for v_1 in Fig. 5.8(a) and v_2 in Fig. 5.8(b), we have

$$v_1 = 6 \text{ V}, \qquad v_2 = 9 \text{ V}$$

and therefore $v = 15$ V. The power is then given by

$$p = \frac{(15)^2}{3} = 75 \text{ W}$$

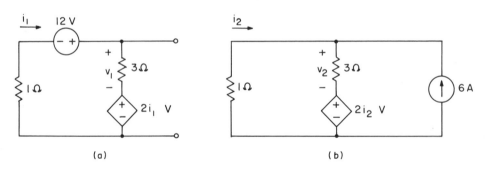

(a)   (b)

FIGURE 5.8 *Circuit of Fig. 5.7 with (a) the current source killed and (b) the voltage source killed*

There are at least three things that are illustrated by the last example. First, as already mentioned, we cannot use superposition directly to calculate power, but we may use it to get current or voltage, from which power is then found. Second, in applying superposition, *only* the independent sources are killed, *never* the dependent sources. Finally, superposition is often a poor method for solving circuits with dependent sources, because each of the individual single-input circuits frequently is almost as difficult to analyze as the original circuit. For example, there are two meshes in Fig. 5.7, the original circuit, and one mesh current is known. Therefore only one application of KVL is needed, around the left mesh, to solve for the current in the 1-Ω resistor. This is also exactly the case in the circuit of Fig. 5.8(b).

EXERCISES

5.2.1 Solve Ex. 4.2.3 using superposition.

5.2.2 Solve Prob. 3.9 using superposition.

5.2.3 Find the power delivered to the 3-Ω resistors in Figs. 5.8 (a) and (b), and show that their sum is *not* equal to the total power delivered to the 3-Ω resistor in Fig. 5.7.   *Ans.* 12 W, 27 W

5.3 THEVENIN'S AND NORTON'S THEOREMS

In the previous section of this chapter we saw that the analysis of some circuits could be greatly simplified by applying the principle of superposition. However, as we saw in the last example of Sec. 5.2, superposition alone may not reduce the complexity of the problems, but its use may lead to additional work. In this section we shall consider *Thevenin's* and *Norton's theorems*, which will in many cases be applicable and greatly simplify the circuit to be analyzed.

We shall assume that the circuit to be considered can be separated into two parts, as shown in Fig. 5.9. The part denoted as circuit A is a linear circuit containing resistors, dependent sources, or independent sources. Circuit B may contain nonlinear elements. We also add the constraint that any dependent source in either circuit must have its controlling element in the same circuit. That is, no dependent source in circuit A can be controlled by a voltage or current associated with an element in circuit B and vice versa. The reason for this will become clear in the development which follows.

We shall show that we can replace circuit A by an *equivalent* circuit containing a source and a resistor in such a way that the voltage-current relations at terminals *a-b* remain the same. Since our aim is to maintain the same terminal relations at *a-b*, clearly, from the standpoint of circuit A, we may obtain the same effect if we replace circuit B by a voltage source of v volts with the proper polarity, as shown in Fig. 5.10. Insofar as circuit A is concerned, it has the same terminal voltage, and since circuit A itself is unchanged, the same terminal current must flow. We have now obtained a linear circuit and can consequently make use of all the properties we have established for such circuits.

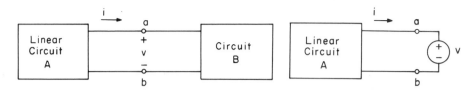

FIGURE 5.9 *Partitioned circuit*   **FIGURE 5.10** *Replacement of circuit B by a voltage source*

In particular, applying superposition to the linear circuit which we have now obtained, we see that the current i will be given by

$$i = i_1 + i_{sc} \tag{5.17}$$

where i_1 is produced by the voltage source v with the network A dead (all its independent sources killed) and i_{sc} is the *short-circuit* current produced by any sources inside circuit A with v killed (replaced by a short circuit). These two cases are shown in Figs. 5.11(a) and (b).

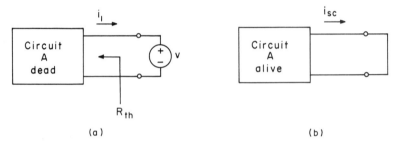

(a)   (b)

FIGURE 5.11 *Circuits obtained for applying superposition*

Since the independent sources are dead in circuit A [Fig. 5.11(a)], from the terminals of the source v we see only a resistive circuit, the equivalent resistance of which we shall call R_{th}. By Ohm's law, we therefore have

$$i_1 = -\frac{v}{R_{th}} \tag{5.18}$$

By (5.17) the expression for the current i is then

$$i = -\frac{v}{R_{th}} + i_{sc} \tag{5.19}$$

Since (5.19) describes network A in the general case, it must hold for any condition at the terminals. Suppose the terminals are open. In this case, $i = 0$, and we shall denote the voltage by $v = v_{oc}$, the *open-circuit* voltage. Substituting these values into

(5.19), we have

$$0 = -\frac{v_{oc}}{R_{th}} + i_{sc}$$

or

$$v_{oc} = R_{th}i_{sc} \qquad\qquad (5.20)$$

Eliminating i_{sc} in (5.19) and (5.20), we have

$$v = -R_{th}i + v_{oc} \qquad\qquad (5.21)$$

The relations given in (5.19) and (5.21) may be used to obtain two very useful circuits which are equivalent to circuit *A*. Historically, the first of these was *Thevenin's equivalent circuit*, named in honor of the French telegraph engineer Charles Leon Thevenin (1857–1926), who published his results in 1883. [The circuit might have been more appropriately named for the great German physicist H. S. F. Helmholtz (1821–1894), who gave a restricted case of it in 1853.]

The Thevenin equivalent circuit is simply one which is described by (5.21) with terminal voltage v and terminal current i, oriented as in Fig. 5.9. To draw the circuit we note that v is the sum of two terms, which therefore must represent two elements in series whose terminal voltages add up to v. The first term evidently corresponds to a resistance R_{th}, called the *Thevenin* resistance, and the second term corresponds to a voltage source with terminal voltage v_{oc}. The result is shown in Fig. 5.12, where the dashed lines represent the connections to the external circuit *B* of Fig. 5.9. Analysis of the circuit will show that (5.21) is satisfied. The statement that Fig. 5.12 is equivalent at the terminals to Fig. 5.9 is known as *Thevenin's theorem*.

Another circuit that is equivalent to circuit *A* of Fig. 5.9 is obtained from (5.19). This circuit is the dual of the Thevenin circuit and is called the *Norton equivalent circuit* in honor of the American engineer E. L. Norton (1898–), whose work was published some 50 years after Thevenin's. From (5.19) we see that i is the sum of two terms, which must then represent two parallel elements whose currents add up to i. The first term evidently arises from the Thevenin resistance R_{th}, and the second term corresponds to a current source i_{sc}. The result is shown in Fig. 5.13, and the statement

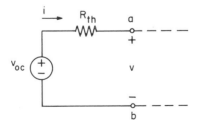

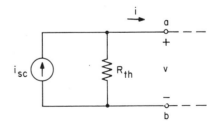

FIGURE 5.12 *Thevenin equivalent circuit of Circuit A of Fig. 5.9*

FIGURE 5.13 *Norton equivalent circuit of Fig. 5.9*

of equivalence of Figs. 5.9 and 5.13 is *Norton's theorem*. Again, the dashed lines represent the connections to the external circuit *B*.

As an example, let us find the Thevenin and Norton equivalent circuits for the network to the left of terminals *a-b* in Fig. 5.14. Then using the results, let us obtain the current *i*, as shown, in terms of the load resistance *R*.

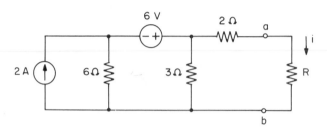

FIGURE 5.14 *Circuit with a variable load resistor*

To obtain the Thevenin circuit we need to find R_{th} and v_{oc}. The Thevenin resistance R_{th} is found from the dead circuit (the two independent sources made zero), shown in Fig. 5.15(a), from which we have

$$R_{\text{th}} = 2 + \frac{(3)(6)}{3 + 6} = 4\,\Omega$$

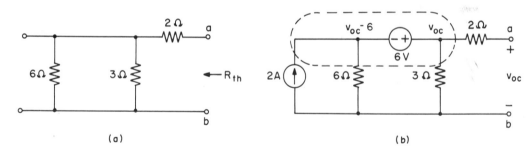

FIGURE 5.15 *Circuits for obtaining the Thevenin circuit of Fig. 5.14*

The open-circuit voltage v_{oc} is obtained from Fig. 5.15(b). Since the terminals *a-b* are open, the voltage v_{oc} is across the 3-Ω resistor. Labeling the nodes as shown, with node *b* as reference, and writing a nodal equation at the generalized node, shown dashed, we have

$$\frac{v_{\text{oc}} - 6}{6} + \frac{v_{\text{oc}}}{3} = 2$$

or

$$v_{\text{oc}} = 6\,\text{V}$$

The Thevenin equivalent circuit, with the load *R* connected, is shown in Fig. 5.16.

We note that the polarity for $v_{oc} = 6$ V is such that the correct voltage polarity results at terminals *a-b* when they are opened.

The current *i* in Fig. 5.14 is the same as that of Fig. 5.16, which in the latter case is readily seen to be

$$i = \frac{6}{R + 4}$$

We may use this result to find the load current for any load *R* that we choose.

To obtain the Norton equivalent circuit we use $R_{th} = 4\ \Omega$, as before, and calculate i_{sc}. We may short terminals *a* and *b* and calculate i_{sc} from the resulting circuit, or we may use the v_{oc} we already have and get i_{sc} from (5.20). In the latter case we have

$$i_{sc} = \tfrac{6}{4} = 1.5 \text{ A}$$

The Norton equivalent circuit, with the load *R* connected, is shown in Fig. 5.17. Using current division, we have, as before,

$$i = \left(\frac{4}{R + 4}\right)(1.5) = \frac{6}{R + 4}$$

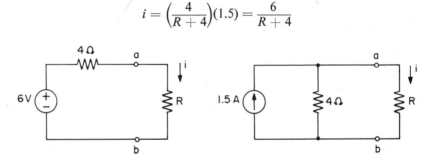

FIGURE 5.16 *Loaded Thevenin equivalent of the circuit of Fig. 5.14*   **FIGURE 5.17** *Loaded Norton equivalent of the circuit of Fig. 5.14*

The direction of the 1.5-A source is such that when *R* is replaced by a short, $i_{sc} = 1.5$ A has the correct direction. In this case and in the Thevenin case of Fig. 5.16, it is a simple matter to place the sources correctly, but in a complex example some care needs to be exercised so that the polarities are correct.

Let us now consider an example containing a dependent source, such as the circuit of Fig. 5.18(a). Suppose we want the Norton equivalent circuit at terminals *a-b*. We shall need, in this case, R_{th} defined for the dead circuit of Fig. 5.18(b) and i_{sc} shown in Fig. 5.18(c).

We may find i_{sc} from Fig. 5.18(c) by noticing that

$$i_2 = 10 - i_1 - i_{sc}$$

and writing the two mesh equations,

$$-4(10 - i_1 - i_{sc}) - 2i_1 + 6i_1 = 0$$
$$-6i_1 + 3i_{sc} = 0$$

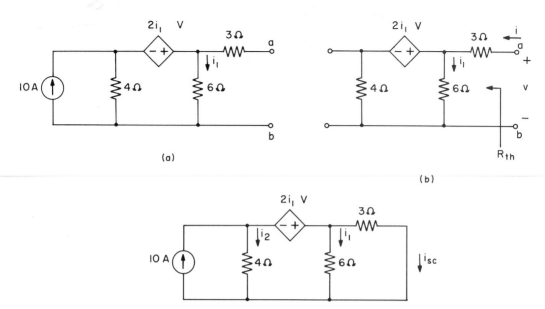

(a)

(b)

(c)

FIGURE 5.18 *(a) Circuit to be analyzed, with (b) its source killed and (c) its terminals short-circuited*

Eliminating i_1, we have

$$i_{sc} = 5 \text{ A} \tag{5.22}$$

We cannot find R_{th} from Fig. 5.18(b) simply by calculating equivalent resistance, because of the dependent source. However, we could excite the circuit with a voltage v (or a current i) at its terminals and calculate the resulting i (or v). Then $R_{th} = v/i$. Alternatively, we could find v_{oc}, as for the Thevenin circuit, and obtain R_{th} from (5.20). This is the method we shall use.

To find v_{oc} we refer to Fig. 5.19, where we have

$$v_{oc} = 6i_1$$

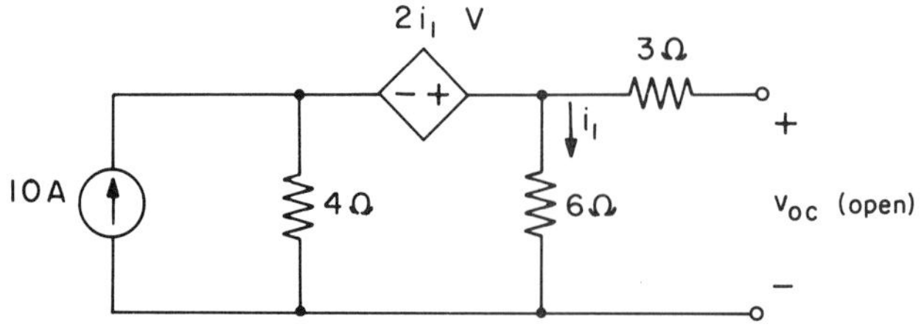

FIGURE 5.19 *Circuit of Fig. 5.18(a) with its terminals opened*

and around the center mesh,

$$-4(10 - i_1) - 2i_1 + 6i_1 = 0$$

From these equations we find $v_{oc} = 30$ V, and thus

$$R_{th} = \frac{v_{oc}}{i_{sc}} = \frac{30}{5} = 6\,\Omega$$

As a final example, let us find the Thevenin equivalent of the circuit of Fig. 5.20. To begin with, we see by inspection that since there is no independent source present, we must have

$$v_{oc} = i_{sc} = 0 \qquad\qquad (5.23)$$

Also, the dead circuit is the given circuit itself, so that R_{th} is simply the resistance seen at the terminals of Fig. 5.20. In view of (5.23), we cannot use $v_{oc} = R_{th}i_{sc}$ to get R_{th}, as we did in the previous example. The only recourse we have is to excite the circuit at its terminals and calculate R_{th} from the results.

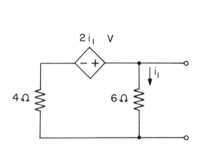

FIGURE 5.20 *Circuit with a dependent source*

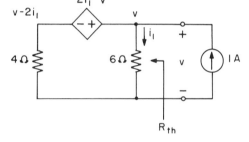

FIGURE 5.21 *Circuit of Fig. 5.20 excited by a current source*

For example, suppose we excite the circuit with a 1-A current source, as shown in Fig. 5.21. Then we have

$$R_{th} = \frac{v}{1} = v$$

where v is the resulting terminal voltage. Taking the bottom node as reference, the nonreference node voltages are as shown. A nodal analysis yields

$$\frac{v - 2i_1}{4} + \frac{v}{6} = 1$$

where

$$v = 6i_1$$

From these equations we have $v = 3$ V and thus $R_{th} = 3\,\Omega$. The Thevenin equivalent (as well as the Norton equivalent) is shown in Fig. 5.22.

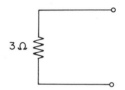

FIGURE 5.22 *Thevenin equivalent of Fig. 5.20*

EXERCISES

5.3.1 Replace the circuit to the left of terminals *a-b* by its Thevenin equivalent and find *i*.   *Ans.* $v_{oc} = 24$ V, $R_{th} = 6$ Ω, $i = 3$ A

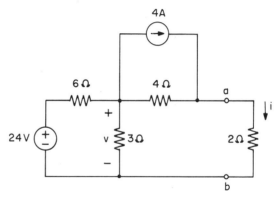

EXERCISE 5.3.1

5.3.2 Replace everything in the figure for Ex. 5.3.1 except the 3-Ω resistor by its Norton equivalent and find *v*.   *Ans.* $i_{sc} = \frac{4}{3}$ A, $R_{th} = 3$ Ω, $v = 2$ V

5.3.3 Find R_{th} in Fig. 5.18 (b) by exciting the circuit with a voltage (or current) and finding the current (or voltage) which results.   *Ans.* 6 Ω

5.4 PRACTICAL SOURCES

In Chapter 1 we defined independent sources and pointed out that they were *ideal* elements. An ideal 12-V battery, for example, supplies 12 V between its terminals regardless of the load connected to the terminals. However, a real, or *practical*, 12-V battery supplies 12 V when its terminals are open-circuited and supplies less than 12 V as current is drawn through the terminals. A practical voltage source thus appears to have an internal drop in voltage when current flows through its terminals, and this internal drop diminishes the voltage at the terminals.

We may represent a practical voltage source by the mathematical model of Fig. 5.23, consisting of an ideal source v_g in series with an *internal* resistance R_g. The voltage v seen at the terminals of the source now depends on the current i drawn from the source. The relationship is easily seen to be

$$v = v_g - R_g i \qquad (5.24)$$

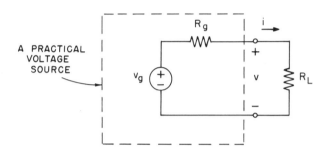

FIGURE 5.23 *Practical voltage source connected to a load R_L*

Thus under open-circuit conditions ($i = 0$), we have $v = v_g$, and under short-circuit conditions ($v = 0$), we have $i = v_g/R_g$. If $R_g > 0$, as it is for a practical source, the source can never deliver an infinite current, as an ideal source can.

For a given practical voltage source (fixed values of v_g and R_g in Fig. 5.23), the load resistance R_L determines the current drawn from the terminals. For example, in Fig. 5.23 the load current is

$$i = \frac{v_g}{R_g + R_L} \qquad (5.25)$$

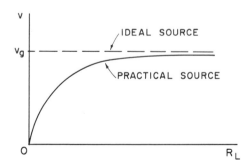

FIGURE 5.24 *Practical and ideal voltage source characteristics*

Also, by voltage division we have

$$v = \frac{R_L v_g}{R_g + R_L} \tag{5.26}$$

Therefore as we vary R_L both i and v vary. A sketch of v vs. R_L is shown in Fig. 5.24, along with the ideal case, which is dashed. For large values of R_L relative to R_g, v is very nearly equal to the ideal value of v_g. (If R_L is infinite, corresponding to an open circuit, then v is v_g.)

We may replace the practical voltage source of Fig. 5.23 by a practical current source by means of Norton's theorem. The result is shown in Fig. 5.25 and is seen to be an ideal current source in parallel with the internal resistance. The two sources in Figs. 5.23 and 5.25 are equivalent at the terminals, of course, if R_g is the same in both cases, and

$$v_g = R_g i_g \tag{5.27}$$

By current division, we find, in Fig. 5.25,

$$i = \frac{R_g i_g}{R_g + R_L} \tag{5.28}$$

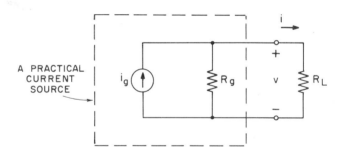

FIGURE 5.25 *Practical current source connected to a load R_L*

Therefore, for a given current source (fixed values of i_g and v_g), the load current depends on R_L. A sketch of i vs. R_L is shown in Fig. 5.26, along with the ideal case, which is dashed.

Very often network analysis can be greatly simplified by changing practical voltage sources to practical current sources, and vice versa, by means of Norton's and Thevenin's theorems. For example, suppose we wish to find the current i shown in Fig. 5.27. We could solve the problem in a number of ways, such as replacing everything except the 4-Ω resistor by its Thevenin equivalent and then finding i. However, we shall illustrate instead the method of successive transformation of sources.

Let us begin by replacing the 32-V source and internal 3-Ω resistance by a practical current source of a 3-Ω internal resistance and a $\frac{32}{3}$-A ideal source. Then let us replace the 4-A source and the internal 2-Ω resistance by a voltage source of $2(4) = 8$ V and a

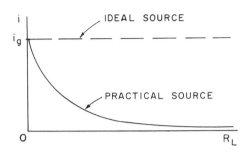

FIGURE 5.26 *Practical and ideal current source characteristics*

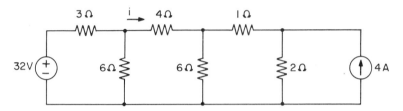

FIGURE 5.27 *Circuit with two practical sources*

2-Ω internal resistance. We are applying, respectively, Norton's and Thevenin's theorems, or equivalently, (5.27). The results of these two source transformations are shown in Fig. 5.28.

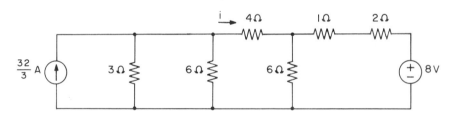

FIGURE 5.28 *Result of two transformations applied to Fig. 5.27*

We may now combine the parallel 3- and 6-Ω resistances and the series 1- and 2-Ω resistances, as shown in Fig. 5.29(a), and repeat the source transformation procedure. Continuing the process, as shown in Figs. 5.29(b), (c), and (d), we finally arrive at an equivalent circuit (insofar as i is concerned) which can be analyzed by inspection. In this case, from Fig. 5.29(d), the answer is

$$i = \frac{\frac{64}{3} - \frac{16}{3}}{2 + 4 + 2} = 2 \text{ A}$$

This procedure may seem unduly long, but it should be observed that most of the steps may be carried out mentally.

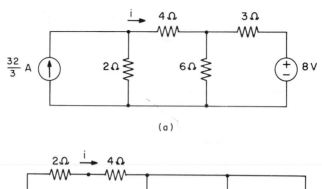

(a)

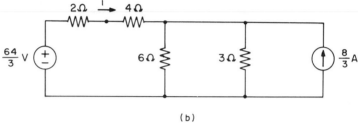

(b)

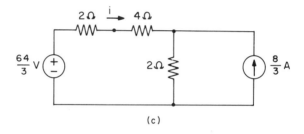

(c)

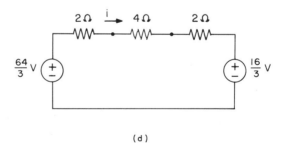

(d)

FIGURE 5.29 *Steps in obtaining i in Fig. 5.28*

As a final note in this section, we often may combine sources as we do resistors to obtain equivalent sources. For example, if we are interested only in *i* in Fig. 5.29(d), we may combine the three series resistors, as we know, but we may also combine the series sources. They represent a net source of $\frac{64}{3} - \frac{16}{3} = 16$ V with a polarity like that of the larger source. Thus an equivalent circuit insofar as i is concerned is that of Fig. 5.30. Similarly, we may combine parallel current sources to obtain an equivalent source.

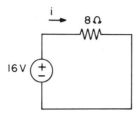

FIGURE 5.30 *Circuit equivalent to Fig. 5.29 for finding i*

EXERCISES

5.4.1 By source transformations, replace the network to the left of terminals *a-b* in the figure for Ex. 5.3.1 by an equivalent network having only one resistor R. From this network, find i as shown.   *Ans. $R = 6 \Omega$, $i = 3$ A*

5.4.2 Convert all the sources in the figure for Ex. 4.2.3 to voltage sources and find v_3.
          Ans. 5 V

5.4.3 In the figure for Prob. 4.5, convert the network to the left of terminals *a-b* to a single practical current source by source transformations and find v.   *Ans.* 10 V

5.5 MAXIMUM POWER TRANSFER

There are many applications in circuit theory where it is desirable to obtain the maximum possible power that a given practical source can deliver. It is very easy, using Thevenin's theorem, to see what maximum power a source is capable of delivering and how to load the source so as to obtain this maximum power. This will be the subject of this section.

Let us begin with the practical voltage source shown earlier in Fig. 5.23 with a load resistance R_L. The power p_L delivered to the resistor R_L is given by

$$p_L = \left(\frac{v_g}{R_g + R_L}\right)^2 R_L \tag{5.29}$$

and it is this quantity which we wish to maximize.

Since the source is assumed to be given, v_g and R_g are fixed, and thus p_L is a function of R_L. To maximize p_L we can make $dp_L/dR_L = 0$ and solve for R_L. From (5.29) we obtain

$$\frac{dp_L}{dR_L} = v_g^2 \left[\frac{(R_g + R_L)^2 - 2(R_g + R_L)R_L}{(R_g + R_L)^4} \right]$$

$$= \frac{(R_g - R_L)v_g^2}{(R_g + R_L)^3} = 0 \tag{5.30}$$

which results in

$$R_L = R_g \tag{5.31}$$

It may be readily shown that

$$\left. \frac{d^2 p_L}{dR_L^2} \right|_{R_L = R_g} = -\frac{v_g^2}{8R_g^3} < 0$$

and therefore (5.31) is the condition which maximizes p_L. We see, therefore, that the maximum power is delivered by a given practical source when the load R_L is equal to the internal resistance of the source. This statement is sometimes called the *maximum power transfer theorem*. We have developed it for a voltage source, but in view of Norton's theorem it also holds for a practical current source.

The maximum power that the practical voltage source is capable of delivering to the load is given by (5.29) and (5.31) to be

$$p_{L_{\text{max}}} = \frac{v_g^2}{4R_g} \tag{5.32}$$

In the case of the practical current source, the maximum deliverable power is

$$p_{L_{\text{max}}} = \frac{R_g i_g^2}{4} \tag{5.33}$$

This may be seen from Fig. 5.25 and (5.31) or by (5.32) and Norton's theorem.

We may extend the maximum power transfer theorem to a linear circuit rather than a single source by means of Thevenin's theorem. That is, the maximum power is obtained from a linear circuit at a given pair of terminals when the terminals are loaded by the Thevenin resistance of the circuit. This is obviously true since by Thevenin's theorem the circuit is equivalent to a practical voltage source with internal resistance R_{th}.

As an example, we may draw the maximum power from the circuit of Fig. 5.18(a) if we load terminals *a-b* with the Thevenin resistance

$$R_L = R_{\text{th}} = 6 \, \Omega$$

Since $i_{\text{sc}} = 5$ A by (5.22), we may draw the Norton equivalent circuit with the required R_L as shown in Fig. 5.31. The power supplied to the load is given by

$$p = \frac{900R_L}{(R_L + 6)^2}$$

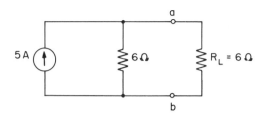

FIGURE 5.31 *Equivalent circuit of Fig. 5.18(a) loaded for maximum power transfer*

which for $R_L = 6$ yields

$$p_{\max} = 37.5 \text{ W}$$

Any other value of R_L will result in a lower value of p. For example, if $R_L = 5 \ \Omega$, then we have $p = 37.19$ W, and for $R_L = 7 \ \Omega$, we have $p = 37.28$ W.

EXERCISES

5.5.1 Find the maximum power that can be transferred from the circuit to the left of terminals *a-b* in the figure for Ex. 5.3.1.   *Ans.* 24 W

5.5.2 Show that the two networks given are equivalent at terminals *a-b* and find the power dissipated in the 3-Ω resistor in each case.

Ans. (a) 12 W, (b) 3 W

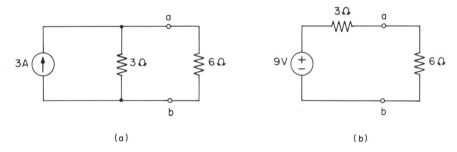

(a)   (b)

EXERCISE 5.5.2

5.5.3 Find the maximum power delivered to the load R_L in Fig. 5.23 if v_g and $R_L > 0$ are fixed and R_g is variable.   *Ans.* v_g^2/R_L when $R_g = 0$

PROBLEMS

5.1 Solve Prob. 2.16 using the method applied to Fig. 5.3.

5.2 Solve Prob. 2.17 using the method of Prob. 5.1.

5.3 Solve Prob. 3.13 using the method of Prob. 5.1. (Suggestion: Assume $i = 1$ mA, etc.)

5.4 Find v in Prob. 4.5 using superposition.

5.5 Solve Prob. 4.10 using superposition.

5.6 Using superposition, find the power delivered to the 3-kΩ resistor.

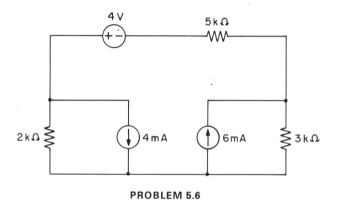

PROBLEM 5.6

5.7 Find v using superposition.

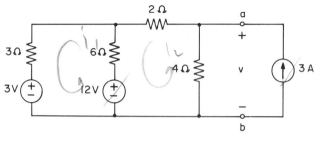

PROBLEM 5.7

5.8 Find v using superposition.

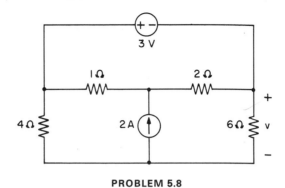

PROBLEM 5.8

5.9   Find v in Prob. 4.5 by replacing the network to the left of terminals a-b by its Thevenin equivalent.

5.10 Find the power delivered to the 4-Ω resistor in Prob. 4.9 by replacing the network to the left of terminals a-b by its Norton equivalent.

5.11 Solve Prob. 4.21 by replacing the network to the left of terminals a-b by its Thevenin equivalent.

5.12 Replace everything except the 6-kΩ resistor in Prob. 3.13 by its Thevenin equivalent and find i.

5.13 Find v by replacing the network to the left of terminals a-b by its Norton equivalent.

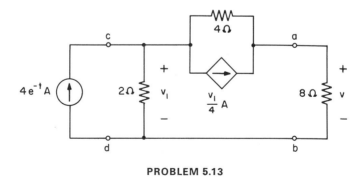

PROBLEM 5.13

5.14 In the figure for Prob. 5.13, replace the network to the right of terminals c-d by its Thevenin equivalent and find v_1.

5.15 By successive source transformations, change the circuit shown to an equivalent one at terminals a-b containing only one resistor. From this equivalent circuit, find v.

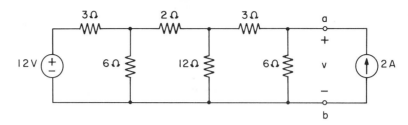

PROBLEM 5.15

5.16 Solve Prob. 4.19 by using source transformations to obtain an equivalent circuit, insofar as the 4-kΩ resistor is concerned, containing only one source and one resistor, in addition to the 4-kΩ resistor.

5.17 Find the maximum power that can be transferred to R by the circuit to the left of terminals a-b. (Suggestion: Use successive source transformations to find the Thevenin or Norton circuit.)

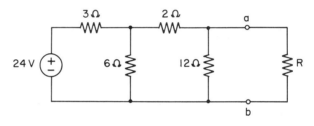

PROBLEM 5.17

5.18 In the figure for Prob. 4.22, find the resistance R to replace the 1-Ω resistance between terminals a-b which will draw the maximum power from the rest of the circuit. Find this maximum power.

5.19 Show that the 1-Ω resistance between terminals c-d in the figure for Prob. 4.22 has the correct value to draw the maximum power from the rest of the circuit. Find this maximum power.

5.20 Solve Prob. 4.23 by *assuming* $v_2 = 1$ V and using the proportionality principle.

6

INDEPENDENCE
OF EQUATIONS*

An electric network is determined by the type of elements it contains and the manner in which the elements are connected. We have spent considerable time in the previous chapters considering the elements themselves and their volt-ampere characteristics. In this chapter we shall consider the manner in which the network elements are connected, or, as it is sometimes called, the network *topology*. As we shall see, a study of the topology of the network provides us with a systematic way of determining how many equations are required in the analysis, which ones are independent, and the best set of equations to select for the most straightforward analysis.

6.1 GRAPH OF A NETWORK

To illustrate the problems involved in the analysis of more complicated networks, let us consider the circuit of Fig. 6.1. The resistors are numbered 1, 2, . . . , 9, with values of resistance, say, $R_1, R_2, \ldots, R_9$. Suppose we are required to perform a loop analysis, in which case we need to write a set of independent KVL equations. (We note that the circuit is nonplanar, and thus we cannot perform a mesh analysis. Anyone doubting this is welcome to try redrawing the circuit in a planar fashion.)

There are 15 loops in the circuit, as may be verified by several means, one of which is trial and error (much trial and more error). For the curious reader the 15 loops are (1, 3, 4, 5), (1, 3, 7, 9), (2, 3, 5, 6), (1, 2, 8, 9), (1, 2, 4, 6), (4, 5, 7, 9), (2, 3, 4, 5, 8, 9), (1, 2, 5, 6, 7, 9), (1, 3, 5, 6, 8, 9), (2, 3, 7, 8), (2, 3, 4, 6, 7, 9), (4, 6, 8, 9), (5, 6, 7, 8), (1, 3, 4, 6, 7, 8), and (1, 2, 4, 5, 7, 8). That is, resistors 1, 3, 4, and 5 form a loop; 1, 3, 7, and 9 (with the source v_g) form a loop; etc.

To undertake a loop analysis of Fig. 6.1 we need to know which of these loops are independent and how many are required. To answer these questions we need consider only how the elements are connected; it is unimportant what kind of elements are involved. To facilitate matters, then, we may retain the nodes of the network and replace its elements by lines. The network topology thus is preserved in a much simpler configuration.

The configuration of lines and nodes obtained by replacing the elements of a network by lines is called the *graph* of the network. The lines of the graph are called its *branches*, and the *nodes* of the graph are, of course, the nodes of the network. As an example, the graph of the network of Fig. 6.1 is shown in Fig. 6.2. It has nine branches and six nodes. (We could consider a node between resistor 9 and v_g, but inasmuch as this will not affect the number of loops, we have chosen to consider these two series elements as one element, namely a nonideal voltage source.)

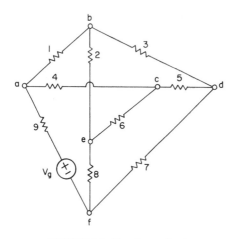

FIGURE 6.1 *Nonplanar circuit*

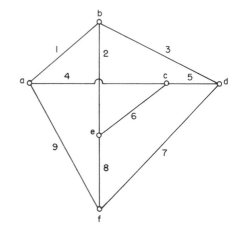

FIGURE 6.2 *Graph of a network*

We shall say that a graph is *connected* if there is a path of one or more branches between any two nodes. The graph of Fig. 6.2 is evidently connected. An example of a graph that is not connected is shown in Fig. 6.3. There is, for instance, no path between nodes *a* and *d*. For the present we shall consider only connected graphs.

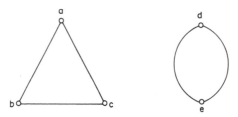

FIGURE 6.3 *Unconnected graph*

EXERCISES

6.1.1   Show that the given graph is planar.

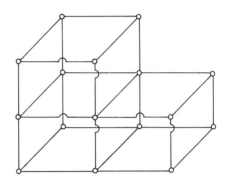

EXERCISE 6.1.1

6.1.2   Show that the given graph is planar.

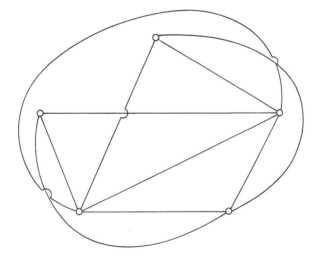

EXERCISE 6.1.2

6.2 TREES AND LINKS

We define a *tree* of a graph as a connected portion of the graph that contains all the nodes but no loops. As an example, Fig. 6.4(b) is a tree of the graph of Fig. 6.4(a). The tree is connected, has no loops, and contains all the nodes of the graph.

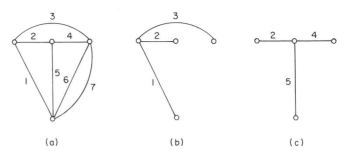

(a)   (b)   (c)

FIGURE 6.4 *Graph and two of its trees*

Generally, a graph has many trees. The configuration of Fig. 6.4(c) is evidently another tree of the graph of Fig. 6.4(a), since it satisfies all the requirements. This particular graph has 24 trees, which the reader may wish to try to discover. It will help in enumerating the trees to notice that each one has exactly three branches, since it takes at least three lines to connect four nodes and more than three lines will form a loop. There are 35 ways to select seven branches three at a time, but 11 of these combinations are not trees.

In the general case, let B be the number of branches and N be the number of nodes in a given graph. Then any tree of the graph contains N nodes and $N - 1$ branches. The number of nodes follows from the definition of a tree, and the number of tree branches may be established by the following construction argument. Let us build the tree starting with one branch and the two nodes to which it is connected. Each additional branch connected to build the tree adds one additional node. The number of nodes is, therefore, one more than the number of branches, and since there are N nodes, there must be $N - 1$ branches.

As examples, the graph of Fig. 6.4(a) has $N = 4$, and thus the number of tree branches in both Figs. 6.4(b) and (c) is $N - 1 = 3$.

Once the branches of a tree are designated, the remaining branches of the graph are called *links*. The number of links in a graph is then $B - (N - 1)$ or $B - N + 1$. As an example, the tree of Fig. 6.4(b) is redrawn in Fig. 6.5, with the tree branches shown as solid lines and the links as dashed lines. The number of links in this case, since $B = 7$, is $7 - 4 + 1 = 4$.

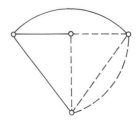

FIGURE 6.5 *Tree branches and links of a graph*

EXERCISES

6.2.1 Find the number of tree branches and links in the graph of Fig. 6.2. *Ans.* 5, 4

6.2.2 Find all the trees of the graph shown.
Ans. (1, 2, 4), (1, 2, 5), (1, 3, 4), (1, 3, 5), (1, 4, 5), (2, 3, 4), (2, 3, 5), (2, 4, 5)

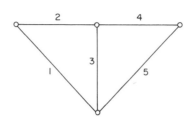

EXERCISE 6.2.2

6.2.3 In the figure for Ex. 6.2.2, let i_2 and i_4 be link currents in elements 2 and 4, directed to the right. Let i_1, i_3, and i_5 be tree currents directed downward. Find the tree currents in terms of the link currents.   *Ans.* $-i_2$, $i_2 - i_4$, i_4

6.3 INDEPENDENT VOLTAGE EQUATIONS

In this section we shall consider the nodal analysis of a circuit whose graph has N nodes and B branches. A tree of the graph then has $N - 1$ branches and, of course, $N - 1$ tree branch voltages. There are also $B - N + 1$ link voltages in the circuit.

Let us imagine that all the branch voltages of any tree are made zero by short-circuiting the branches (i.e., the branches are replaced by short circuits). Then all the nodes of the circuit, being in the tree, are at the same potential, and thus the link voltages are all zero also. Therefore the link voltages depend on the tree branch voltages, for if a link voltage were independent of the tree voltages, it could not be forced to zero by short-circuiting the tree branches. On the other hand, if one tree branch were not short-circuited, there would be one node at a different potential and thus one tree branch voltage not dependent on the others. We may conclude, therefore, that the $N - 1$ branch voltages of any tree are independent and may be used to find the link voltages.

Another way to see that any link voltage may be expressed in terms of tree voltages is to note that adding the link to the tree completes a circuit whose only other elements are tree branches. Thus by KVL the link voltage is an algebraic sum of tree branch voltages.

As an example, the graph of Fig. 6.6 has tree branch voltages v_1, v_2, and v_3, indicated by the solid lines. The dashed lines are the links whose voltages are v_4, v_5, and v_6. If the link labeled v_4 is added to the tree, the circuit of v_2, v_3, v_4 is formed. By KVL around this circuit, we may obtain

$$v_4 = v_3 - v_2$$

In like manner, adding first link v_5 and then link v_6, we obtain

$$v_5 = v_1 - v_2$$
$$v_6 = v_1 - v_3$$

Thus the link voltages may be found from the tree branch voltages.

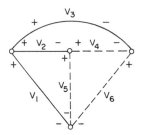

FIGURE 6.6 *Tree and link voltages*

A systematic way of writing equations involving tree branch voltages is to imagine opening a tree branch, noting that this separates the tree into two parts. Currents flow between the two parts through the tree branch imagined to be open and through links. Thus by KCL the algebraic sum of these currents in a given direction is zero. This procedure may then be repeated for the other tree branches.

To illustrate the use of tree branch voltages, let us consider the circuit in Fig. 6.7(a). A graph of the circuit is shown in Fig. 6.7(b) with the tree branches shown as solid lines and the links as dashed lines. Since the tree branch voltages are independent, we have included in the tree the 20-V source. Thus the number of unknowns is reduced

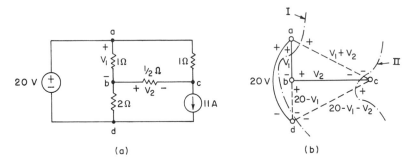

(a)   (b)

FIGURE 6.7 *Circuit and its graph*

by one. By KVL we may find the link voltages in terms of the tree voltages, with the results shown in Fig. 6.7(b).

If we imagine the tree branch (a, b) labeled v_1 as open, then the tree is separated into two parts. These two parts are connected by branch (a, b) and links (a, c), (b, d), and (d, c), as indicated by the line marked I. Summing the currents across the line in the direction of the arrow, we have

$$v_1 + v_2 + v_1 - \frac{20 - v_1}{2} - 11 = 0$$

Repeating the procedure for tree branch (b, c), labeled v_2, leads to line II and the equation

$$-(v_1 + v_2) - 2v_2 + 11 = 0$$

Solving these equations, we have $v_1 = 8$ V and $v_2 = 1$ V. All the link voltages, and consequently all the link and tree currents, may now be found.

In any circuit analysis procedure where the unknowns are voltages, we need to find only the $N - 1$ tree branch voltages which constitute an independent set. This means that only $N - 1$ independent voltage equations are required in the analysis. Since *any* independent set of equations will suffice, then *any* independent set of $N - 1$ voltages constitutes a solution.

Another independent set of $N - 1$ voltages, other than the tree branch voltages, is the set of nondatum node voltages, considered in the nodal method of Chapter 4. To see this, we note that any nondatum node is in the tree and is connected through tree branches to the datum node. Thus every nondatum node voltage is an algebraic sum of tree branch voltages (the tree branches between the nondatum and datum nodes). On the other hand, every tree branch voltage is the difference between its two node voltages. In summary, the node voltages may be determined from the tree voltages and vice versa. Thus the nondatum node voltages are also an independent set.

In the example of Fig. 6.7 we see that if d is the datum node, then the nondatum node voltages v_a, v_b, and v_c are related to the tree voltages v_1, v_2, and 20 by

$$v_a = 20$$
$$v_b = 20 - v_1$$
$$v_c = 20 - v_1 - v_2$$

Conversely, we have

$$v_1 = v_a - v_b$$
$$v_2 = v_b - v_c$$
$$20 = v_a$$

The nodal method, in many cases, is easier to apply than the loop method because the nodes are easy to find. In the loop method, as exemplified by the example of Fig. 6.1, the appropriate loops may be difficult to identify. In the next section we shall consider a method based on graph theory of finding sets of independent loop equations.

EXERCISES

6.3.1 In the graph of the circuit in the figure for Ex. 4.3.1, select the tree of the voltage sources and the 3-Ω resistor. Using the method of this section, write one KCL equation and determine v_1. *Ans.* −4 V

6.3.2 Select the tree of the voltage source, and the 3- and 6-Ω resistors, and use the method of this section to find v. *Ans.* 3 V

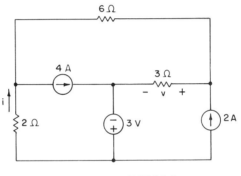

EXERCISE 6.3.2

6.3.3 Using an appropriate tree and the methods of this section, find v in Prob. 4.18. (Note: The tree should contain the voltage v and the three sources as well as one other branch.) *Ans.* 20 V

6.4 INDEPENDENT CURRENT EQUATIONS

As we have seen in the example of Fig. 6.1, it is not always easy to identify the independent loops for a loop analysis of a circuit. To develop a systematic means of writing loop equations, let us consider a general network with B branches and N nodes. Corresponding to a given tree there are $B - N + 1$ links.

Suppose that all the link currents are made zero by open-circuiting the links (replacing the links by open circuits). Since the tree contains no loops, then all the tree branch currents are zero also. The tree currents thus depend on the link currents, i.e., they may be expressed in terms of the link currents, for if a tree current were independent of the link currents it could not be forced to zero by open-circuiting the links.

Moreover, if one link is not open-circuited, a loop is left in the graph, and a current will flow in the link. A link current thus is not dependent on the other link currents. In summary, the $B - N + 1$ link currents are an independent set, and the loop analysis of the circuit requires $B - N + 1$ independent equations.

One systematic way to find $B - N + 1$ independent loops is to start with the tree and add one of the links. This determines the loop containing that link, since adding the link to the tree closes a loop. Remove this link and add another link to the tree, determining a second loop. Continue the process until the $B - N + 1$ loops are found. The set is independent because each loop contains a different link.

As an example, let us reconsider the circuit of Fig. 6.1. One tree consists of branches 1, 5, 7, 8, and 9, with corresponding links 2, 3, 4, and 6, as shown in Fig. 6.8. Closing the links one at a time results in the four independent loops I, II, III, and IV shown. Loop I contains link 2 and tree branches 8, 9, and 1; loop II contains 3, 7, 9, and 1; loop III contains 4, 5, 7, and 9; and loop IV contains 6, 5, 7, and 8. These four loops are sufficient for performing a loop analysis.

To illustrate the use of link currents in circuit analysis, let us return to the example of Fig. 6.7(a). The graph is redrawn in Fig. 6.9, showing the link currents i_1, i_2, and 11 A. We have chosen the current source as a link because the link currents are an independent set. This reduces the number of unknowns by one. Generally, for this reason, one should place voltage sources in the tree and current sources in the links.

The tree branch currents, as in the general case, may be found from the link currents, as shown in Fig. 6.9. Closing the links labeled i_1 and i_2 forms loops 1 and 2, as indicated. Applying KVL to these loops yields, from Figs. 6.7(a) and 6.9,

$$2i_1 - 20 + i_1 - i_2 + 11 = 0$$

$$i_2 - \frac{11 - i_2}{2} - i_1 + i_2 - 11 = 0$$

the solution of which is $i_1 = 6$ A and $i_2 = 9$ A.

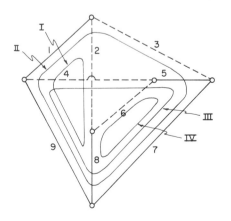

FIGURE 6.8 *Circuits of Fig. 6.1*

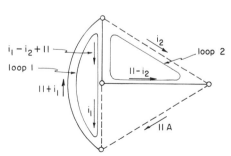

FIGURE 6.9 *Graph of Fig. 6.7(a)*

Since the link current 11 A is known, we needed only two loops involving link currents i_1 and i_2. Incidentally, in this simple example the links were chosen so that the link currents are also mesh currents. This is, of course, not the case in general.

The results obtained thus far in this chapter are valid for general networks, which may be either planar or nonplanar. In the special case of planar networks, as we saw in Chapter 4, a mesh analysis is possible. In the circuits of that chapter the mesh currents were independent and were sufficient in number to perform the analysis. We shall now show that this is the case in general for planar networks.

Let us begin by taking apart the planar circuit with M meshes and reconstructing it one mesh at a time. The first mesh in the reconstruction has the same number, say k_1, of nodes and branches, for the first branch has two nodes, each additional branch adds one new node, and the last branch adds no nodes since it is tied back to a node of the first branch. This is illustrated by the graph of four meshes in Fig. 6.10(a). The first mesh constructed, shown in Fig. 6.10(b), has the same number of branches and nodes, namely four in this case.

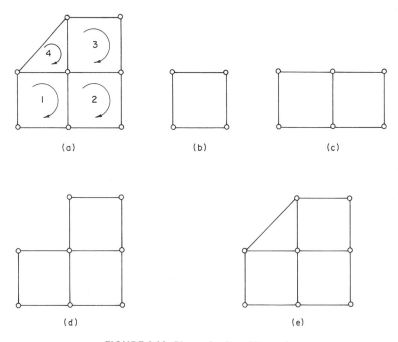

FIGURE 6.10 *Planar circuit and its meshes*

After the first mesh, each subsequent mesh is formed by connecting branches and nodes to previous meshes. Each time the number of nodes added is one less than the number of branches, because each added branch adds one node, except for the last branch, which is connected to a node of a previously added mesh. This process is illustrated in Figs. 6.10(c), (d), and (e).

Thus, if the second mesh adds k_2 branches, then it adds only $k_2 - 1$ nodes. Similarly, the third mesh adds k_3 branches and $k_3 - 1$ nodes, and so forth. The last mesh, the Mth, adds k_M branches and $k_M - 1$ nodes. If in the completed graph the number of branches is B and the number of nodes is N, we have

$$k_1 + k_2 + \ldots + k_M = B \tag{6.1}$$

and

$$k_1 + (k_2 - 1) + \ldots + (k_M - 1) = N \tag{6.2}$$

The latter equation may be written

$$k_1 + k_2 + \ldots + k_M - (M - 1) = N$$

which by (6.1) becomes

$$B - (M - 1) = N$$

Solving for the number of meshes, we have

$$M = B - N + 1 \tag{6.3}$$

which is also the number of links in the graph.

The mesh currents, therefore, constitute an appropriate set of currents to completely describe a planar network. They are the same in number as the independent set of link currents, and they are independent since each new mesh contains at least one branch not in the previous meshes.

EXERCISE

6.4.1 Show that (6.3) holds for the circuits of Probs. 4.19, 4.20, and 4.22.

6.5 A CIRCUIT APPLICATION

As a final illustration in this chapter we shall analyze a moderately complicated circuit, shown in Fig. 6.11(a). Its graph is shown in Fig. 6.11(b), where we have selected a tree shown by the solid lines.

From the graph we see that there are four tree branches labeled v_1, v_2, $3v_2$, and 10. Thus if the tree branch voltage method is used, there will be only two unknowns, v_1 and v_2, requiring two equations. There are five link currents in the graph, one of which is known (the 6-A source) and another, the $2v_1$ source, which may be expressed in

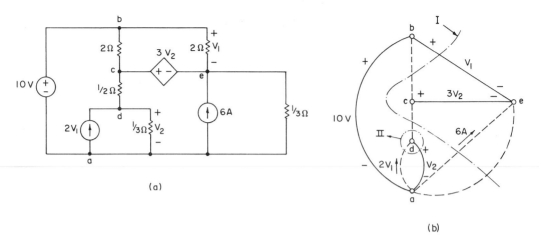

FIGURE 6.11 *Network and its graph*

terms of the tree branch current i_{be}, and subsequently in terms of other link currents. Thus if the link current method is used, we must have three equations. Accordingly, we shall analyze the circuit using tree branch voltages.

To write the two necessary equations we first imagine tree branch (b, e) opened, which separates the tree into two parts. These parts are connected by branch (b, e) and links (b, c), (c, d), and (a, e) of the 6-A source and link (a, e) of the $\frac{1}{3}$-Ω resistor. These elements are identified by line I. A similar procedure using branch (a, d) leads to line II. Applying KCL to the currents crossing lines I and II in the direction indicated yields

$$\frac{v_1}{2} + \frac{v_{bc}}{2} + 2v_{dc} + 6 + 3v_{ae} = 0 \tag{6.4}$$

and

$$2v_{dc} - 2v_1 + 3\,v_2 = 0 \tag{6.5}$$

Inspection of the graph shows that

$$v_{bc} = v_1 - 3v_2$$
$$v_{dc} = v_1 - 2v_2 - 10$$
$$v_{ae} = v_1 - 10$$

Substituting these values into (6.4) and (6.5) and solving for the tree branch voltages, we have

$$v_1 = -11 \text{ V}, \qquad v_2 = -20 \text{ V}$$

We may note that nodal analysis is equally easy to apply. The node voltages may be expressed in terms of the two unknowns v_1 and v_2, and thus only two nodal equations are required. We shall leave the details to the problems (Prob. 6.4).

EXERCISES

6.5.1 Using the graph described for the figure for Ex. 6.3.2, write one KVL equation and find i, using the method of Sec. 6.4.   *Ans.* 3 A

6.5.2 Using an appropriate tree for the graph of Prob. 4.16, find the current i flowing to the right in the 4-Ω resistor. (An appropriate tree should *not* contain the current sources or the current i. Thus, using the methods of Sec. 6.4, only one KVL equation is required.)   *Ans.* −9.5 A

6.5.3 Solve Prob. 4.9 using the method of Sec. 6.4.   *Ans.* 4 W

PROBLEMS

6.1 Find a tree, if possible, that contains all the voltage sources and the branches whose voltages control dependent sources but does not contain current sources or branches whose currents control dependent sources. Use this tree with an appropriate graph theory method to find v_1.

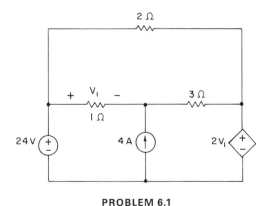

PROBLEM 6.1

6.2 Select a tree as described in Prob. 6.1 and use an appropriate graph theory method to find v_1.

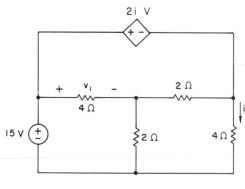

PROBLEM 6.2

6.3   Solve Prob. 4.21 selecting an appropriate tree and using graph theory methods.

6.4   Find v_1 and v_2 in Fig. 6.11 using nodal analysis.

6.5   In the figure shown, all the resistances are 1 Ω, element w is also a 1-Ω resistor, elements x and y are independent 1-A current sources directed upward, and element z is an independent 3-A current source directed to the left. Selecting an appropriate tree and using graph theory methods, find i. (Note that this circuit is similar to that of Fig. 6.1 and is thus nonplanar. However, only one loop equation is required in this case.)

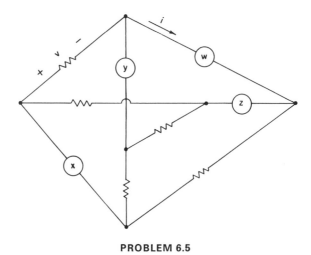

PROBLEM 6.5

6.6   Find v in Prob. 6.5 using graph theory methods if element x is a 4-V source with positive terminal at the top, y is a 2-V source with positive terminal at the bottom, z is a 6V source with positive terminal at the left, and w is a 4-V source with positive terminal at the left.

6.7   The figure for Ex. 6.2.2 and Figs. (a) and (b) shown are examples of graphs of ladder networks. There is a theorem that states that the number of trees in a ladder graph of n branches is the *Fibonacci number* a_n, defined by $a_0 = a_1 = 1$, $a_2 = a_0 + a_1 = 2$, $a_3 = a_1 + a_2 = 3$, $a_4 = a_2 + a_3 = 5$, etc. That is, except for a_0 and a_1, each Fibonacci number is obtained by adding the previous two. Verify that the theorem holds for the ladder graphs shown and also for the graph in the figure for Ex. 6.2.2.

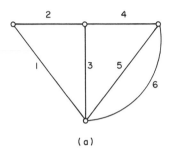 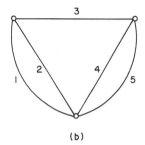

(a)   (b)

PROBLEM 6.7

6.8   The graphs shown in (a) and (b) are two basic nonplanar graphs [the one in (b) is that in the figure for Prob. 6.5 redrawn]. Branch *a-b* in (a) is an ideal 6-V voltage source with its positive terminal at the top and in (b) is a 4-A ideal current source directed upward. All the other branches in both figures are 1-Ω resistors. Find i shown in each figure using graph theory methods.

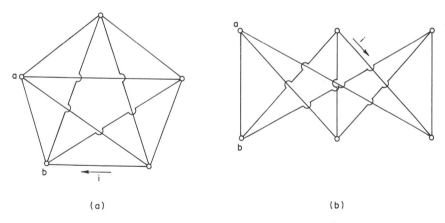

(a)   (b)

PROBLEM 6.8

6.9   Find v_1 and v_2 in Fig. 6.11 (a) using link currents as the unknowns.

6.10 Solve Prob. 4.22 using graph theory methods.

7

ENERGY-STORAGE
ELEMENTS

Up to now we have considered only resistive circuits, that is, circuits containing resistors and sources. The terminal characteristics of these elements are simple algebraic equations which lead to circuit equations that are algebraic. In this chapter we shall introduce two important dynamic circuit elements, the capacitor and the inductor, whose terminal equations are differential equations rather than algebraic equations. These elements are referred to as *dynamic* because, in the ideal case, they store energy which can be retrieved at some later time. Another term which is used, for this reason, is *storage* elements.

We shall first describe the property of capacitance and discuss the mathematical model of an ideal device. The terminal characteristics and energy relations will then be given, followed by derivations for parallel and series connections of two or more capacitors. We shall then repeat this procedure for the inductor.

The chapter will be concluded with a discussion of practical capacitors and inductors and their equivalent circuits.

7.1 CAPACITORS

A *capacitor* is a two-terminal device that consists of two conducting bodies that are separated by a nonconducting material. Such a nonconducting material is known as a *dielectric*. As a result of the dielectric, charges are not permitted to move from one conducting body to the other within the device. They must therefore be transported between the conducting bodies via external circuitry connected to the terminals of the capacitor. One very simple type, called a parallel-plate capacitor, is shown in Fig. 7.1. The conducting bodies are flat, rectangular conductors that are separated by the dielectric material.

To describe the charge-voltage relationship for the device, let us transfer charge from one plate to the other. Suppose, for instance, by means of some external circuit, that we take a small number of electrons, say $-\Delta q$, from the upper plate to the lower plate. This, of course, leaves a charge of $+\Delta q$ on the top plate and deposits a charge of $-\Delta q$ on the bottom plate. Since moving these charges requires the separation of unlike charges (recall that unlike charges attract one another), a small amount of work is performed, and the top plate is raised to a potential of say $+\Delta v$ with respect to the bottom plate.

Each increment of charge $-\Delta q$ that we transfer increases the potential difference between the plates by Δv. Therefore the potential difference between the plates is proportional to the charge being transferred. This suggests that a change in the terminal voltage by an amount Δv causes a corresponding change in the charge on the upper plate by an amount Δq. Thus the charge is proportional to the potential difference, and we may write

$$q = Cv \tag{7.1}$$

where C is the constant of proportionality, known as the *capacitance* of the device, in coulombs per volt. The unit of 1 C/V is known as the *farad* (F), named for the famous British physicist Michael Faraday (1791–1867). Capacitors which satisfy (7.1) are called *linear capacitors* since their charge-voltage relationship is the equation of a straight line having a slope of C.

It is interesting to note in the above example that the net charge within the capacitor is always zero. Charges removed from the top plate always appear on the lower one so that the total charge remains zero. We should also observe that charges leaving one terminal enter the other. This fact is consistent with the requirement that current entering one terminal must exit the other in a two-terminal device.

Since the current is defined as the rate of change of charge, differentiating (7.1), we find that

$$i = C\frac{dv}{dt} \tag{7.2}$$

which is the current-voltage relation for a capacitor.

The circuit symbol for the capacitor and the current-voltage convention which satisfy (7.2) are shown in Fig. 7.2. It is apparent that moving a charge of $-\Delta q$ in Fig. 7.1 from the upper to the lower conductor represents a current flowing into the upper

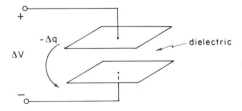

FIGURE 7.1 *Parallel-plate capacitor*

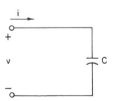

FIGURE 7.2 *Circuit symbol for a capacitor*

terminal. The movement of this charge causes the upper terminal to become more positive than the lower one by an amount Δv. Hence the current-voltage convention of Fig. 7.2 is satisfied. Obviously, if v is assigned so that the lower terminal is positive or if the current is assumed to enter the lower terminal, but not both, then a minus sign must be used in the right-hand side of (7.2). We recall that this is also necessary in the case of a resistor.

In (7.2) we see that if v is constant, then the current i is zero. Therefore a capacitor acts like an open circuit to a dc voltage. On the other hand, the more rapidly v changes, the larger is the current flowing through its terminals. Consider, for example, a voltage which increases linearly from 0 to 1 V in a^{-1} s, given by

$$
\begin{aligned}
v &= 0, & t &\leq 0 \\
&= at, & 0 &\leq t \leq a^{-1} \\
&= 1, & t &\geq a^{-1}
\end{aligned}
$$

If this voltage is applied to the terminals of a 1-F capacitor (an unusually large value which is convenient for illustrative purposes), the resulting current is

$$
\begin{aligned}
i &= 0, & t &< 0 \\
&= a, & 0 &< t < a^{-1} \\
&= 0, & t &> a^{-1}
\end{aligned}
$$

Plots of v and i are shown in Fig. 7.3. We see that i is zero when v is constant and that it is equal to a when v increases linearly. If a is made larger, then v changes more rapidly and i increases. It is apparent that if $a^{-1} = 0$ (a is infinite), v changes abruptly (in zero time) from 0 to 1 V.

In general any abrupt or instantaneous changes in voltage, such as in the above example, require that an infinite current flow through the capacitor. An infinite current, however, requires that an infinite power exist at the capacitor terminals, which

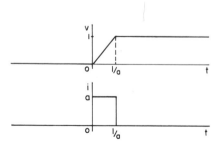

FIGURE 7.3 *Voltage and current waveforms for a 1-F capacitor*

is a *physical* impossibility. Thus abrupt or instantaneous changes in the voltage across a capacitor are *not* possible, and the voltage is continuous even though the current may be discontinuous. (It is possible, of course, to draw on paper circuits that contradict this statement. These circuits are mathematical models that do not sufficiently describe the entire physical picture, as we shall see in Sec. 7.9.) An alternative statement concerning abrupt changes in voltages in circuits containing more than one capacitor is that *the total charge cannot change instantaneously* (conservation of charge).

Let us now find $v(t)$ in terms of $i(t)$ by integrating both sides of (7.2) between times t_o and t. The result is

$$v(t) = \frac{1}{C} \int_{t_0}^{t} i \, dt + v(t_0) \tag{7.3}$$

where $v(t_0) = q(t_0)/C$ is the voltage on C at time t_0. In this equation, the integral term represents the voltage that accumulates on the capacitor in the interval from t_0 to t, whereas $v(t_0)$ is that which accumulates from $-\infty$ to t_0. The voltage $v(-\infty)$, of course, is taken to be zero. Thus an alternative form of (7.3) is

$$v(t) = \frac{1}{C} \int_{-\infty}^{t} i \, dt$$

In applying this result, we obviously are obtaining the area associated with a plot of i from $-\infty$ to t. In Fig. 7.3, for example, since $v(-\infty) = 0$, we have

$$v = \frac{1}{C} \int_{-\infty}^{t} (0) \, dt + v(-\infty) = 0, \qquad t \leq 0$$

Therefore $v(0) = 0$, and

$$v = \frac{1}{C} \int_{0}^{t} a \, dt + v(0) = at, \qquad 0 \leq t \leq a^{-1}$$

Similarly, $v(1/a) = 1$, which yields

$$v = \frac{1}{C} \int_{1/a}^{t} (0) \, dt + v\left(\frac{1}{a}\right) = 1, \qquad t \geq a^{-1}$$

(Recall that $C = 1$ F in this example.)

In this example, we see that v and i do not necessarily have the same shape. Specifically, the maximum and minimum values of v and i do not necessarily occur at at the same time, unlike the case for the resistor. In fact, inspection of Fig. 7.3 reveals that the current can be discontinuous even though the voltage is continuous, as stated previously.

EXERCISES

7.1.1   If a 100-μF capacitor has $v(t) = f(t)$ V, as shown, find (a) $i(-3)$, (b) $i(-1)$, (c) $i(1.5)$, and (d) $i(2.5)$. (e) Sketch $i(t)$ for all t.

Ans. (a) 50 μA, (b) -50 μA, (c) 0, (d) 0.1 mA

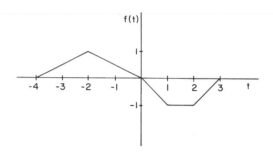

EXERCISE 7.1.1

7.1.2   If a 1000-μF capacitor has $i(t) = f(t)$ mA, where $f(t)$ is given in Ex. 7.1.1, find (a) $v(-2)$, (b) $v(0)$, and (c) $v(2)$. (d) Sketch $v(t)$ for all t. [Note: $v(-4) = 0$.]

Ans. (a) 1 V, (b) 2 V, (c) 0.5 V

7.1.3   A 1-μF capacitor has a terminal voltage of 100 cos 1000t V. Find $i(t)$.

Ans. -0.1 sin 1000t A

7.1.4   A 10-μF capacitor has a terminal current of $10e^{-100t}$ mA. If $v(0) = -10$ V, find $v(t)$ for $t > 0$.

Ans. $-10e^{-100t}$ V

7.2 ENERGY STORAGE IN CAPACITORS

The terminal voltage across a capacitor is accompanied by a separation of charges between the capacitor plates. These charges have electrical forces acting on them. An *electric field*, a basic quantity in electromagnetic theory, is defined as the force acting on a unit positive charge. Thus the forces acting on the charges within the capacitor can be considered to result from an electric field. It is for this reason that the energy stored or accumulated in a capacitor is said to be stored in the electric field.

The energy stored in a capacitor, from (1.6) and (7.2), is given by

$$w_C(t) = \int_{-\infty}^{t} vi \, dt = \int_{-\infty}^{t} v\left(C\frac{dv}{dt}\right) dt$$

$$= C \int_{-\infty}^{t} v \, dv = \tfrac{1}{2}Cv^2(t)\Big|_{t=-\infty}^{t}$$

Since $v(-\infty) = 0$, we may write

$$w_C(t) = \tfrac{1}{2}Cv^2(t) \text{ J} \tag{7.4}$$

From this result, we see that $w_C(t) \geq 0$. Therefore, from (1.7), the capacitor is a passive circuit element. In terms of the charge on the device, (7.1) and (7.4) yield

$$w_C(t) = \frac{1}{2}\frac{q^2(t)}{C} \text{ J} \tag{7.5}$$

The ideal capacitor, unlike the resistor, *cannot* dissipate any energy. The energy which is stored in the device can thus be recovered. Consider, for instance, a 1-F capacitor which has a voltage of 10 V. The energy stored is

$$w_C = \tfrac{1}{2}Cv^2 = 50 \text{ J}$$

Suppose the capacitor is not connected in a circuit; then no current can flow, and the charge, voltage, and energy remain constant. If we now connect a resistor across the capacitor, a current flows until all the energy (50 J) is absorbed as heat by the resistor and the voltage across the combination is zero. The analysis of such a network is found in the next chapter.

As has been pointed out earlier, the voltage on a capacitor is a continuous function. Thus by (7.4) we see that the energy stored in a capacitor is also continuous. This is not surprising since otherwise energy would have to be transported from one place to another in zero time, which is an impossibility.

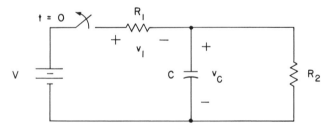

FIGURE 7.4 *Circuit illustrating continuity of capacitor voltage*

To illustrate continuity of capacitor voltage, let us consider Fig. 7.4, which contains a switch that is opened at $t = 0$, as indicated. (Ideally, a switch transforms a pair of terminals from an open circuit to a short circuit, or vice versa, in zero time.) To discuss the effect of the switching action we first need to consider two different types of time $t = 0$. We shall denote $t = 0^-$ as the time just before the switching action and $t = 0^+$ as the time just after the switching action. Theoretically, of course, no time has elapsed between 0^- and 0^+, but the two times represent radically different states of the circuit. Thus $v_C(0^-)$ is the voltage on the capacitor just before the switch is moved and $v_C(0^+)$ is the voltage immediately after switching. Mathematically, $v_C(0^-)$ is the limit of $v_C(t)$ as t approaches zero through negative ($t < 0$) values and $v_C(0^+)$ is the limit as t

approaches zero through positive $(t > 0)$ values. The same notation applies to v_1 across the resistor R_1.

Suppose that in Fig. 7.4 we have $V = 6$ V and $v_C(0^-) = 4$ V. Just prior to the switching action $(t = 0^-)$ we have $v_1(0^-) = V - v_C(0^-) = 2$ V. Immediately after the switch is opened we have $v_1(0^+) = 0$, since no current is flowing in R_1. However, since v_C is continuous we have

$$v_C(0^+) = v_C(0^-) = 4\,\text{V}$$

Thus the voltage on the resistor has changed abruptly, but that on the capacitor has not.

Obviously, one could consider circuits on paper in which capacitor voltages are forced to change abruptly. For example, if two capacitors having different voltages are suddenly connected in parallel by a switching action, their resulting common voltage cannot be the same as both their previous, different voltages. We shall consider such *singular* circuits in Sec. 7.9, where we shall see that stored energy has *appeared* to change abruptly. The apparent change cannot be accounted for in the lumped circuit models we are using, but it is a remarkable fact that the lumped models are valid before and after (though not during) the switching action. Physical circuits, however, have resistance associated with the capacitor (such as in the leads and the dielectric) which precludes the infinite currents that must accompany discontinuous capacitor voltages. These are the type circuits we shall be concerned with, in general.

EXERCISES

7.2.1 A 1-μF capacitor is charged to 100 V. Find the charge and energy.

Ans. 10^{-4} C, 5 mJ

7.2.2 In Fig. 7.4, let $C = \frac{1}{3}$ F, $R_1 = R_2 = 3\,\Omega$, and $V = 9$ V. If the current in R_2 at $t = 0^-$ is 1 A directed downward, find at $t = 0^-$ and at $t = 0^+$ (a) the charge on the capacitor, (b) the current in R_1 directed to the right, and (c) the current in C directed downward.   *Ans.* (a) 1, 1 C, (b) 2, 0 A, (c) 1, -1 A

7.3 SERIES AND PARALLEL CAPACITORS

In this section we shall determine the equivalent capacitance for series and parallel connections of capacitors. As we shall see, equivalent capacitance is in direct analogy with equivalent conductance.

Let us first consider the series connection of N capacitors, as shown in Fig. 7.5(a). Applying KVL, we find

$$v = v_1 + v_2 + \ldots + v_N \tag{7.6}$$

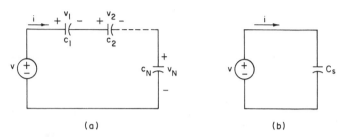

(a)   (b)

FIGURE 7.5 *(a) Series connection of N capacitors; (b) equivalent circuit*

From (7.3), this equation can be written

$$v = \frac{1}{C_1} \int_{t_0}^{t} i \, dt + v_1(t_0) + \frac{1}{C_2} \int_{t_0}^{t} i \, dt + v_2(t_0)$$

$$+ \ldots + \frac{1}{C_N} \int_{t_0}^{t} i \, dt + v_N(t_0)$$

$$= \left(\frac{1}{C_1} + \frac{1}{C_2} + \ldots + \frac{1}{C_N} \right) \int_{t_0}^{t} i \, dt + v_1(t_0) + v_2(t_0) + \ldots + v_N(t_0)$$

or, by (7.6),

$$v(t) = \left(\sum_{n=1}^{N} \frac{1}{C_n} \right) \int_{t_0}^{t} i \, dt + v(t_0)$$

In Fig. 7.5(b), we see that

$$v = \frac{1}{C_s} \int_{t_0}^{t} i \, dt + v(t_0)$$

where $v(t_0)$ is the voltage on C_s at $t = t_0$.

Suppose we require that the circuit of Fig. 7.5(b) be an equivalent circuit for that of Fig. 7.5(a). Comparing the last two equations, we see that

$$\frac{1}{C_s} = \frac{1}{C_1} + \frac{1}{C_2} + \ldots + \frac{1}{C_N} = \sum_{n=1}^{N} \frac{1}{C_n} \tag{7.7}$$

from which we may find the equivalent capacitance C_s.

As an example of the utility of (7.6) and (7.7), consider the series connection of 1- and $\frac{1}{3}$-F capacitors having initial voltages of 4 and 6 V, respectively. Then

$$\frac{1}{C_s} = 1 + 3$$

or

$$C_s = 0.25 \text{ F}$$

and

$$v(t_0) = 4 + 6 = 10 \text{ V}$$

Let us now consider the parallel connection of N capacitors, as shown in Fig. 7.6(a). Application of KCL gives

$$i = i_1 + i_2 + \ldots + i_N$$

Substituting from (7.2), we have

$$i = C_1 \frac{dv}{dt} + C_2 \frac{dv}{dt} + \ldots + C_N \frac{dv}{dt}$$

$$= (C_1 + C_2 + \ldots + C_N) \frac{dv}{dt} = \left(\sum_{n=1}^{N} C_n \right) \frac{dv}{dt}$$

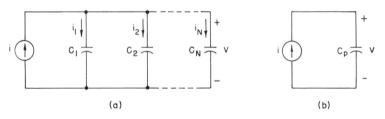

(a)

(b)

FIGURE 7.6 *Parallel connection of N capacitors; (b) equivalent circuit*

In the circuit of Fig. 7.6(b), the current is

$$i = C_p \frac{dv}{dt}$$

If we now require that this circuit be an equivalent circuit for that of Fig. 7.6(a), the above equations give

$$C_p = C_1 + C_2 + \ldots + C_N = \sum_{n=1}^{N} C_n \tag{7.8}$$

Thus the equivalent capacitance of N parallel capacitors is simply the sum of the individual capacitances. An initial voltage, of course, would be equal to that which is present across the parallel combination.

It is interesting to notice that the equivalent capacitance of series and parallel capacitors is analogous to the equivalent conductance of series and parallel conductances.

EXERCISES

7.3.1 Find the equivalent capacitance and the initial voltage.   *Ans.* 0.1 F, −2 V

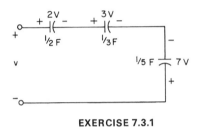

EXERCISE 7.3.1

7.3.2 Find the equivalent capacitance.   *Ans.* 6 F

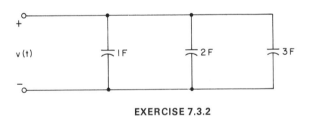

EXERCISE 7.3.2

7.3.3 Find the equivalent capacitance.   *Ans.* 3 μF

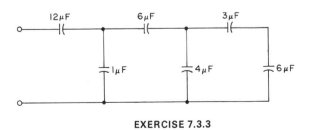

EXERCISE 7.3.3

7.4   INDUCTORS

In the previous sections, we found that the electrical characteristics of the capacitor are the result of forces that exist between electric charges. Just as static charges exert forces upon one another, it is found that moving charges, or currents, also influence one another. The force which is experienced by two neighboring current-carrying wires was experimentally determined by Ampere in the early nineteenth century. These forces can be characterized by the existence of a *magnetic field*. The magnetic field, in turn, can be thought of in terms of a *magnetic flux* that forms closed loops about electric currents. The origin of the flux, of course, is the electric currents. The study of magnetic fields, like that of electric fields, comes in a later course on electromagnetic theory.

An inductor is a two-terminal device that consists of a coiled conducting wire. A current flowing through the device produces a magnetic flux ϕ which forms closed loops encircling the coils making up the inductor, as shown by the simple model of Fig. 7.7. Suppose that the coil contains N turns and that the flux ϕ passes through each turn. In this case, the total flux linked by the N turns of the coil is

$$\lambda = N\phi$$

This total flux is commonly referred to as the *flux linkage*. The unit of magnetic flux is the *weber* (Wb), named for the German physicist Wilhelm Weber (1804–1891).

In a linear inductor, the flux linkage is directly proportional to the current flowing through the device. Therefore, we may write

$$\lambda = Li \tag{7.9}$$

where L, the constant of proportionality, is the *inductance* in webers per ampere. The unit of 1 Wb/A is known as the *henry* (H), named for the American physicist Joseph Henry (1797–1878).

In (7.9) we see that an increase in i produces a corresponding increase in λ. This increase in λ produces a voltage in the N-turn coil. The fact that voltages occur with changing magnetic flux was first discovered by Henry. Henry, however, repeating the mistake of Cavendish with the resistor, failed to publish his findings. As a result, the famous Michael Faraday is credited with discovering the law of electromagnetic induction. This law states that the voltage is equal to the time rate of change of the total magnetic flux. In mathematical form, the law is

$$v = \frac{d\lambda}{dt}$$

Thus, differentiating (7.9), we find

$$v = L\frac{di}{dt} \tag{7.10}$$

Clearly, as i increases, a voltage is developed across the terminals of the inductor, the polarity of which is shown in Fig. 7.7. This voltage opposes an increase in i, for if this were not the case, that is, if the polarity were reversed, the induced voltage would "aid" the current. This physically cannot be true because the current would increase indefinitely.

The circuit symbol and the current-voltage convention for the inductor is shown in Fig. 7.8. Just as in the cases of the resistor and the capacitor, if either the current direction or the voltage assignment, but not both, are reversed, then a negative sign must be employed in the right-hand side of (7.10).

An examination of (7.10) shows that if i is constant, then the voltage v is zero. Therefore an inductor acts like a short circuit to a dc current. On the other hand, the more rapidly i changes, the greater is the voltage that appears across its terminals. Consider, for instance, a current that decreases linearly from 1 to 0 A in b^{-1} s, defined by

$$i = 1, \qquad t \leq 0$$
$$= 1 - bt, \qquad 0 \leq t \leq b^{-1}$$
$$= 0, \qquad t \geq b^{-1}$$

A 1-H inductor having this terminal current has a terminal voltage given by

$$v = 0, \qquad t < 0$$
$$= -b, \qquad 0 < t < b^{-1}$$
$$= 0, \qquad t > b^{-1}$$

Plots of i and v for this case are shown in Fig. 7.9. We see that v is zero when i is constant and is equal to $-b$ when i decreases linearly. If b is made larger, i changes

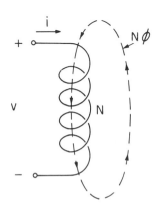

FIGURE 7.7 *Simple model of an inductor*

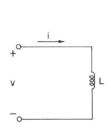

FIGURE 7.8 *Circuit symbol for the inductor*

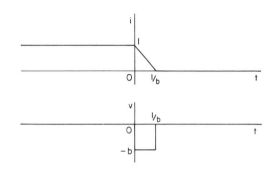

FIGURE 7.9 *Current and voltage waveforms for a 1-H inductor*

more rapidly and v becomes more negative. Clearly, if $b^{-1} = 0$ (b infinite), then i changes abruptly from 1 to 0 A.

In general, abrupt changes in the current, such as in the above example, require that an infinite voltage appear across the terminals of the inductor. As described in the case of the capacitor, this requires that an infinite power exist at the terminals of the inductor, a physical impossibility. Thus instantaneous changes in the current through an inductor are *not* possible. We observe that the current is continuous even though the voltage may be discontinuous.

An alternative statement concerning abrupt changes of the currents flowing in circuits containing more than one inductor is that *the total flux linkage cannot change instantaneously*. That is, for a circuit containing inductors $L_1, L_2, \ldots, L_N$, the sum $\lambda_1 + \lambda_2 + \ldots + \lambda_N$ cannot change instantaneously. If we compare (7.1) and (7.9), we see that the flux linkage in an inductor is analogous to the charge on a capacitor. Thus the sum of the flux linkages given above (conservation of flux linkage) is analogous to conservation of charge. An example employing the conservation of flux linkage is given in Sec. 7.9.

Let us now find the current $i(t)$ in terms of the voltage $v(t)$. Integrating (7.10) from time t_0 to t and solving for $i(t)$, we have

$$i(t) = \frac{1}{L} \int_{t_0}^{t} v(t)\, dt + i(t_0) \tag{7.11}$$

In this equation, the integral term represents the current buildup from time t_0 to t, whereas $i(t_0)$ is the current at t_0. Obviously, $i(t_0)$ is the current which accumulates from $t = -\infty$ to t_0, where $i(-\infty) = 0$. Thus an alternative expression is

$$i(t) = \frac{1}{L} \int_{-\infty}^{t} v(t)\, dt$$

In the application of (7.11), we are obtaining the net area under the graph of v from $-\infty$ to t, since $i(t_0)$ represents the area from $-\infty$ to t_0. In Fig. 7.9, for instance, since $i(0) = 1$, we have, for $L = 1$ H,

$$i(t) = \frac{1}{L} \int_{0}^{t} (-b)\, dt + i(0) = -bt + 1, \qquad 0 \leq t \leq b^{-1}$$

Thus $i(1/b) = 0$, and

$$i(t) = \frac{1}{L} \int_{1/b}^{t} (0)\, dt + i\left(\frac{1}{b}\right) = 0, \qquad b^{-1} \leq t$$

In the above example, we see that v and i, just as in the case of the capacitor, do not necessarily have the same variation in time. Inspection of Fig. 7.9, for example, shows that the voltage can be discontinuous even though the current is continuous, as previously mentioned.

EXERCISES

7.4.1 If a 100-mH inductor has $i(t) = f(t)$ mA, where $f(t)$ is given in Ex. 7.1.1, find (a) $v(-3)$, (b) $v(-1)$, (c) $v(1.5)$, and (d) $v(2.5)$. (e) Sketch $v(t)$ for all t.
Ans. (a) 50, (b) -50, (c) 0, (d) 100 μV

7.4.2 If a 10-mH inductor has $v(t) = f(t)$ mV, where $f(t)$ is given in Ex. 7.1.1, find (a) $i(-2)$, (b) $i(0)$, and (c) $i(2)$. (d) Sketch $i(t)$ for all t. [Note: $i(-4) = 0$.]
Ans. (a) 0.1 A, (b) 0.2 A, (c) 50 mA

7.4.3 A 1-mH inductor has a terminal current of 10 sin 100t A. Find $v(t)$.
Ans. cos 100t V

7.4.4 Find the current for $t > 0$ in a 100-mH inductor having a terminal voltage of $10e^{-100t}$ V if $i(0) = 0$ A.
Ans. $1 - e^{-100t}$ A

7.5 ENERGY STORAGE IN INDUCTORS

A current i flowing through an inductor causes a total flux linkage λ to be produced that passes through the turns of the coils making up the device. Just as work was performed in moving charges between the plates of a capacitor, a similar work is necessary to establish the flux ϕ in the inductor. The work or energy required in this case is said to be stored in the magnetic field.

The energy stored in an inductor, employing (1.6) and (7.10), is given by

$$w_L(t) = \int_{-\infty}^{t} vi\, dt = \int_{-\infty}^{t} \left(L \frac{di}{dt} \right) i\, dt$$

$$= L \int_{-\infty}^{t} i\, di = \tfrac{1}{2} L i^2(t) \Big|_{t=-\infty}^{t}$$

Recalling that $i(-\infty) = 0$, we have

$$w_L(t) = \tfrac{1}{2} L i^2(t) \text{ J} \tag{7.12}$$

Inspection of this equation reveals that $w_L(t) \geq 0$. Therefore, from (1.7), we see that the inductor is a passive circuit element.

The ideal inductor, like the ideal capacitor, does not dissipate any power. Therefore the energy stored in the inductor can be recovered. Consider, for example, a 2-H inductor that is carrying a current of 5 A. The energy stored is

$$w_L = \tfrac{1}{2}Li^2 = 25 \text{ J}$$

Suppose the inductor, by means of an external circuit, is connected in parallel with a resistor. In this case, a current flows through the inductor-resistor combination until all the energy previously stored in the inductor (25 J) is absorbed by the resistor and the current is zero. Solutions of circuits of this type are found in the next chapter.

Since inductor currents are continuous, it follows that the energy stored in an inductor, like that stored in a capacitor, is also continuous. To illustrate this let us consider the circuit of Fig. 7.10, which contains a switch that is closed at $t = 0$, as indicated. Suppose that $i_L(0^-) = 2$ A and $I = 3$ A. Then by KCL, $i_1(0^-) = 3 - 2 = 1$ A. After the switch is closed ($t = 0^+$), we have $i_1(0^+) = 0$ since a short circuit is placed across R_1. However, we have

$$i_L(0^+) = i_L(0^-) = 2 \text{ A}$$

Thus the resistor current has changed abruptly but the inductor current has not.

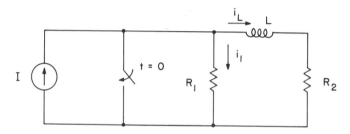

FIGURE 7.10 *Circuit illustrating continuity of inductor current*

An example of a *singular* circuit for which inductor currents *appear* to be discontinuous is given in Sec. 7.9. As in the case of singular capacitive circuits, the apparent discontinuity in the energy stored in the inductors cannot be accounted for in the lumped circuit model. However, lumped circuit theory is valid before and after (though not during) the switching action. Physical circuits having inductors contain associated resistance which does not permit the infinite inductor voltages that must accompany abrupt changes in inductor currents. We shall be concerned primarily with circuits of this type.

EXERCISES

7.5.1 Derive an expression for the energy stored in an inductor in terms of the flux linkage λ and the inductance L.
Ans. $\lambda^2/2L$

7.5.2 A 10-mH inductor has a current of 100 mA. Find the flux linkage and the energy.
Ans. 1 mWb, 50 μJ

7.6 SERIES AND PARALLEL INDUCTORS

In this section we shall determine the equivalent inductance for series and parallel connections of inductors. Let us first consider a series connection of N inductors, as shown in Fig. 7.11(a). Applying KVL, we see that

$$v = v_1 + v_2 + \ldots + v_N$$

from which we may write

$$v = L_1 \frac{di}{dt} + L_2 \frac{di}{dt} + \ldots + L_N \frac{di}{dt}$$

$$= (L_1 + L_2 + \ldots + L_N) \frac{di}{dt}$$

$$= \left(\sum_{n=1}^{N} L_n \right) \frac{di}{dt}$$

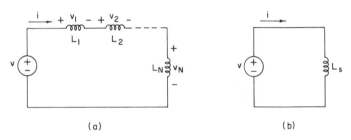

(a)   (b)

FIGURE 7.11 *(a) Series connection of N inductors; (b) equivalent circuit*

For the circuit of Fig. 7.11(b), the voltage is

$$v = L_s \frac{di}{dt}$$

If we now require that this circuit be an equivalent circuit for the series connection, then the above equations yield

$$L_s = L_1 + L_2 + \ldots + L_N = \sum_{n=1}^{N} L_n \tag{7.13}$$

Therefore the equivalent inductance of N series inductors is simply the sum of the individual inductances. In addition, an initial current would clearly be equal to that flowing in the series connection.

Let us now consider the parallel connection of N inductors, as shown in Fig. 7.12(a). Application of KCL gives

$$i = i_1 + i_2 + \ldots + i_N \tag{7.14}$$

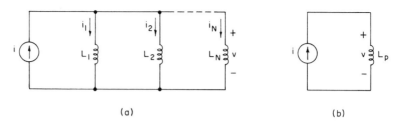

(a) (b)

FIGURE 7.12 *(a) Parallel connection of N inductors; (b) equivalent circuit*

Substituting from (7.11), we have

$$i = \frac{1}{L_1}\int_{t_0}^{t} v\,dt + i_1(t_0) + \frac{1}{L_2}\int_{t_0}^{t} v\,dt + i_2(t_0) + \ldots + \frac{1}{L_N}\int_{t_0}^{t} v\,dt + i_N(t_0)$$

$$= \left(\frac{1}{L_1} + \frac{1}{L_2} + \ldots + \frac{1}{L_N}\right)\int_{t_0}^{t} v\,dt + i_1(t_0) + i_2(t_0) + \ldots + i_N(t_0)$$

or, by (7.14),

$$i(t) = \left(\sum_{n=1}^{N}\frac{1}{L_n}\right)\int_{t_0}^{t} v\,dt + i(t_0)$$

In Fig. 7.12(b), we see that

$$i = \frac{1}{L_p}\int_{t_0}^{t} v\,dt + i(t_0)$$

where $i(t_0)$ is the current in L_p at $t = t_0$. If this circuit is an equivalent network for the parallel connection, then the above equations require that the equivalent parallel inductance be given by

$$\frac{1}{L_p} = \frac{1}{L_1} + \frac{1}{L_2} + \ldots + \frac{1}{L_N} = \sum_{n=1}^{N}\frac{1}{L_n} \tag{7.15}$$

As an example of the application of (7.14) and (7.15), suppose we have two parallel inductors of 2 and 3 H carrying initial currents of 2 and 1 A, respectively. The parallel combination could be replaced by a single inductance L_p satisfying

$$\frac{1}{L_p} = \frac{1}{2} + \frac{1}{3}$$

or

$$L_p = \tfrac{6}{5}\,\text{H}$$

carrying an initial current of

$$i(t_0) = 2 + 1 = 3\,\text{A}$$

In the case of inductors, it is interesting to observe that the equivalent inductance for series and parallel connections is analogous to the equivalent resistance of series and parallel resistors.

EXERCISES

7.6.1 Find the equivalent inductance.   *Ans.* 9 H

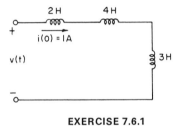

EXERCISE 7.6.1

7.6.2 Find the equivalent inductance and the initial current. The currents shown are initial values.   *Ans.* 0.1 H, −2 A

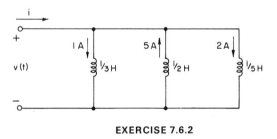

EXERCISE 7.6.2

7.6.3 Find the equivalent inductance.   *Ans.* 3 H

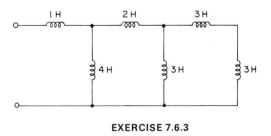

EXERCISE 7.6.3

7.7 PRACTICAL CAPACITORS
AND INDUCTORS*

Commercially available capacitors are manufactured in a wide variety of types, values, and voltage ratings. The capacitor type is generally classified by the kind of dielectric used, and its value is determined by the type of dielectric and the physical geometry of the device. The voltage rating, or *working voltage*, is the maximum voltage that can safely be applied to the capacitor. Voltages exceeding this value may permanently damage the device by destroying or breaking down the dielectric.

Simple capacitors are often constructed employing two sheets of metal foil which are separated by a dielectric material. The foil and dielectric are pressed together into a laminar form and are then rolled or folded into a compact package. Electrical conductors attached to each metal-foil sheet constitute the terminals of the capacitor.

Practical capacitors, unlike ideal capacitors, generally dissipate a small amount of power. This is due primarily to *leakage* currents that occur within the dielectric material in the device. Practical dielectrics have a nonzero conductance which allows an *ohmic* current to flow between the capacitor plates. This current is easily included in an equivalent circuit for the device by placing a resistance in parallel with an ideal capacitance, as shown in Fig. 7.13. In this figure, R_c represents the ohmic losses of the dielectric and C the capacitance. The leakage resistance R_c is inversely proportional to the capacitance C. Therefore the product of the leakage resistance and capacitance R_cC, a quantity often given by manufacturers, is useful in specifying the capacitor loss.

Common types of capacitors include ceramic (barium titanate), Mylar, Teflon, and polystyrene. These types are available in capacitance values ranging typically from 100 pF to 1 μF having tolerances of 3, 10, and 20%. Resistance-capacitance products for these types range from 10^3 Ω-F (ceramic) to 2×10^6 Ω-F (Teflon).

Another type of capacitor which gives larger values of C is the electrolytic capacitor. This capacitor is constructed of polarized layers of aluminum oxide or tantalum oxide and has values of 1 to 100,000 μF. Resistance-capacitance products, however, range from 10 to 10^3 Ω-F, which indicates that electrolytics are more *lossy* than nonelectrolytic types. Also, since electrolytic capacitors are polarized, they must be connected into a circuit with the proper voltage polarity. If the incorrect polarity is used, the oxide will be reduced, and heavy conduction will occur between the plates.

Practical inductors, like practical capacitors, usually dissipate a small amount of power. This dissipation results from ohmic losses associated with the wire making up the inductor coil and *core* losses due to induced currents arising in the core on which the coil is wound. An equivalent circuit for an inductor can be realized by placing a resistance in series with an ideal inductor, as shown in Fig. 7.14, where R_L represents the ohmic losses and L the inductance.

Inductors are available with values ranging from 1 μH to 100 H. Large inductance values are obtained by employing many turns and ferrous (iron) core materials; hence as the inductance increases, the series resistance generally increases.

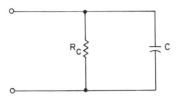
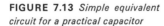
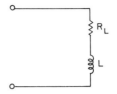

FIGURE 7.13 *Simple equivalent circuit for a practical capacitor*

FIGURE 7.14 *Simple equivalent circuit for a practical inductor*

Like the resistor and the operational amplifier, the capacitor can be fabricated in integrated-circuit form. However, attempts at integrating the inductor have not been very successful because of geometry constraints and because semiconductors do not exhibit the necessary magnetic properties. For this reason, in many applications, circuits are designed using only resistors, capacitors, and electronic devices, such as op amps.

EXERCISE

7.7.1 Mylar capacitors have a resistance-capacitance product of 10^5 Ω-F. Find the equivalent parallel resistor in Fig. 7.13 for the following capacitors: (a) 100 pF, (b) 0.1 μF, and (c) 1 μF.   *Ans.* (a) 10^{15} Ω, (b) 10^{12} Ω, (c) 10^{11} Ω

7.8 DUALITY AND LINEARITY

Let us now determine the dual relationships for the capacitor and the inductor. This is easily done by considering the current-voltage relations of (7.2) and (7.10) for the elements. Repeating these equations for convenience, we have

$$i = C \frac{dv}{dt} \qquad (7.16)$$

and

$$v = L \frac{di}{dt} \qquad (7.17)$$

Comparing the equations, we see that replacing i by v, v by i, and C by L in the first equation yields the second equation. Therefore it is clear that the capacitor and the inductor are dual elements. A similar comparison of the equations for charge and flux, given by (7.1) and (7.9), shows that these are also dual quantities. A summary of the dual quantities that we shall consider in the book is given in Table 7.1.

TABLE 7.1 *Dual quantities*

Voltage	Current
Charge	Flux
Resistance	Conductance
Inductance	Capacitance
Short circuit	Open circuit
Impedance	Admittance*
Nonreference node	Mesh
Reference node	Outer mesh
Tree branch	Link*
Series	Parallel
KCL	KVL

* Tree branch and link were considered in Chapter 6, which the reader may have omitted, and impedance and admittance will be considered in Chapter 11.

We are now able to construct dual circuits for networks containing the dual quantities listed in Table 7.1. Consider, for example, the two-mesh network of Fig. 7.15(a). Using the geometric method of Sec. 4.7, the dual circuit is shown dashed in Fig. 7.15(a) and is redrawn in Fig. 7.15(b). The solutions for the currents and voltages

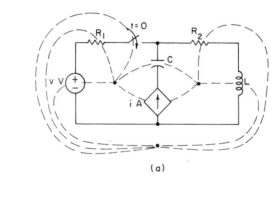

(a)

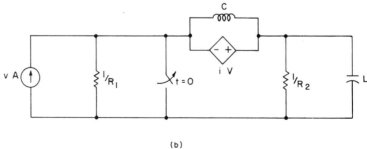

(b)

FIGURE 7.15 *(a) Two-mesh network; (b) dual network*

in networks including combinations of resistors, capacitors, and inductors will be considered in the coming chapters.

Let us now consider the property of linearity for capacitors and inductors, which are defined by (7.16) and (7.17), respectively. Comparing these equations to that of (5.2), we see that the elements satisfy the proportionality property and that they are therefore linear elements. Thus circuits containing any combination of independent sources, and linear dependent sources, resistors, capacitors, and inductors are linear, and superposition and Thevenin's or Norton's theorems are applicable. These topics will be considered in later chapters.

EXERCISE

7.8.1   Construct dual circuits for the networks of (a) Fig. 7.4, (b) Fig. 7.10, (c) Fig. 7.12, and (d) Fig. 7.14.

7.9   SINGULAR CIRCUITS*

A circuit in which a switching action takes place that *appears* to produce discontinuities in capacitor voltages or inductor currents is sometimes called a *singular* circuit. In this section we shall consider two such circuits, one containing capacitors and one containing inductors.

Let us consider first Fig. 7.16, where C_1 and C_2 have voltages of 1 and 0 V, respectively, prior to the closing of the switch. That is, $v_1(0^-) = 1$ V and $v_2(0^-) = 0$ V. We shall now determine $w_1(0^+)$ and $w_2(0^+)$, the energies stored in C_1 and C_2 at $t = 0^+$.

The energy stored in the circuit prior to closing the switch is

$$w_1(0^-) = \tfrac{1}{2}C_1 v_1^2(0^-) = \tfrac{1}{2} \text{ J}$$

and

$$w_2(0^-) = \tfrac{1}{2}C_2 v_2^2(0^-) = 0 \text{ J}$$

The charge on each capacitor at this time is

$$q_1(0^-) = C_1 v_1(0^-) = 1 \text{ C}$$

and

$$q_2(0^-) = C_2 v_2(0^-) = 0 \text{ C}$$

After the switch closes, we see from KVL that

$$v_1(0^+) = v_2(0^+)$$

Therefore, from (7.1), the charge on each capacitor satisfies

$$q_1(0^+) = q_2(0^+)$$

In addition, when the switch is closed, conservation of charge requires that the total charge remain constant; hence

$$q_1(0^-) + q_2(0^-) = q_1(0^+) + q_2(0^+)$$

Combining these equations, we find that

$$q_1(0^+) + q_2(0^+) = 2q_1(0^+) = 2q_2(0^+) = 1 \text{ C}$$

or

$$q_1(0^+) = q_2(0^+) = \tfrac{1}{2} \text{ C}$$

Substituting into (7.1), we find

$$v_1(0^+) = v_2(0^+) = \tfrac{1}{2} \text{ V}$$

Therefore the energy stored in each capacitor at $t = 0^+$ is

$$w_1(0^+) = w_2(0^+) = \tfrac{1}{2}C_1 v_1^2(0^+) = \tfrac{1}{8} \text{ J}$$

Let us now compare the energy in the system before and after closing the switch. At $t = 0^-$,

$$w_1(0^-) + w_2(0^-) = \tfrac{1}{2} \text{ J}$$

At $t = 0^+$,

$$w_1(0^+) + w_2(0^+) = \tfrac{1}{4} \text{ J}$$

We know that capacitors do not dissipate power. What, then, has happened to the $\tfrac{1}{4}$ J from $t = 0^-$ to $t = 0^+$? Looking back over our work, we see that v_1 changes abruptly from 1 to $\tfrac{1}{2}$ V at $t = 0$. As pointed out in Sec. 7.1, instantaneous changes in the voltage are not possible. Therefore during the infinitesimal time from $t = 0^-$ to $t = 0^+$, our mathematical model is not valid. In reality, what has happened is the following. When the switch closes at time $t = 0$, a large current is produced as charges are transferred from C_1 to C_2. This rapidly changing current gives rise to an electromagnetic wave which *radiates* $\tfrac{1}{4}$ J of energy. The voltage v_1 changes in a short, but nonzero, time from 1 to $\tfrac{1}{2}$ V. Our network during this interval does not behave as a lumped-parameter circuit, and concepts from electromagnetic theory (a later course) are required for the solution we have described.

Although our circuit model is not valid at the instant the switch closes, the solutions for the voltages and energies before and after the closing of the switch *are correct*. This is due entirely to the fact that the total charge did not change during this time interval.

As pointed out earlier, most circuit models do not permit an infinite current in a capacitance. Physical circuits normally have a finite value of resistance and inductance which limit such currents. As a result, the capacitor voltages and energies are continuous functions. If, for example, a series resistance is included in the circuit of Fig. 7.16, the voltage on each capacitor is continuous. That is,

$$v_1(0^-) = v_1(0^+)$$

and

$$v_2(0^-) = v_2(0^+)$$

Analyses for circuits of this type are given in Chapter 8.

For a second example, consider the circuit of Fig. 7.17. Inductors L_1 and L_2 have currents of 1 and 0 A, respectively, before the switch is closed at time $t = 0$. Therefore $i_1(0^-) = 1$ A and $i_2(0^-) = 0$ A. We shall now determine $w_1(0^+)$ and $w_2(0^+)$, the energy stored in the inductors at $t = 0^+$.

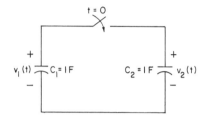

FIGURE 7.16 *Circuit containing two capacitors which are switched at time $t = 0$*

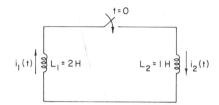

FIGURE 7.17 *Circuit containing two inductors which are switched at time $t = 0$*

The energy stored in the network prior to the closing of the switch is

$$w_1(0^-) = \tfrac{1}{2}L_1 i_1^2(0^-) = 1 \text{ J}$$

and

$$w_2(0^-) = \tfrac{1}{2}L_2 i_2^2(0^-) = 0 \text{ J}$$

The flux linkage of each inductor at this time is

$$\lambda_1(0^-) = L_1 i_1(0^-) = 2 \text{ Wb}$$

and

$$\lambda_2(0^-) = L_2 i_2(0^-) = 0 \text{ Wb}$$

After the switch is closed, we see from KCL that

$$i_1(0^+) = i_2(0^+)$$

Also, when the switch is closed, conservation of flux linkage requires that the total flux remain constant; hence

$$L_1 i_1(0^-) + L_2 i_2(0^-) = L_1 i_1(0^+) + L_2 i_2(0^+)$$
$$= (L_1 + L_2)i_1(0^+)$$

or

$$2(1) + 1(0) = 3i_1(0^+)$$

Therefore

$$i_1(0^+) = i_2(0^+) = \tfrac{2}{3} \text{ A}$$

Thus the energy stored in each inductor at $t = 0^+$ is

$$w_1(0^+) = \tfrac{1}{2}L_1 i_1^2(0^+) = \tfrac{4}{9} \text{ J}$$

and

$$w_2(0^+) = \tfrac{1}{2}L_2 i_2^2(0^+) = \tfrac{2}{9} \text{ J}$$

If we now compare the total energy stored in the network, we see at $t = 0^-$ that

$$w_1(0^-) + w_2(0^-) = 1 \text{ J}$$

and at $t = 0^+$ that

$$w_1(0^+) + w_2(0^+) = \tfrac{2}{3} \text{ J}$$

which indicates that $\tfrac{1}{3}$ J has been lost by the circuit even though ideal inductors can dissipate no power.

Looking back over the problem, we see that $i_1(t)$ changes abruptly from 1 to $\tfrac{2}{3}$ A at $t = 0$. We know, however, that abrupt changes in the current are not possible. Therefore during the infinitesimal time from $t = 0^-$ to $t = 0^+$, our mathematical model is once again not valid. As pointed out previously, a rapidly changing current gives rise to an electromagnetic wave which radiates energy. In this case $\tfrac{1}{3}$ J is radiated, and the current changes from 1 to $\tfrac{2}{3}$ A in a short, but nonzero, time. Our circuit, of course, does not behave as a lumped-parameter network during this interval of time.

As in the case of a capacitive circuit having infinite currents, most inductive circuit models do not permit infinite voltages to occur across an inductor as a result of abrupt currents. As discussed in the case of the capacitor, physical circuits having inductors contain a finite value of resistance and capacitance which limit such voltages. The currents and energies in these circuits are continuous functions. If, for instance, a parallel resistance is included in Fig. 7.17, the currents are continuous at $t = 0$. Circuits of this type are studied in Chapter 8.

EXERCISES

7.9.1 If in Fig. 7.16 C_2 is changed to $\tfrac{1}{2}$ F and $v_1(0^-)$ to 1 V, find the voltage and the total stored energy at $t = 0^+$. *Ans.* $\tfrac{2}{3}$ V, $\tfrac{1}{3}$ J

7.9.2 If $L_2 = 3$ H in Fig. 7.17, find the current and the total energy after the switch is closed if $i_1(0^-) = 1$ A. *Ans.* $\tfrac{2}{3}$ A, $\tfrac{2}{3}$ J

PROBLEMS

7.1 Find the charge residing on each plate of a 2-μF capacitor that is charged to 100 V. If the same charge resides on a 1-μF capacitor, what is the voltage?

7.2 Determine the voltage required to store 50 μC on a 0.25-μF capacitor. What time will be required for a constant current of 25 mA to deliver this charge?

7.3 The voltage on a 0.01-μF capacitor increases linearly from 0 V at $t = 0$ to 100 V at $t = 10$ ms. It then decreases linearly to 0 V at $t = 30$ ms. Find (a) $q(5)$, (b) $i(15)$, and (c) $p(15)$ (arguments in milliseconds). (d) Sketch $q(t)$, $i(t)$, and $p(t)$.

7.4 Determine the current flowing in a 0.1-μF capacitor having the following terminal voltages: (a) 10, (b) 100t, (c) $50e^{-10t}$, (d) 150 cos 100t, and (e) $100e^{-t}$ sin t V.

7.5 The current in a 10-μF capacitor decreases linearly from 0 mA at $t = 0$ to -10 mA at 10 ms. It then increases linearly to 0 mA at $t = 20$ ms. If $v(0) = 10$ V, find (a) $v(15)$ and (b) $p(15)$ (arguments in milliseconds). (c) Sketch $v(t)$ and $p(t)$.

7.6 Find the work required to charge a 0.1-μF capacitor to 100 V.

7.7 Determine $w(t)$ for each case of Prob. 7.4.

7.8 Determine $v_1(0^+)$ if $v_1(0^-) = 2v_2(0^-) = 2$ V. Find the current at $t = 0^+$ flowing to the right through R.

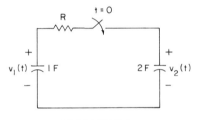

PROBLEM 7.8

7.9 What are the maximum and minimum values of capacitance that can be obtained from 10 1-μF capacitors?

7.10 The capacitances shown are all in microfarads. Find C_{eq} at terminals a-b (see figure below).

7.11 Derive an equation for current division between two parallel capacitors.

7.12 Derive an equation for voltage division between two series capacitors.

7.13 Determine the flux linkage of a 100-mH inductor carrying a current of 0.1 A.

7.14 The current in a 10-H inductor increases linearly from 0 A at $t = 0$ to 0.1 A at $t = 0.1$ s. It then decreases linearly to 0 A at $t = 0.5$ s. Find (a) $\lambda(0.05)$, (b) $v(0.30)$, and (c) $p(0.30)$. (d) Sketch $\lambda(t)$, $v(t)$, and $p(t)$.

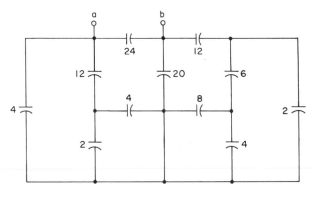

PROBLEM 7.10

7.15 Determine the terminal voltage for a 10-mH inductor having the following terminal currents: (a) 10 A, (b) $50t^2$ A, (c) 10 sin 377t A, and (d) $100te^{-t}$ A.

7.16 The terminal voltage of a 100-mH inductor increases linearly from 0 V at $t = 0$ to 10 V at $t = 1$ ms. It then returns immediately to 0 V and repeats this pattern indefinitely. If $i(0) = 0$ A, find (a) $i(0.5)$, (b) $\lambda(2)$, and (c) $p(0.5)$ (arguments in milliseconds). (d) Sketch $i(t)$ and $p(t)$ for $0 \leq t \leq 2$ ms.

7.17 Determine the terminal current for a 10-mH inductor having a terminal voltage of $10e^{-100t}$ V if $i(0) = 0$ A.

7.18 Find the work required to establish a current of 100 mA in a 10-mH inductor.

7.19 Determine $w(t)$ for each case of Prob. 7.15.

7.20 Determine $i_1(0^+)$ and $v(0^+)$.

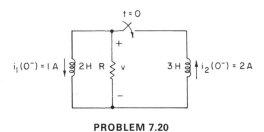

PROBLEM 7.20

7.21 Find the maximum and minimum values of inductance which can be obtained from five 10-mH inductors.

7.22 Determine L_{eq} (values shown are in henrys).

7.23 Derive an equation for voltage division between two series inductors.

7.24 Derive an equation for current division between two parallel inductors.

*7.25 A 400- and a 600-pF ceramic capacitor are connected in parallel. The resistance-capacitance product for a ceramic capacitor is 10^3 Ω-F. What is the equivalent capacitance and parallel resistance for the combination?

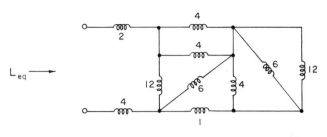

PROBLEM 7.22

7.26 Determine a dual circuit for the network of (a) Fig. 7.13, (b) the figure for Prob. 7.10, and (c) the figure for Prob. 7.22.

*7.27 (a) Make R = 0 in Prob. 7.8 and find the energy lost during the switching action due to radiation. (b) Replace R by an open circuit in Prob. 7.20 and determine $i_1(0^+)$.

8

SIMPLE *RC* AND *RL* CIRCUITS

In this chapter we shall consider simple circuits containing resistors and capacitors or resistors and inductors, which we shall refer to for brevity as *RC* or *RL* circuits, respectively. The application of Kirchhoff's laws to these networks gives rise to *differential equations* that, in general, are more difficult to solve than the algebraic equations encountered in the previous chapters. Several methods of solving these equations will be presented.

We shall first concern ourselves with source-free *RC* and *RL* circuits, so-called because they contain no independent sources. As we shall see, the source-free responses result from energies stored in the dynamic circuit elements and are characterized by the nature of the circuit itself. For this reason the response is known as the natural response of the circuit.

Following our study of source-free circuits, we shall consider driven *RC* and *RL* circuits in which the forcing or driving functions are constant independent sources that are suddenly applied to the networks. We shall find that the responses in these networks consist of two parts, a natural response, similar in form to that of the source-free case, and a forced response, which is characterized by the forcing function.

8.1 SOURCE-FREE *RC* CIRCUIT

We shall begin our study of a source-free network by considering the series connection of a capacitor and a resistor, as shown in Fig. 8.1. In this circuit, the capacitor is charged to a voltage of V_0 at time $t = t_0$. Since there are no current or voltage sources in the network, the circuit response is due entirely to the energy which is stored in the capacitor. The energy in this case, at time $t = t_0$, is

$$w(t_0) = \frac{1}{2}CV_0^2 \text{ J} \tag{8.1}$$

Let us now determine $v(t)$ and $i(t)$ for $t \geq t_0$. Applying KCL at the top node, we find that

$$C\frac{dv}{dt} + \frac{v}{R} = 0$$

or

$$\frac{dv}{dt} + \frac{1}{RC}v = 0 \tag{8.2}$$

which is a *first-order* differential equation. (The *order* of a differential equation is the order of the highest-order derivative in the equation.)

Numerous methods are available for solving differential equations of the form of (8.2). One straightforward method is to rearrange the terms in the equation so as to *separate* the variables v and t. Then simply integrating the result leads to the solution. In (8.2) the variables may be separated by first writing

$$\frac{dv}{dt} = -\frac{1}{RC}v$$

from which we obtain

$$\frac{dv}{v} = -\frac{1}{RC}dt \tag{8.3}$$

We can now obtain the indefinite integral of both sides of this equation, given by

$$\int \frac{dv}{v} = -\frac{1}{RC}\int dt + K \tag{8.4}$$

where K is a constant of integration. Completing the integration, we find that

$$\ln v = -\frac{t}{RC} + K$$

For the solution to be valid for $t \geq t_0$, we see that K must be selected such that the initial condition of $v(t_0) = V_0$ is satisfied. Therefore, at $t = t_0$, we have

$$\ln v(t_0) = \ln V_0 = -\frac{t_0}{RC} + K$$

or

$$K = \ln V_0 + \frac{t_0}{RC}$$

Substituting the value of K into the solution yields

$$\ln v - \ln V_0 = \ln \frac{v}{V_0} = -\frac{t - t_0}{RC}$$

If we now recall the relationship

$$e^{\ln x} = x$$

it is apparent that

$$v(t) = V_0 e^{-(t-t_0)/RC} \qquad (8.5)$$

In Fig. 8.1, we see that this is the voltage across R; therefore the current is

$$i(t) = \frac{v(t)}{R} = \frac{V_0}{R} e^{-(t-t_0)/RC}$$

Another method of solving the separated equation of (8.3) is performed by integrating each side of the equation between its appropriate limits. In our case v has a value of V_0 at time t_0, and thus

$$\int_{V_0}^{v} \frac{dv}{v} = -\frac{1}{RC} \int_{t_0}^{t} dt \qquad (8.6)$$

where the integrals in this equation are *definite* integrals. Performing the integrations, we have

$$\ln v - \ln V_0 = -\frac{t - t_0}{RC}$$

which is equivalent to (8.5).

A graph of (8.5) for $t_0 = 0$ is shown in Fig. 8.2. We see that the voltage is initially V_0 and that it decays exponentially toward zero as t becomes large. The rate at which the voltage decays is determined solely by the product of the resistance and the capacitance of the network. Since the response is characterized by the circuit elements and not by an external voltage or current source, the response is called the *natural response* of the circuit.

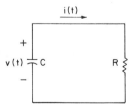

FIGURE 8.1 *Source-free RC circuit*

FIGURE 8.2 *Graph of the voltage response for $t_0 = 0$ in the simple RC circuit of Fig. 8.1*

The energy stored in the network at $t = t_0$ is given in (8.1). As time increases, the voltage across the capacitor, and hence the energy stored in the capacitor, decreases. From a physical standpoint, it is apparent that all the energy stored in the capacitor at $t = t_0$ must be dissipated by the resistor as time becomes infinite. The instantaneous power absorbed by the resistor is

$$p_R(t) = \frac{v^2(t)}{R} = \frac{V_0^2}{R} e^{-2(t-t_0)/RC} \text{ W}$$

Therefore the energy absorbed by the resistor as time becomes infinite is

$$\begin{aligned} w_R(\infty) &= \int_{t_0}^{\infty} p_R(t)\, dt \\ &= \int_{t_0}^{\infty} \frac{V_0^2}{R} e^{-2(t-t_0)/RC}\, dt \\ &= -\tfrac{1}{2} C V_0^2 e^{-2(t-t_0)/RC} \Big|_{t_0}^{\infty} \\ &= \tfrac{1}{2} C V_0^2 \text{ J} \end{aligned}$$

which is, indeed, equal to the energy initially stored in the network.

EXERCISES

8.1.1 In Fig. 8.1, let $t_0 = 0$, $V_0 = 10$ V, $R = 1$ kΩ, and $C = 1$ μF. Find (a) $v(10^{-3})$, (b) $i(10^{-4})$, and (c) $w_C(10^{-5})$.   *Ans.* (a) 3.68 V, (b) 9.05 mA, (c) 49.01 μJ

8.1.2 The capacitor is charged to a voltage of 100 V prior to the closing of the switch. For $t > 0$, find (a) $v(t)$, (b) $i(t)$, (c)$w_C(t)$, and (d) the time at which $v(t) = 50$ V.
Ans. (a) $100e^{-100t}$ V, (b) $-0.01e^{-100t}$ A, (c) $5e^{-200t}$ mJ, (d) 6.93 ms

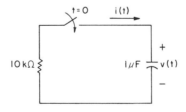

EXERCISE 8.1.2

8.1.3 The switch in the network opens at $t = 1$. For $t > 1$, find (a) $v(1^+)$, (b) $v(t)$, and (c) $w_C(t)$. (d) Sketch $i(t)$.
Ans. (a) 10 V, (b) $10e^{-(t-1)}$ V, (c) $0.5e^{-2(t-1)}$ J

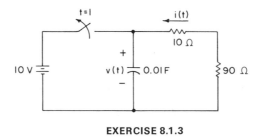

EXERCISE 8.1.3

8.2 TIME CONSTANTS

In networks that contain energy-storage elements it is very useful to characterize with a single number the rapidity with which the natural response decreases. To describe such a number, let us consider the network of Fig. 8.1 and for convenience take $t_0 = 0$ in (8.5). The voltage response, in this case, is given by

$$v(t) = V_0 e^{-t/RC}$$

where V_0 is now the voltage at $t = 0$.

Graphs of $v(t)$ for $RC = k =$ a constant, $RC = 2k$, and $RC = 3k$ are shown in Fig. 8.3. We see that the smaller the RC product, the more rapidly the exponential function $v(t)$ decreases. In fact, the voltage for $RC = k$ decays to a specific value in one-half the time of that required for $RC = 2k$ and in one-third the time of that required for $RC = 3k$. It is also clear that the voltage response remains unchanged if R is increased and C is decreased, or vice versa, such that the product RC is the same. For instance, if we double R and halve C, the voltage response is unchanged.

The current in the network of this example is

$$i(t) = \frac{V_0}{R} e^{-t/RC}$$

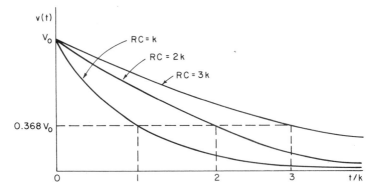

FIGURE 8.3 *Graphs of v(t) for various values of RC*

Clearly, the current decreases in the same manner as the voltage. It should be noticed that changing R and C such that the product of RC remains constant causes a change in the initial current V_0/R. The current response, however, still decreases in the same fashion because $e^{-t/RC}$ is unchanged.

The time required for the natural response to decay by a factor of $1/e$ is defined as the *time constant* of a circuit, which we shall denote by τ. In our case, this requires that

$$V_0 e^{-(t+\tau)/RC} = \frac{V_0}{e} e^{-t/RC} = V_0 e^{-(t+RC)/RC}$$

which yields

$$\tau = RC$$

The units of τ are $\Omega\text{-F} = (\text{V/A})(\text{C/V}) = \text{C/A} = \text{s}$. In terms of the time constant, the voltage response is

$$v(t) = V_0 e^{-t/\tau} \tag{8.7}$$

The response at the end of one time constant is reduced to $e^{-1} = 0.368$ of its initial value. At the end of two time constants it is equal to $e^{-2} = 0.135$ of its initial value, and at the end of five time constants it has become $e^{-5} = 0.0067$ of its initial value. Therefore, after four or five time constants, the response is essentially zero.

An interesting property of exponential functions is shown in Fig. 8.4. A tangent to the curve at $t = 0$ intersects the time axis at $t = \tau$. This is easily verified by considering the equation of a straight line tangent to the curve at $t = 0$, given by

$$v_1 = mt + V_0$$

where m is the slope of the line. Differentiating $v(t)$, we have

$$\frac{dv}{dt} = -\frac{V_0}{\tau} e^{-t/\tau}$$

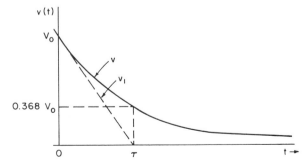

FIGURE 8.4 *Graph illustrating the relation between a line tangent to v(t) at t = 0 and τ*

Therefore

$$m = \frac{dv}{dt}\Big|_{t=0} = -\frac{V_0}{\tau}$$

and

$$v_1 = -\frac{V_0}{\tau}t + V_0$$

The line intersects the time axis at $v_1 = 0$, which requires that $t = \tau$. It can also be shown that a tangent to the curve at a time t_1 intersects the time axis at $t_1 + \tau$ (see Prob. 8.6).

From Fig. 8.4, we see that an alternative definition for the time constant is the time required for the natural response to become zero if it decreases at a constant rate equal to the initial rate of decay.

As a second example, let us consider the more general *RC* circuit of Fig. 8.5(a). Prior to the opening of the switch, the circuit is in a *dc steady-state* condition, by which we mean that the currents and voltages are constant. As we shall see in Sec. 8.4 on driven circuits, a dc steady-state condition is established when a switch, such as that in Fig. 8.5(a), has been closed for a long period of time.

We see by inspection that the capacitor voltage is equal to the voltage across the equivalent resistance to the left of the capacitor terminals. The equivalent resistance is given by

$$R_{eq} = 8 + \frac{3(2+4)}{3+(2+4)} = 10\,\Omega$$

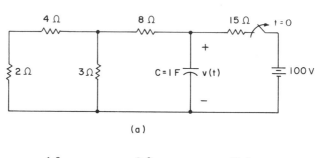

(a)

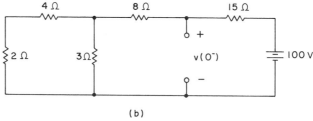

(b)

FIGURE 8.5 *(a) More general RC circuit; (b) its equivalent at $t = 0^-$*

Since the capacitor is an open circuit to steady-state dc, we see in Fig. 8.5(b) that at $t = 0^-$ the capacitor voltage (by voltage division between R_{eq} and the 15-Ω resistor) is simply

$$v(0^-) = \left(\frac{10}{10 + 15}\right) \times 100 = 40 \text{ V}$$

The capacitor voltage is continuous at $t = 0$, so that

$$V_0 = v(0^+) = v(0^-) = 40 \text{ V}$$

For $t > 0$ the time constant for the network is simply the product of the capacitance and the equivalent resistance, given by

$$\tau = R_{eq}C = 10 \text{ s}$$

Therefore, from (8.7), with $t_0 = 0$ and $V_0 = 40$ V, the voltage is

$$v(t) = 40e^{-t/10}$$

EXERCISES

8.2.1 In a series RC circuit, determine (a) τ for $R = 10$ kΩ and $C = 10$ μF, (b) C for $R = 100$ kΩ and $\tau = 10$ μs, and (c) R for $v(t)$ to halve every 10 ms on a 1-μF capacitor.   *Ans.* (a) 0.1 s, (b) 100 pF, (c) 14.43 kΩ

8.2.2 A series RC circuit consists of a 50-kΩ resistor and a 0.02-μF capacitor. It is desired to increase the current in the network by a factor of 4 without changing the capacitor voltage. Find the necessary values of R and C.

Ans. 12.5 kΩ, 0.08 μF

8.2.3 The circuit is in a dc steady-state condition at $t = 0^-$. For $t > 0$, find (a) $v(0^+)$, (b) the time constant τ, and (c) $v_1(t)$.

Ans. (a) 30 V, (b) 20 ms, (c) $10e^{-50t}$ V

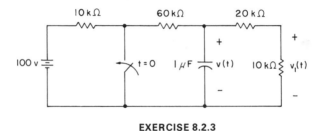

EXERCISE 8.2.3

8.3 SOURCE-FREE *RL* CIRCUIT

In this section we shall study the series connection of an inductor and a resistor, as shown in Fig. 8.6. The inductor, in this case, is carrying a current I_0 at time $t = t_0$. As in the case of the source-free *RC* circuit, there are no current or voltage sources in the network, and the current and voltage responses are due entirely to the energy stored in the inductor. The stored energy at $t = t_0$ is given by

$$w_L(t_0) = \tfrac{1}{2}LI_0^2 \text{ J} \tag{8.8}$$

Summing the voltages around the circuit, we have

$$L\frac{di}{dt} + Ri = 0$$

or

$$\frac{di}{dt} + \frac{R}{L}i = 0 \tag{8.9}$$

This equation is of the same form as that of (8.2) for the *RC* circuit. We may therefore solve it by separating the variables. Instead, however, let us introduce a second, very powerful method, which we shall generalize in the next chapter. The method consists of assuming or guessing (a perfectly legitimate mathematical technique) a general form of the solution based on an inspection of the equation to be solved. In guessing a solution, we shall include several unknown constants and determine their values so that our assumed solution satisfies the differential equation and the initial conditions for the network.

A close inspection of (8.9) reveals that *i* must be a function that does not change its form upon differentiation; that is, di/dt is a multiple of *i*. The only function which satisfies this requirement is an exponential function of *t*. Realizing also that the function must be multiplied by a constant to allow for a proper amplitude in the response, we make an obvious guess that the solution is of the form

$$i(t) = Ae^{st} \tag{8.10}$$

where *A* and *s* are constants to be determined. Substituting the solution into (8.9), we obtain

$$\left(s + \frac{R}{L}\right)Ae^{st} = 0$$

From this result, we see that our solution is valid if $Ae^{st} = 0$ or if $s = -R/L$. The first case is disregarded since, by (8.10), it results in $i = 0$ for all *t* and cannot satisfy $i(t_0) = I_0$. Thus we take the case $s = -R/L$, and (8.10) becomes

$$i(t) = Ae^{-Rt/L}$$

The constant A can now be determined from the initial condition $i(t_0) = I_0$. This condition requires that

$$i(t_0) = I_0 = Ae^{-Rt_0/L}$$

or

$$A = I_0 e^{Rt_0/L}$$

Therefore the solution becomes

$$i(t) = I_0 e^{-R(t-t_0)/L} \qquad\qquad (8.11)$$

In terms of the time constant τ, we may write this result as

$$i(t) = I_0 e^{-(t-t_0)/\tau}$$

where $\tau = L/R$. Evidently, the units of τ are H/Ω = (V-s/A)/(V/A) = s. Increasing L, like increasing C in the RC circuit, increases the time constant. However, an increase in R, in contrast to the RC circuit, lowers the value of the time constant. A graph of a typical current response for $t_0 = 0$ is shown in Fig. 8.7.

The instantaneous power delivered to the resistor in Fig. 8.6 is

$$p(t) = Ri^2(t) = RI_0^2 e^{-2R(t-t_0)/L}$$

Therefore the energy absorbed by the resistor as time becomes infinite is given by

$$
\begin{aligned}
w(\infty) &= \int_{t_0}^{\infty} p(t)\, dt \\
&= \int_{t_0}^{\infty} RI_0^2 e^{-2R(t-t_0)/L}\, dt \\
&= \tfrac{1}{2} L I_0^2 \text{ J}
\end{aligned}
$$

Comparing this result to (8.8), we see that the energy initially stored in the inductor is dissipated by the resistor, as expected.

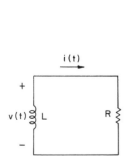

FIGURE 8.6 *Source-free RL circuit*

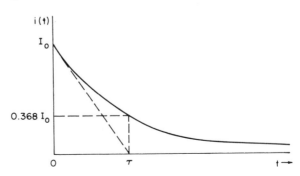

FIGURE 8.7 *Current response of a simple RL circuit for $t_0 = 0$*

Suppose we had chosen to find the inductor voltage v in the circuit instead of the current i. Applying KCL, we find, for $t_0 = 0$,

$$\frac{v}{R} + \frac{1}{L} \int_0^t v \, dt + i(0) = 0$$

which is an *integral* equation. Differentiating this equation with respect to time, we see that

$$\frac{1}{R} \frac{dv}{dt} + \frac{1}{L} v = 0$$

or

$$\frac{dv}{dt} + \frac{R}{L} v = 0$$

This equation is a differential equation that we can solve using one of the methods discussed previously. It is also interesting to note that if we replace v by iR, we have

$$\frac{di}{dt} + \frac{R}{L} i = 0$$

which is equivalent to (8.9), that was obtained using KVL.

Let us now determine $i(t)$ and $v(t)$ in the more general *RL* circuit of Fig. 8.8. The circuit is in a dc steady-state condition at $t = 0^-$; therefore, recalling that an inductor is a short circuit to dc, we have

$$i(0^-) = \tfrac{100}{50} = 2 \text{ A}$$

Since the current in the inductor is continuous at $t = 0$, we have

$$i(0^+) = i(0^-) = 2 \text{ A}$$

The time constant for the network for $t > 0$ is clearly the ratio of the inductance and the equivalent resistance as seen from the terminals of the inductor. The equivalent resistance is

$$R_{eq} = 50 + \frac{(75)(150)}{75 + 150} = 100 \, \Omega$$

and hence the time constant is

$$\tau = \frac{L}{R_{eq}} = 0.1 \text{ s}$$

Therefore, since $I_0 = i(0^+) = 2$ A and $t_0 = 0$, we have

$$i(t) = 2e^{-10t} \text{ A}$$

Summing the voltages of the inductor and the 50-Ω resistor, the voltage $v(t)$ is given by

$$v(t) = 10\frac{di}{dt} + 50i$$
$$= -100e^{-10t}V$$

As a final example, consider the network of Fig. 8.9, which contains a dependent voltage source. The initial current is $i(0) = I_0$. Summing the voltages around the loop, we find that

$$L\frac{di}{dt} + Ri + ki = 0$$

or

$$\frac{di}{dt} + \left(\frac{R+k}{L}\right)i = 0$$

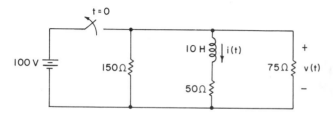

FIGURE 8.8 *More general RL circuit*

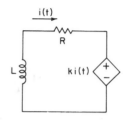

FIGURE 8.9 *RL circuit containing a dependent voltage source*

Comparing this equation to (8.9), we see that the equations are identical if R in (8.9) is replaced by $R + k$. Thus, from (8.11) with $t_0 = 0$, we have

$$i(t) = I_0 e^{-(R+k)t/L}$$

The time constant in this case, which is modified by the presence of the dependent source, is given by

$$\tau = \frac{L}{R+k}$$

EXERCISES

8.3.1 In a series *RL* circuit, determine (a) τ for $R = 200\ \Omega$ and $L = 10$ mH, (b) L for $R = 1$ kΩ and $\tau = 100$ ms, and (c) R for the stored energy in a 0.1-H inductor to halve every 10 ms.   *Ans.* (a) 50 μs, (b) 100 H, (c) 3.47 Ω

8.3.2 A series *RL* circuit has a time constant of 1 ms with an inductance of 1 H. It is desired to halve the inductor voltage without changing the current response. Find the new values of inductance and resistance required.

Ans. 0.5 H, 500 Ω

8.3.3 The circuit is in a dc steady-state condition at $t = 0^-$. Determine $i(0^+)$ and $v(t)$ for $t > 0$.   *Ans.* -2 A, $100e^{-100t}$ V

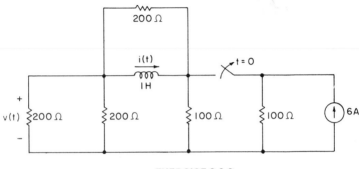

EXERCISE 8.3.3

8.3.4 The circuit shown is in a dc steady-state condition at $t = 0^-$. Find $v(t)$ for $t > 0$.

Ans. $-20e^{-10t}$ V

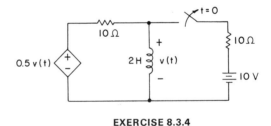

EXERCISE 8.3.4

8.4 RESPONSE TO A CONSTANT FORCING FUNCTION

In the preceding sections we have considered source-free circuits whose responses have been the result of initial energies stored in capacitors and inductors. All independent current or voltage sources were removed or switched out of the circuits prior to finding the natural responses. It was shown that these responses, when arising in circuits containing a single capacitor or inductor and an equivalent resistor, die out with increasing time.

In this section we shall examine circuits which, in addition to having initial stored energies, are *driven* by constant independent current or voltage sources, or *forcing functions*. For these circuits we shall obtain solutions which are the result of inserting or switching sources into the networks. We shall find that the responses in these cases, unlike those of source-free circuits, consist of two parts, one of which is always a constant.

Let us begin by considering the circuit of Fig. 8.10. The network consists of the parallel connection of a constant current source and a resistor which is switched at time $t = 0$ across a capacitor having a voltage $v(0^-) = V_0$ V. For $t > 0$, a nodal equation at the upper node is given by

$$C\frac{dv}{dt} + \frac{v}{R} = I_0$$

or

$$\frac{dv}{dt} + \frac{1}{RC}v = \frac{I_0}{C} \tag{8.12}$$

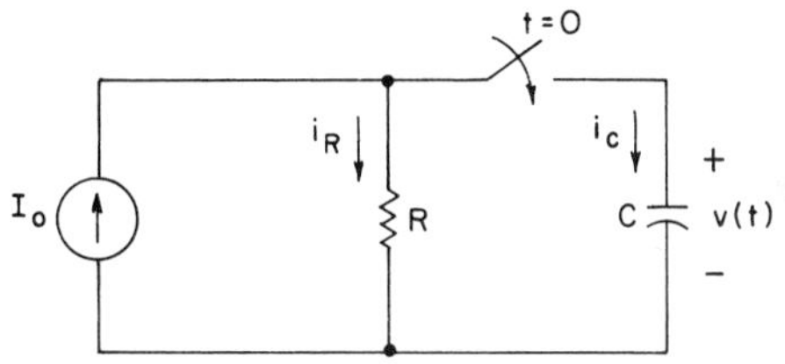

FIGURE 8.10 *Driven RC network*

Equations of this type that have constant forcing functions (I_0 in this case) can be solved by the method of separation of variables. We may first write (8.12) in the form

$$\frac{dv}{dt} = -\frac{v - RI_0}{RC}$$

Multiplying both sides by $dt/(v - RI_0)$ and forming indefinite integrals, we have

$$\int \frac{dv}{v - RI_0} = -\frac{1}{RC}\int dt + K$$

where K is a constant of integration. Thus, performing the integrations, we obtain

$$\ln(v - RI_0) = -\frac{t}{RC} + K$$

This result can be written as

$$v - RI_0 = e^{-(t/RC)+K}$$

or, solving for v, we find

$$v = Ae^{-t/RC} + RI_0 \tag{8.13}$$

where we have taken $A = e^K$, a constant to be determined by the initial condition of the circuit.

From (8.13) we see that the general solution for the voltage response consists of two parts, an exponential function and a constant function. The exponential function is of the identical form as that of the natural response in a source-free circuit composed of R and C. Since this part of the solution is characterized entirely by the RC time constant, we shall refer to it as the *natural response* v_n of the driven circuit. As in the case of the source-free circuit, this response approaches zero as time increases.

The second part of the solution, given by RI_0, bears a close resemblance to the forcing function I_0. In fact, as time increases, the natural response disappears, and the solution is simply RI_0. This component is due entirely to the forcing function, and we shall call it the *forced response* v_f of the driven circuit.

Let us now evaluate the constant A in (8.13). As in the case of the source-free circuit, its value must be selected so that the initial voltage in the circuit is satisfied. At $t = 0^+$, we see that

$$v(0^+) = v(0^-) = V_0$$

Therefore, at $t = 0^+$, (8.13) requires that

$$V_0 = A + RI_0$$

or

$$A = V_0 - RI_0$$

Substituting this value of A back into our solution yields

$$v(t) = RI_0 + (V_0 - RI_0)e^{-t/RC} \tag{8.14}$$

We should observe in this solution that the constant A is now determined not only by the initial voltage (or energy) on the capacitor but also by the forcing function I_0.

Graphs of v_n, v_f, and v are shown in Figs. 8.11(a) and (b). In (a) the natural response v_n for $V_0 - RI_0 > 0$ and the forced response v_f are shown. In (b) the complete response is shown.

The current in the capacitor for $t > 0$ is

$$i_C = C\frac{dv}{dt} = -\frac{V_0 - RI_0}{R}e^{-t/RC}$$

whereas the current in the resistor is

$$i_R = I_0 - i_C = I_0 + \frac{V_0 - RI_0}{R}e^{-t/RC}$$

It is interesting to note that the resistor voltage has changed abruptly from RI_0 at $t = 0^-$ to V_0 at $t = 0^+$. The capacitor voltage, as previously pointed out, is continuous.

The solutions that we have encountered so far in this chapter are often referred to in other more descriptive terms. Two such terms that are very popular are the *transient response* and the *steady-state response*. The transient response is the transitory portion of the complete response which approaches zero as time increases. The steady-state response, on the other hand, is that part of the complete response which remains after the transient response has become zero.

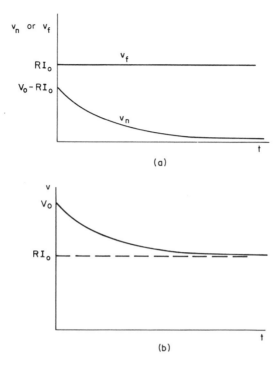

FIGURE 8.11 *Graphs of voltage response for the driven RC network of Fig. 8.10. (a) Natural and forced responses; (b) complete response*

In our example, we see that the transient response and the natural response are identical, as are those for the source-free circuits of the previous sections. The steady-state response is therefore identical to the forced response. In our example this response is RI_0, a dc value which we defined as a dc steady-state condition in Sec. 8.2.

We should not conclude from the above discussion that the natural and forced responses are always the transient and steady-state responses. If the forcing function is a transitory function, for instance, the steady-state response is zero, as we shall see in the next chapter. In this case the complete response is the transient response.

EXERCISES

8.4.1 In Fig. 8.10 $I_0 = 100$ mA and $R = 1$ kΩ. For $t > 0$, determine (a) V_0 such that the natural response is zero, (b) $i_C(10^{-4})$ for $V_0 = 50$ V and $C = 1$ μF, and (c) C for $V_0 = -50$ V such that $v(10^{-3}) = 0$.

Ans. (a) 100 V, (b) 45.2 mA, (c) 2.47 μF.

8.4.2   Find $v(t)$ for $t > 0$ if $v(0^-) = 5$ V.   *Ans.* $10 - 5e^{-100t}$ V

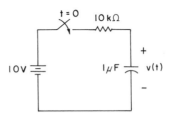

EXERCISE 8.4.2

8.5 THE GENERAL CASE

The equations describing the networks of the previous sections are all special cases of a general expression given by

$$\frac{dy}{dt} + Py = Q \tag{8.15}$$

where y is the unknown variable, such as v or i, and P and Q are constants. If, for instance, we compare this equation to that of (8.2) for the source-free circuit of Sec. 8.1, we see that $y = v$, $P = 1/RC$, and $Q = 0$. The same relations are valid for the forced RC circuit of Sec. 8.4 except that $Q = I_0/C$.

A solution of (8.15) can be found by separation of variables. However, let us introduce another method which also is applicable when Q is a function of time, an important case in later chapters. This method, known as the *integrating factor method*, consists of multiplying the equation by a factor which makes its left-hand side a perfect derivative and simply integrating both sides.

Let us begin by considering the derivative of a product, given by

$$\frac{d}{dt}(ye^{Pt}) = \frac{dy}{dt}e^{Pt} + Pye^{Pt}$$

$$= \left(\frac{dy}{dt} + Py\right)e^{Pt}$$

We see that if we multiply both sides of (8.15) by e^{Pt}, the left-hand side becomes a perfect derivative. Thus performing this multiplication, we have

$$\frac{d}{dt}(ye^{Pt}) = Qe^{Pt}$$

Integrating both sides of the equation, we find

$$ye^{Pt} = \int Qe^{Pt}\,dt + A$$

where A is a constant of integration. Solving for y, we have

$$y = e^{-Pt} \int Q e^{Pt} \, dt + A e^{-Pt} \qquad (8.16)$$

which is valid, of course, if Q is a function of time or a constant. Taking Q to be a constant, however, we obtain

$$y = A e^{-Pt} + \frac{Q}{P} \qquad (8.17)$$

$$= y_n + y_f$$

where $y_n = A e^{-Pt}$ and $y_f = Q/P$ are the natural and forced responses. We observe that y_n has the same mathematical form as the source-free natural response and that y_f is *always* a constant which is proportional to Q. In addition, $1/P$ is the time constant in the natural response.

To illustrate the method, let us find i_2 for $t > 0$ in the circuit of Fig. 8.12, given that $i_2(0) = 1$ A. Although the circuit is a somewhat complex combination of elements, the solution of (8.17) is valid since the network contains a constant forcing function and a *single* energy-storage element (the inductor). The loop equations for the circuit are

$$8i_1 - 4i_2 = 10$$

$$-4i_1 + 12i_2 + \frac{di_2}{dt} = 0$$

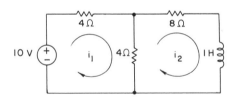

FIGURE 8.12 *Driven RL circuit*

Eliminating i_1 from these equations, we find that

$$\frac{di_2}{dt} + 10i_2 = 5$$

Comparing this equation with (8.15), we see that $P = 10$ and $Q = 5$. Hence (8.17) yields

$$i_2 = A e^{-10t} + \tfrac{1}{2}$$

Applying the initial condition, we have

$$i_2(0) = A + \tfrac{1}{2} = 1$$

Therefore $A = \frac{1}{2}$, and the solution is given by

$$i_2 = \frac{1}{2}e^{-10t} + \frac{1}{2}$$

EXERCISES

8.5.1 Find $v(t)$ for $t > 0$ if $v(0) = 0$.   *Ans.* $12(1 - e^{-10t})$ V

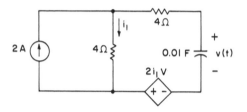

EXERCISE 8.5.1

8.5.2 Find $i_2(t)$ in Fig. 8.12 if the source voltage is $10e^{-t}$ V and $i_2(0) = 1$ A.

Ans. $\frac{5}{9}e^{-t} + \frac{4}{9}e^{-10t}$ A

8.5.3 Find $i(t)$ for $t > 0$ if $v(0) = 12$ V.   *Ans.* $2 - e^{-5t}$ A

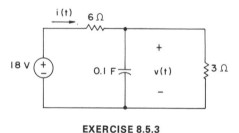

EXERCISE 8.5.3

8.6 A SHORTCUT PROCEDURE

Let us now introduce a shortcut procedure that is very useful for finding the currents and voltages in many circuits. The technique involves formulating the solution by merely inspecting the circuit. Consider, for instance, the example of the previous section (Fig. 8.12). We know that

$$i_2 = i_{2n} + i_{2f}$$

where i_{2n} and i_{2f} are the natural and forced responses, respectively. Since i_{2n} has the same form as the source-free response, we can look at the network in the absence of the forcing function (i.e., make the 10-V source zero by replacing it by a short circuit), as shown in Fig. 8.13(a). The natural response is then

$$i_{2n} = Ae^{-10t}$$

The forced response is constant; therefore, insofar as the forced response is concerned, it does not matter at what time we look at the circuit. We may choose then to look at the circuit in the steady state when i_{2n} is zero. At this time the inductor is a short circuit, as shown in Fig. 8.13(b), and

$$i_{2f} = \tfrac{1}{2}$$

Therefore

$$i_2 = Ae^{-10t} + \tfrac{1}{2}$$

The constant A is now determined as before from the initial condition, $i_2(0) = 1$.

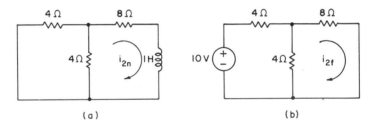

FIGURE 8.13 *Circuits for finding the response of Fig. 8.12. (a) Circuit for finding i_{2n}; (b) circuit for finding i_{2f}*

A word of caution is appropriate at this point. When evaluating the constant A, the student should *always* apply the initial condition to the complete response—never to the natural response alone—because the initial condition is always given for the current, not for a part of it.

As a second example, let us find $i(t)$ for $t > 0$ in Fig. 8.14, given $v(0) = 10$ V. The current is given by

$$i = i_n + i_f$$

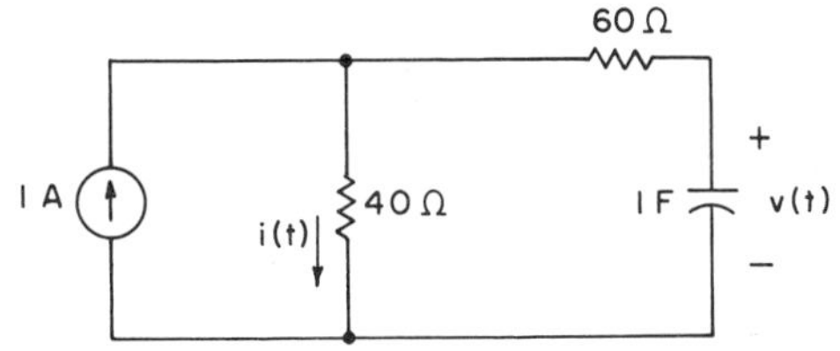

FIGURE 8.14 *Driven RC circuit*

To obtain i_n we note that it has the same *form* as v_n, the natural response of the capacitor voltage. In fact, the natural response of *every* current or voltage in the circuit has the same form as v_n. This is true because all the other currents and voltages in the source-free circuit may be obtained from v_n by applying one or more of the operations of addition, subtraction (in KCL and KVL), differentiation, and integration, none of which changes the nature of the exponential $e^{-t/\tau}$. Examining the source-free circuit (the current source open circuited), we see that the time constant for the capacitor voltage is $\tau = 100$ s. Therefore

$$i_n = Ae^{-t/100}$$

In the steady state the capacitor is an open circuit, and the forced response is, by inspection,

$$i_f = 1 \text{ A}$$

Therefore

$$i(t) = Ae^{-t/100} + 1$$

To evaluate A, we must find the value of $i(0^+)$. Since $v(0) = v(0^+) = 10$ V, summing the voltages around the right-hand mesh, for $t = 0^+$, we have

$$-40i(0^+) + 60[1 - i(0^+)] + 10 = 0$$

or

$$i(0^+) = 0.7$$

Substituting this initial current into our solution, we find that

$$0.7 = 1 + A$$

Therefore $A = -0.3$ and

$$i(t) = 1 - 0.3e^{-t/100}$$

Before concluding this section let us, as a final example, determine $i(t)$ and $v(t)$ in the circuit of Fig. 8.15(a). The network is in a dc steady-state condition at $t = 0^-$ with the switch open; therefore the inductor and capacitor are a short circuit and an open circuit, respectively, at this time. The capacitor voltage is equal to the voltage that appears across the 20-Ω resistor, and the inductor current is equal to the current in the 15-Ω resistor. By current division, the currents in the 15- and 20-Ω resistors are easily shown to be 2 and 3 A, respectively. Thus

$$i(0^-) = 2 \text{ A}$$

and

$$v(0^-) = 60 \text{ V}$$

When the switch closes at $t = 0$, we observe that nodes a and b are short-circuited together, and we can redraw the network as shown in Fig. 8.15(b). It should be noticed that the 30-Ω resistor need not be included in this circuit because the switch is a short circuit across its terminals. The combination is equivalent to a 30-Ω resistor in parallel with a 0-Ω resistor, which, of course, is 0 Ω or a short circuit.

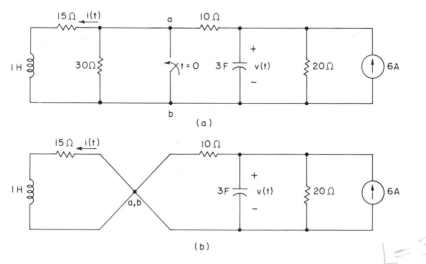

FIGURE 8.15 *(a) Circuit containing an inductance and a capacitance; (b) equivalent circuit for t > 0*

Let us next consider the current $i(t)$ leaving a in Fig. 8.15(b) through the 15-Ω resistor. From KCL, this same current must enter a through the 1-H inductor; hence no current flowing in the circuit to the left of a can enter the other part of the circuit to the right of a, and vice versa. Thus after the switch is closed, the network reduces to two independent circuits, each of which can be solved individually.

The first circuit, consisting of the 1-H inductor and the 15-Ω resistor, is simply a source-free *RL* network having $i(0^+) = i(0^-) = 2$ A. Therefore

$$i(t) = 2e^{-15t}$$

The second circuit, composed of all the elements to the right of a, is simply a driven *RC* network with $v(0^+) = v(0^-) = 60$ V. From our shortcut procedure, we find

$$v(t) = 40 + 20e^{-t/20}$$

The shortcut procedure presented in this section is applicable also to circuits containing dependent sources. However, no savings in time or effort usually result because the circuit equations still have to be written for the source-free and dc steady-state cases.

EXERCISES

8.6.1   Find $v(t)$ for $t > 0$ if the circuit is in dc steady state at $t = 0^-$.

Ans. $2.5(1 - e^{-10^5 t})$ V

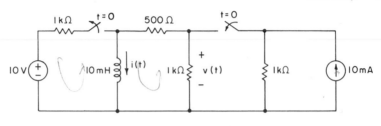

EXERCISE 8.6.1

8.6.2   Find $i(t)$ and $v(t)$ for $t > 0$ if the circuit is in dc steady state at $t = 0^-$.

Ans. $-0.45e^{-10t}$ mA, $-45e^{-10^4 t}$ V

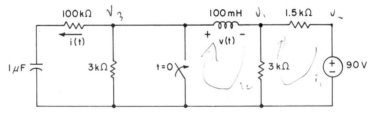

EXERCISE 8.6.2

8.7   THE UNIT STEP FUNCTION

In the previous sections we have analyzed circuits in which energy sources have been
suddenly inserted into the networks. At the instant these sources are applied the
voltages or currents, at the points of application, change abruptly. Forcing functions
whose values change in this manner are called *singularity functions*. There are many
singularity functions that are useful in circuit analysis. One of the most important is
the unit step function, so named by the English engineer Oliver Heaviside (1850–1925).

The *unit step function* is a dimensionless function which is equal to zero for all
negative values of its argument and which is equal to one for all positive values of its
argument. If we denote the unit step function by the symbol $u(t)$, then a mathematical
description is

$$u(t) = 0, \quad t < 0$$
$$= 1, \quad t > 0$$

(8.18)

From a graph of (8.18), shown in Fig. 8.16, we see that at $t = 0$, $u(t)$ changes abruptly from 0 to 1. Some authors define $u(0)$ to be 1, but we are leaving $u(t)$ undefined at $t = 0$.

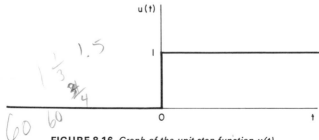

FIGURE 8.16 *Graph of the unit step function u(t)*

Since the unit step function is a dimensionless quantity, a voltage or current source can be obtained by multiplying $u(t)$ by a voltage or a current. A voltage step of V volts is represented by the product $Vu(t)$. Clearly, this voltage is 0 for $t < 0$ and V volts for $t > 0$. A voltage step source of V volts is shown in Fig. 8.17(a). A circuit which is equivalent to this source is shown in Fig. 8.17(b). A short circuit exists for $t < 0$, and the voltage is, of course, zero. For $t > 0$, a voltage V appears at the terminals. We have assumed in our model that the switching action occurs in zero time.

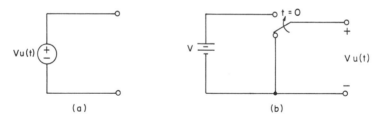

(a)   (b)

FIGURE 8.17 *(a) Voltage step source of V volts; (b) equivalent circuit*

Equivalent circuits for a current step source of I amperes are shown in Fig. 8.18. An open circuit exists for $t < 0$, and the current is zero. For $t > 0$, the switching action causes a terminal current of I amperes to flow.

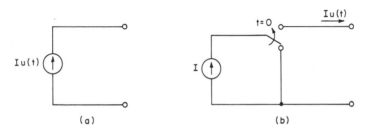

(a)   (b)

FIGURE 8.18 *(a) Current step source of I amperes; (b) equivalent circuit*

The switching action shown in Fig. 8.17 can only be approximated in actual circuits. However, in many cases, it is not necessary to require that the voltage source be a short circuit for $t < 0$, as we shall see in the next section. If the terminals of a network to which the source is to be connected remain at 0 V for $t < 0$, then a series connection of a source V and a switch is equivalent to the voltage step generator, as shown in Fig. 8.19.

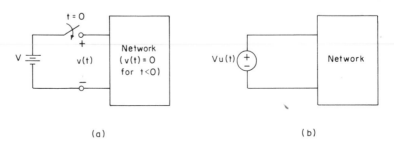

(a)   (b)

FIGURE 8.19 *(a) Network with V applied at t = 0; (b) equivalent circuit*

Equivalent circuits for a current step generator in a network are shown in Fig. 8.20. In each case the current in the network terminals must be zero for $t < 0$.

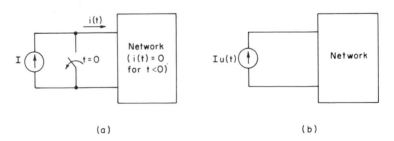

(a)   (b)

FIGURE 8.20 *(a) Network with I applied at t = 0; (b) equivalent circuit*

Let us now return to our definition of the unit step function given in (8.18). We may generalize this definition by replacing t by $t - t_0$ in the three places that it occurs, which results in

$$u(t - t_0) = 0, \qquad t < t_0$$
$$= 1, \qquad t > t_0$$

(8.19)

The function $u(t - t_0)$ is the function $u(t)$ *delayed* by t_0 seconds, as shown in Fig. 8.21.

Multiplying (8.19) by V or I gives us a voltage step source or a current step source whose value changes abruptly at time t_0. Equivalent networks for these sources are obtained in Figs. 8.17–8.20 by taking all actions related to switching to occur at $t = t_0$.

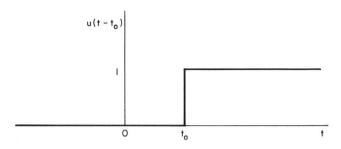

FIGURE 8.21 *Graph of the unit step function u(t − t₀)*

Step functions are very useful in formulating more complex functions. Take, for instance, the rectangular voltage pulse of Fig. 8.22(a). From this figure, we see that

$$v_1(t) = 0, \qquad t < 0$$
$$= V, \qquad 0 < t < t_0$$
$$= 0, \qquad t > t_0$$

Since $u(t)$ becomes 1 for $t > 0$ and $-u(t - t_0)$ becomes -1 for $t > t_0$, it is obvious that

$$v_1(t) = V[u(t) - u(t - t_0)] \tag{8.20}$$

To check this result, we see that, for $t < 0$,

$$v_1(t) = V(0 - 0) = 0$$

for $0 < t < t_0$,

$$v_1(t) = V(1 - 0) = V$$

and for $t > t_0$,

$$v_1(t) = V(1 - 1) = 0$$

Now suppose that we wish to produce a train of these pulses with one occurring every T seconds, where $T > t_0$, as shown in Fig. 8.22(b). Such a wave is called a *square wave*. The first pulse is given by (8.20). The second pulse is simply the first pulse delayed by T seconds. Therefore replacing t by $t - T$ in (8.20). we have

$$\text{Pulse } 2 = V\{u(t - T) - u[t - (T + t_0)]\}$$

The $(n + 1)$th pulse in the pulse train is the first delayed by nT, and therefore

$$\text{Pulse } n + 1 = V\{u(t - nT) - u[t - n(T + t_0)]\}$$

To obtain an expression for the square wave for all $t > 0$, we add the above expressions and obtain

$$v_2(t) = V \sum_{n=0}^{\infty} \{u(t - nT) - u[t - n(T + t_0)]\} \tag{8.21}$$

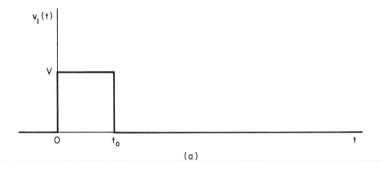

(a)

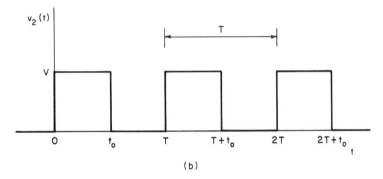

(b)

FIGURE 8.22 *(a) Rectangular pulse; (b) square wave*

Waveforms like those of (8.20) and (8.21) are very common in *digital* circuits such as those in the digital computer.

EXERCISES

8.7.1 Using unit step functions, write an expression for the current $i(t)$ which satisfies

(a) $i(t) = 0$, $t < 0$

          $= -4$ mA, $t > 0$

(b) $i(t) = 0$, $t < -10$ ms

          $= 2$ A, -10 ms $< t < 100$ ms

          $= -1$ A, 100 ms $< t$

(c) $i(t) = 4 \ \mu$A, $t < 1$ s

          $= 0$, $t > 1$ s

Ans. (a) $-4u(t)$ mA, (b) $2u(t + 10^{-2}) - 3u(t - 10^{-1})$ A, (c) $4u(-t + 1) \ \mu$A

8.7.2 Sketch the voltage given by

$$v(t) = (t + 1)u(t - 1) - \tfrac{3}{2}tu(t) + (\tfrac{1}{2}t - 1)u(t - 2)$$

8.7.3   Using unit step functions, write an expression for $v(t)$ for $-\infty < t < \infty$.

Ans. $10 \sin 2\pi t[u(t) - u(t - 1)]$

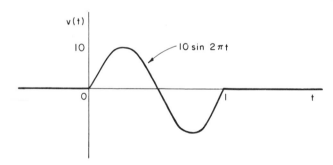

EXERCISE 8.7.3

8.8 THE STEP RESPONSE

The step response is the response of a circuit having only one input which is a unit step function. The response and step input can, of course, be a current or a voltage. The step response is due entirely to the step input since no initial energies are present in the dynamic circuit elements. This is the case because all the currents and voltages in the network are zero at $t = 0^-$ due to the fact that the step function is zero for $-\infty < t < 0$. In this section we shall consider step responses for which the step input may or may not be a unit step.

As an example of a step response, let us find $v(t)$ in the simple *RC* circuit of Fig. 8.23(a) having a voltage step input of $Vu(t)$ volts. Applying KCL, we have

$$C\frac{dv}{dt} + \frac{v - Vu(t)}{R} = 0$$

or

$$\frac{dv}{dt} + \frac{v}{RC} = \frac{V}{RC}u(t)$$

For $t < 0$, this equation becomes

$$\frac{dv}{dt} + \frac{v}{RC} = 0$$

The solution of an equation of this form is

$$v = Ae^{-t/RC}$$

Applying the initial condition $v(0^-) = 0$, we see that $A = 0$, and therefore

$$v(t) = 0, \qquad t < 0$$

which confirms our assertion that the response is zero prior to the change in the input.

For $t > 0$, our differential equation is

$$\frac{dv}{dt} + \frac{v}{RC} = \frac{V}{RC}$$

We know that

$$v = v_n + v_f$$

where

$$v_n = Ae^{-t/RC}$$

and, by inspection,

$$v_f = V$$

Therefore

$$v = V + Ae^{-t/RC}$$

The initial condition $v(0^+) = v(0^-) = 0$ requires that $A = -V$, and therefore our solution for all t is

$$v(t) = 0, \qquad\qquad\qquad t < 0$$
$$= V[1 - e^{-t/RC}], \qquad t > 0$$

This may be written more concisely, using the unit step function, as

$$v(t) = V(1 - e^{-t/RC})u(t)$$

The voltage across the resistor and the capacitor is zero for $t < 0$. Therefore an equivalent circuit for our network is satisfied by the circuit of Fig. 8.23(b) provided we specify that $v(0^-) = 0$.

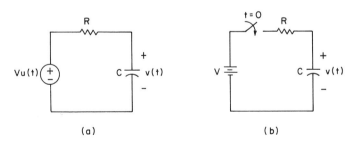

(a)   (b)

FIGURE 8.23 *(a) RC circuit with a voltage step input; (b) equivalent circuit*

As a second example, let us find $v_2(t)$ in the circuit of Fig. 8.24, consisting of a resistor, a capacitor, and an op amp. A nodal equation at the inverting terminal of the op amp is given by

$$\frac{v_1}{R} + C\frac{dv_2}{dt} = 0$$

since the node voltage and the current of the inverting terminal are both zero. Therefore

$$\frac{dv_2}{dt} = -\frac{1}{RC}v_1$$

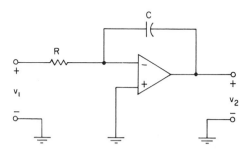

FIGURE 8.24 *Integrator*

Integrating both sides of this equation, we find

$$v_2(t) = -\frac{1}{RC}\int v_1 \, dt + K$$

where K is a constant of integration. The response $v_2(t)$ is proportional to the integral of the input voltage v_1. Thus the circuit is called an *integrator*.

Let us now determine the step response of the network by taking $v_1 = Vu(t)$. In this case,

$$v_2(t) = -\frac{V}{RC}\int u(t) \, dt + K$$

$$= -\frac{V}{RC}tu(t) + K$$

The capacitor voltage at $t = 0^+$ is zero, which requires that $v_2(0^+) = 0$. Thus $K = 0$, and our solution can be written

$$v_2 = -\frac{V}{RC}tu(t)$$

A graph of v_2 is shown in Fig. 8.25. This function is called a *ramp* function with a slope of $-V/RC$.

As a final example, let us find the voltage $v(t)$ in the network of Fig. 8.26, given that

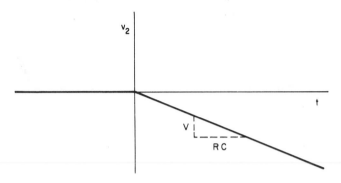

FIGURE 8.25 *Step response of an integrator*

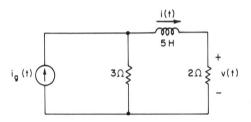

FIGURE 8.26 *RL circuit driven by $i_g(t)$*

$i(0^-) = 0$. The forcing function for the circuit is the current pulse,

$$i_g(t) = 10[u(t) - u(t-1)] \text{ A}$$

which is shown in Fig. 8.27(a).

Examining the circuit, we see that a zero initial condition $[i(0^-) = 0]$ exists and that at $t = 0$ a current step of 10 A is applied. Thus the network response is identical to the step response until time $t = 1$ s. At this time, the forcing function $i_g(t)$ becomes zero, and the response is simply the source-free response resulting from the energy which the inductor has accumulated during the interval of the current pulse. Therefore we see that the solution of the problem involves finding the step response of the circuit for $t < 1$ s and then finding the source-free response for $t > 1$ s.

We know that the step response is of the form

$$v = v_n + v_f$$

where

$$v_n = Ae^{-R_{eq}t/L} = Ae^{-5t/5} = Ae^{-t}$$

and, by current division and Ohm's law, that

$$v_f = 2\left[\frac{(3)(10)}{2+3}\right] = 12$$

Combining these equations, we have

$$v = Ae^{-t} + 12$$

Since $i(0^+) = i(0^-) = 0$, then $v(0^+) = v(0^-) = 0$ and $A = -12$. Therefore

$$v = 0, \qquad\qquad t < 0$$
$$= 12(1 - e^{-t}), \qquad 0 < t < 1$$

For $t > 1$, we know that v is of the form

$$v = Be^{-t}$$

At $t = 1^-$, our solution for the step response gives

$$v(1^-) = 12(1 - e^{-1})$$

Since the inductor current is continuous, $v(1^-) = v(1^+) = 12(1 - e^{-1})$. Therefore

$$v(1^+) = Be^{-1} = 12(1 - e^{-1})$$

or

$$B = 12(1 - e^{-1})e$$

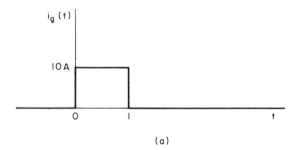

(a)

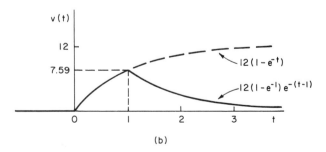

(b)

FIGURE 8.27 *(a) Forcing function $i_g(t)$; (b) response of an RL circuit to $i_g(t)$*

and our solution becomes

$$v = 12(1 - e^{-1})e^{-(t-1)}, \qquad t > 1$$

The solution for all t can be written as

$$v(t) = 12(1 - e^{-t})[u(t) - u(t - 1)] + 12(1 - e^{-1})e^{-(t-1)}u(t - 1) \qquad (8.22)$$

A graph of this response is shown in Fig. 8.27(b).

EXERCISES

8.8.1 Find the step response $i(t)$ to the voltage step $v(t) = 20u(t)$ V.

Ans. $(1 - 0.5e^{-100t})u(t)$ mA

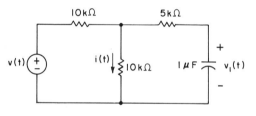

EXERCISE 8.8.1

8.8.2 Find the step response $v_2(t)$ to the voltage step $v_1(t) = 10u(t)$ V. Sketch $v_2(t)$ for $0 < t < 10^{-2}$ s.

Ans. $(10 + 100t)u(t)$

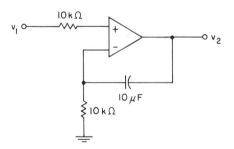

EXERCISE 8.8.2

8.8.3 Find $v_1(t)$ in Ex. 8.8.1 if $v(t) = 20[u(t) - u(t - 0.1)]$ V. Sketch $v_1(t)$.

Ans. $10[1 - e^{-100t}][u(t) - u(t - 0.1)] + 10(e^{10} - 1)e^{-100t}u(t - 0.1)$ V

8.9 APPLICATION OF SUPERPOSITION

In this section we shall consider the use of superposition for obtaining solutions of RC and RL circuits containing two or more independent sources. As a first example, let us consider the circuit of Fig. 8.26 of the previous section. The value of the independent current source is given by

$$i_g = 10u(t) - 10u(t - 1)$$

This source is equivalent to a pair of independent current sources connected in parallel. Thus if we let

$$i_g = i_1 + i_2$$

where $i_1 = 10u(t)$ and $i_2 = -10u(t - 1)$, the circuit of Fig. 8.26 can be redrawn as shown in Fig. 8.28.

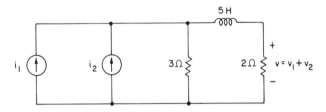

FIGURE 8.28 *Equivalent circuit of Fig. 8.26*

From the principle of superposition, we may write the output voltage as

$$v = v_1 + v_2$$

where v_1 and v_2 are the responses due to i_1 and i_2, respectively. In our previous solution we found that the step response due to the current step i_1 was given by

$$v_1 = 12(1 - e^{-t})u(t)$$

Next, we need the response v_2 due to i_2. We note that i_2 is simply the negative of i_1 delayed 1 s in time. Therefore v_2 is obtained from v_1 by multiplying v_1 by -1 and replacing t by $t - 1$. The result is given by

$$v_2 = -12(1 - e^{-(t-1)})u(t - 1)$$

Our solution is now given by

$$v = 12(1 - e^{-t})u(t) - 12(1 - e^{-(t-1)})u(t - 1) \tag{8.23}$$

which is equivalent to (8.22) (see Ex. 8.9.1).

As a second example, let us consider the RC network of Fig. 8.29, which contains two independent sources and an initial capacitor voltage $v(0) = V_0$. Applying KVL around the left mesh, we find that

$$(R_1 + R_2)i + \frac{1}{C} \int_0^t i \, dt + V_0 = V_1 - R_2 I_1$$

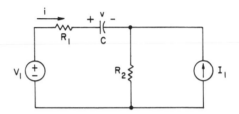

FIGURE 8.29 *RC network*

Multiplying each term in this equation by a constant K, we have

$$(R_1 + R_2)(Ki) + \frac{1}{C} \int_0^t (Ki) \, dt + KV_0 = KV_1 - R_2(KI_1)$$

Clearly, the current response becomes Ki when the independent sources *and* the initial capacitor voltage are multiplied by the factor K, which demonstrates the proportionality property for a linear network. This result is easily extended to any linear circuit containing one or more capacitors. Thus initial capacitor voltages can be treated as independent voltage sources. In a similar manner, it is easily shown that initial inductor currents can be treated as independent current sources.

We shall now employ superposition to determine the voltage v by finding v_1, v_2, and v_3, the responses due to V_1, I_1, and V_0, respectively. The circuit for finding v_1 is shown in Fig. 8.30(a). This is a simple driven RC circuit having a zero initial capacitor voltage. The solution is given by

$$v_1 = V_1(1 - e^{-t/(R_1+R_2)C})$$

The circuit for finding v_2 is shown in Fig. 8.30(b). This again is a simple driven circuit having a zero initial capacitor voltage, for which we find

$$v_2 = -R_2 I_1(1 - e^{-t/(R_1+R_2)C})$$

In the circuit of Fig. 8.30(c), the voltage v_3 is simply the source-free response resulting from the initial capacitor voltage. Since $v_3(0) = V_0$, we find that

$$v_3 = V_0 e^{-t/(R_1+R_2)C}$$

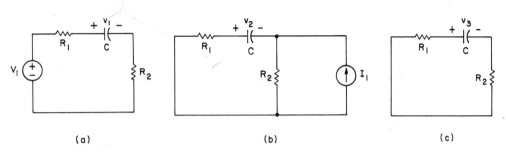

FIGURE 8.30 *Circuits for finding (a) v_1, (b) v_2, and (c) v_3 for an RC network*

Therefore the entire response is given by

$$v = v_1 + v_2 + v_3$$
$$= V_1 - R_2I_1 + (R_2I_1 - V_1 + V_0)e^{-t/(R_1+R_2)C} \qquad (8.24)$$

Inspecting our solution, we see that it consists of a forced response v_f and a natural response v_n, which we could have anticipated. An alternative method of finding the solution is to obtain v_f and v_n, as described previously. Superposition can, of course, be used in finding v_f. In our case, we know that

$$v_f = v_{1f} + v_{2f}$$

From Figs. 8.30(a) and (b) we see, by inspection, that

$$v_{1f} = V_1$$

and

$$v_{2f} = -R_2I_1$$

Therefore the forced response is

$$v_f = V_1 - R_2I_1$$

The natural response, obtained from the source-free circuit [Fig. 8.30(c)], is given by

$$v_n = Ae^{-t/(R_1+R_2)C}$$

Therefore

$$v = V_1 - R_2I_1 + Ae^{-t/(R_1+R_2)C} \qquad (8.25)$$

Since $v(0) = V_0$, we have

$$A = R_2I_1 - V_1 + V_0$$

which substituted into (8.25) gives (8.24).

EXERCISES

8.9.1   Reduce (8.22) to the form of (8.23).

8.9.2   Using superposition, repeat Ex. 8.8.3.

8.9.3   Use superposition to find *i* for *t* > 0. Assume the circuit is in a steady-state condition at *t* = 0⁻.   *Ans.* $5 + 10(1 - e^{-500t})$mA

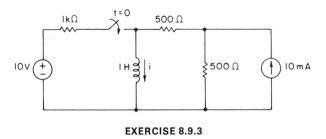

EXERCISE 8.9.3

PROBLEMS

8.1   The current $i(10^{-2}) = 3.68$ mA. Determine (a) $v(0)$, (b) $v(t)$ for $t > 0$, and (c) the energy dissipated by the 10-kΩ resistor as *t* becomes large. (d) Sketch $v(t)$ for $t > 0$.

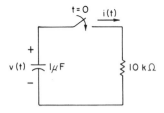

PROBLEM 8.1

8.2   Determine the new value of resistance necessary in Prob. 8.1 for $v(10^{-2}) = 3.68$ V, given $v(0) = 50$ V.

8.3   Find $v(t)$ for $t > 0$, given $v(0) = 15$ V.

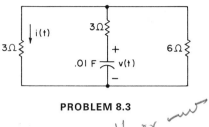

PROBLEM 8.3

8.4 Find $i(t)$ for $t > 0$ in Prob. 8.3. ᴀᴜ ll ᴏᴠ ᴄᴜᴠ

8.5 The circuit is in a steady-state condition at $t = 0^-$. Find $i(t)$ for $t > 0$.

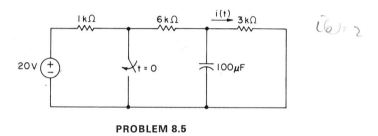

PROBLEM 8.5

8.6 Consider a source-free circuit that has a response $v(t) = V_0 e^{-t/\tau}$. Show that a straight line that is tangent to a graph of this response at time t_1 intersects the time axis (abscissa) at time $t_1 + \tau$.

8.7 The circuit is in a steady-state condition at $t = 0^-$. Find $i(t)$ for $t > 0$.

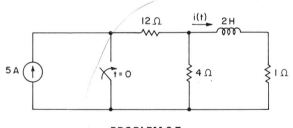

PROBLEM 8.7

8.8 Find $v(t)$ for $t > 0$, given $i(0) = 2$ A.

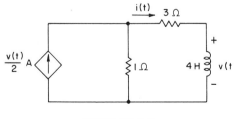

PROBLEM 8.8

8.9   Find $i(t)$ for $t > 0$, given $i(0) = 10$ A.

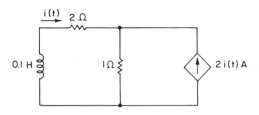

PROBLEM 8.9

8.10   Find $i(t)$ for $t > 0$, given $v(0) = 6$ V.

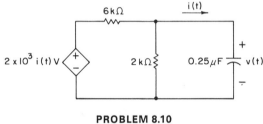

PROBLEM 8.10

8.11   Find $v(t)$ for $t > 0$, given $v(0) = 2$ V.

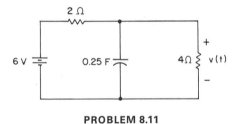

PROBLEM 8.11

8.12   Repeat Prob. 8.11 with the 0.25-F capacitor replaced by a $\frac{1}{3}$-H inductor. Find the inductor current.

8.13   Repeat Prob. 8.11 with the 6-V source replaced by $6e^{-t}$ V.

8.14   Find $i(t)$ for $t > 0$, given $i(0) = 0$.

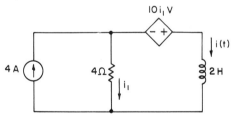

PROBLEM 8.14

8.15 Repeat Prob. 8.14 with the 4-A source replaced by $6e^{-t}$ A.

8.16 The circuit is in a steady-state condition at $t = 0^-$. Find i for $t > 0$.

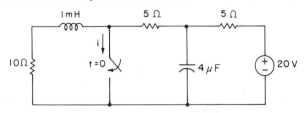

PROBLEM 8.16

$$X = e^{-pt} \int Qe^{pt} + be^{-pt}$$

8.17 Express $v(t)$ in terms of unit step functions, where

$$e^{-pt} + \frac{Q}{p}$$

$$\begin{aligned}
v(t) &= 0 \text{ V}, & t < -10 \\
&= -10 \text{ V}, & -10 < t < 0 \\
&= 20 \text{ V}, & 0 < t < 10 \\
&= 15 \text{ V}, & 10 < t
\end{aligned}$$

8.18 Express $i(t)$ in terms of unit step functions.

$$e^{-pt} \int b e^{pt} dt + V_0 e^{-pt}$$

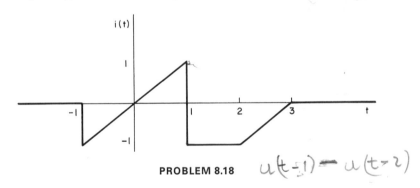

PROBLEM 8.18 $u(t+1) - u(t-2)$

8.19 Sketch the voltage given by

$$v(t) = \sum_{n=0}^{2} (t - n)[u(t - n - 1) - u(t - n - 2)]$$

8.20 Find i for $t > 0$ if $v = 24u(t)$ V.

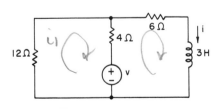

PROBLEM 8.20

8.21 Repeat Prob. 8.20 for $v = 24[u(t) - u(t-1)]$ V.

8.22 Find v for $t > 0$.

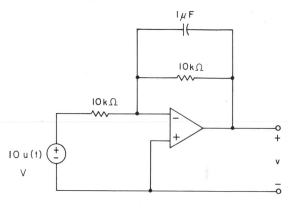

PROBLEM 8.22

8.23 Using superposition, find v for $t > 0$, if $v(0) = 0$.

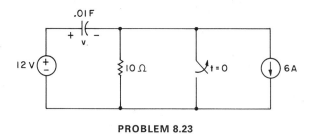

PROBLEM 8.23

8.24 Find v for $t > 0$. Assume the circuit is in a steady-state condition at $t = 0^-$.

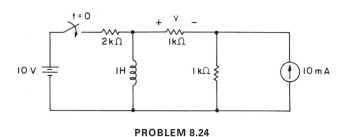

PROBLEM 8.24

9

SECOND-ORDER CIRCUITS

In the case of linear circuits with energy-storage elements the describing equations (those relating the outputs to the inputs) may be expressed as linear differential equations, because the terminal relations of the elements are such that the terms in the loop or nodal equations are derivatives, integrals, or multiples of the unknowns and the source variables. Evidently a single differentiation of an equation will remove any integrals that it may contain, so that in general the loop or nodal equations for a given circuit may be considered to be differential equations. The describing equation then may be obtained from these equations.

The circuits containing storage elements that we have considered so far were first-order circuits. That is, they were described by first-order differential equations. This is always the case when there is only one storage element present or when a switching action converts the circuit into two or more independent circuits each having no more than one storage element.

In this chapter we shall consider second-order circuits, which as we shall see, contain two storage elements and have describing equations that are second-order differential equations. In general, nth-order circuits, containing n storage elements, are described by nth-order differential equations. The results for first- and second-order circuits ($n = 1$ and $n = 2$) may be readily extended to the general case, but we shall not do so here. However, a solution of a third-order differential equation is outlined in Prob. 9.25, which may be used to solve a third-order circuit given in Prob. 9.26. Higher-order circuits will be treated in more detail in Chapter 14.

9.1 CIRCUITS WITH TWO STORAGE ELEMENTS

To introduce the subject of *second-order* circuits, let us begin with the circuit of Fig. 9.1, where the output to be found is the mesh current i_2. The circuit contains two storage elements, the inductors, and as we shall see, i_2 satisfies a second-order differential equation. Methods of solving such equations will be considered in later sections of this chapter.

The mesh equations of Fig. 9.1 are given by

$$2\frac{di_1}{dt} + 12i_1 - 4i_2 = v_g$$

$$-4i_1 + \frac{di_2}{dt} + 4i_2 = 0$$

(9.1)

From the second of these we have

$$i_1 = \frac{1}{4}\left(\frac{di_2}{dt} + 4i_2\right)$$

(9.2)

which differentiated results in

$$\frac{di_1}{dt} = \frac{1}{4}\left(\frac{d^2i_2}{dt^2} + 4\frac{di_2}{dt}\right)$$

(9.3)

Substituting (9.2) and (9.3) into the first of (9.1) to eliminate i_1, we have, after multiplying the resulting equation through by 2,

$$\frac{d^2i_2}{dt^2} + 10\frac{di_2}{dt} + 16i_2 = 2v_g$$

(9.4)

The describing equation for the output i_2 is thus a *second-order* differential equation. That is, it is a differential equation in which the highest derivative is second-order. For this reason we shall refer to Fig. 9.1 as a *second-order* circuit and note that, typically, second-order circuits contain two storage elements.

There are exceptions, however, to the rule that two-storage-element circuits have second-order describing equations. For example, let us consider the circuit of Fig. 9.2, which has two capacitors. With the reference node taken as indicated, nodal equations at the nodes labeled v_1 and v_2 are given by

$$\frac{dv_1}{dt} + v_1 = v_g$$

$$\frac{dv_2}{dt} + 2v_2 = 2v_g$$

(9.5)

The choice of the node voltages v_1 and v_2 as the unknowns has resulted in two first-order differential equations, each containing only one of the unknowns. When this happens, we say that the equations are *uncoupled*, and thus no elimination procedure is required to separate the variables. It was the elimination procedure which, applied to (9.1), gave the second-order equation of (9.4). The equations of (9.5) may be solved separately by the methods of the previous chapter.

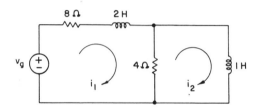

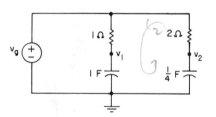

FIGURE 9.1 *Circuit with two inductors*   **FIGURE 9.2** *Circuit with two capacitors*

Evidently Fig. 9.2, although it contains two storage elements, is not a second-order circuit. The same voltage v_g is across each RC combination, and thus the circuit may be redrawn as two first-order circuits. If the source were a practical source rather than an ideal source, then the circuit would be a second-order circuit. (See Prob. 9.5.)

EXERCISES

9.1.1 Find the equation satisfied by the mesh current i_2.

$$Ans. \quad \frac{d^2 i_2}{dt^2} + 7\frac{d i_2}{dt} + 6i_2 = \frac{dv_g}{dt}$$

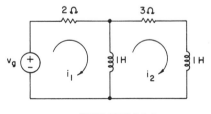

EXERCISE 9.1.1

9.1.2 Let $v_g = 14e^{-2t}$ V, $i_1(0^+) = 6$ A, and $i_2(0^+) = 2$ A in Ex. 9.1.1, and find $di_2(0^+)/dt$ (the value of di_2/dt at $t = 0^+$). *Ans.* -4 A/s

9.1.3 For the values of $i_2(0^+)$, $di_2(0^+)/dt$, and v_g given in Ex. 9.1.2, show that i_2 in Ex. 9.1.1 is given by

$$i_2 = -4e^{-t} + 7e^{-2t} - e^{-6t} \text{ A}$$

(*Suggestion*: Substitute the answer into the differential equation, etc.)

9.2 SECOND-ORDER EQUATIONS

In Chapter 8 we considered first-order circuits in some detail and saw that their describing equations were first-order differential equations of the form

$$\frac{dx}{dt} + a_0 x = f(t) \tag{9.6}$$

In Sec. 9.1 we defined second-order circuits as those having two storage elements with describing equations that were second-order differential equations, given generally by

$$\frac{d^2 x}{dt^2} + a_1 \frac{dx}{dt} + a_0 x = f(t) \tag{9.7}$$

In (9.6) and (9.7) the a's are real constants, x may be either a voltage or a current, and $f(t)$ is a known function of the independent sources.

As an example, for the circuit of Fig. 9.1, the describing equation was (9.4). Comparing this equation with (9.7), we see that $a_1 = 10$, $a_0 = 16$, $f(t) = 2v_g$, and $x = i_2$.

From Chapter 8 we know that the complete response satisfying (9.6) is given by

$$x = x_n + x_f \tag{9.8}$$

where x_n is the natural response obtained when $f(t) = 0$ and x_f is the forced response, which satisfies (9.6). The forced response, in contrast to the natural response, contains no arbitrary constants.

Let us see if this same procedure will apply to the second-order equation (9.7). By a solution to (9.7) we shall mean a function x which satisfies (9.7) identically. That is, when x is substituted into (9.7), the left member becomes identically $f(t)$. We shall also require that x contain *two* arbitrary constants since we must be able to satisfy the two conditions imposed by the initial energy stored in the two storage elements.

If x_n is the natural response, i.e., the response when $f(t) = 0$, then it must satisfy the equation

$$\frac{d^2 x_n}{dt^2} + a_1 \frac{dx_n}{dt} + a_0 x_n = 0 \tag{9.9}$$

Since each term contains x_n to the same degree, namely 1 (the right member may be thought of as $0 = 0x_n$), this equation is sometimes called the *homogeneous* equation.

If x_f is to satisfy the original equation, as it did in the first-order case, then by (9.7) we must have

$$\frac{d^2 x_f}{dt^2} + a_1 \frac{dx_f}{dt} + a_0 x_f = f(t) \tag{9.10}$$

Adding (9.9) and (9.10) and rearranging the terms, we may write

$$\frac{d^2}{dt^2}(x_n + x_f) + a_1 \frac{d}{dt}(x_n + x_f) + a_0(x_n + x_f) = f(t) \tag{9.11}$$

The rearrangement is possible, of course, because the equations involved are linear.

Comparing (9.7) and (9.11) we see that (9.8) is our solution, as it was in the first-order case. That is, x satisfying (9.7) is made up of two components, a natural response

x_n satisfying the homogeneous equation (9.9) and a forced response x_f satisfying the original equation (9.10) or (9.7). As we shall see, the natural response will contain two arbitrary constants and, as in the first-order case, the forced response will have no arbitrary constants. We shall consider methods of finding the natural and forced responses in the next three sections.

Of course, if the driving, or forcing, functions are such that $f(t) = 0$ in (9.7), then the forced response is zero, and the solution of the differential equation is simply the natural response.

EXERCISES

9.2.1 Show that
$$x_1 = A_1 e^{-t}$$
and
$$x_2 = A_2 e^{-2t}$$
are each solutions of
$$\frac{d^2x}{dt^2} + 3\frac{dx}{dt} + 2x = 0$$

regardless of the values of the constants, A_1 and A_2.

9.2.2 Show that
$$x = x_1 + x_2 = A_1 e^{-t} + A_2 e^{-2t}$$

is also a solution of the differential equation of Ex. 9.2.1.

9.2.3 Show that if the right member of the differential equation of Ex. 9.2.1 is changed from 0 to 6, then
$$x = A_1 e^{-t} + A_2 e^{-2t} + 3$$
is a solution.

9.3 THE NATURAL RESPONSE

The natural response x_n of the general solution
$$x = x_n + x_f$$

of (9.7) must satisfy the homogeneous equation, which we repeat as
$$\frac{d^2x}{dt^2} + a_1\frac{dx}{dt} + a_0x = 0 \tag{9.12}$$

Evidently the solution $x = x_n$ must be a function which does not change its form when it is differentiated. That is, the function, its first derivative, and its second derivative must all have the same form, for otherwise the combination in the left member of the equation could not become identically zero for all t.

We are therefore led to *try*

$$x_n = Ae^{st} \tag{9.13}$$

since this is the only function which retains its form when it is repeatedly differentiated. This is, of course, the same function that worked so well for us in the first-order case of Chapter 8. Also, as in the first-order case, A and s are constants to be determined.

Substituting (9.13) for x in (9.12), we have

$$As^2e^{st} + Asa_1e^{st} + Aa_0e^{st} = 0$$

or

$$Ae^{st}(s^2 + a_1s + a_0) = 0$$

Since Ae^{st} cannot be zero [for then by (9.13) $x_n = 0$, and we cannot satisfy any initial energy-storage conditions], we have

$$s^2 + a_1s + a_0 = 0 \tag{9.14}$$

This equation is called the *characteristic equation* and is simply the result of replacing derivatives in (9.12) by powers of s. That is, x, the zeroth derivative, is replaced by s^0, the first derivative by s^1, and the second derivative by s^2.

Since (9.14) is a quadratic equation, we have not one solution, as in the first-order case, but two solutions, say s_1 and s_2, given by the quadratic formula as

$$s_{1,2} = \frac{-a_1 \pm \sqrt{a_1^2 - 4a_0}}{2} \tag{9.15}$$

Therefore we have two natural components of the form (9.13), which we denote by

$$x_{n1} = A_1e^{s_1t}$$
$$x_{n2} = A_2e^{s_2t} \tag{9.16}$$

The coefficients A_1 and A_2 are, of course, arbitrary. Either of the two solutions (9.16) will satisfy the homogeneous equation, because substituting either into (9.12) reduces it to (9.14).

As a matter of fact, because (9.12) is a linear equation, the *sum* of the solutions (9.16) is also a solution. That is,

$$x_n = x_{n1} + x_{n2} \tag{9.17}$$

is a solution of (9.12). To see this, we have only to substitute the expression for x_n into (9.12). This results in

$$\frac{d^2}{dt^2}(x_{n1} + x_{n2}) + a_1 \frac{d}{dt}(x_{n1} + x_{n2}) + a_0(x_{n1} + a_{n2})$$

$$= \left(\frac{d^2 x_{n1}}{dt^2} + a_1 \frac{dx_{n1}}{dt} + a_0 x_{n1}\right) + \left(\frac{d^2 x_{n2}}{dt^2} + a_1 \frac{dx_{n2}}{dt} + a_0 x_{n2}\right)$$

$$= 0 + 0 = 0$$

since both x_{n1} and x_{n2} satisfy (9.12).

By (9.16) and (9.17) we have

$$x_n = A_1 e^{s_1 t} + A_2 e^{s_2 t} \tag{9.18}$$

which is a more general solution (unless $s_1 = s_2$) than either equation of (9.16). In fact, (9.18) is called the *general solution* of the homogeneous equation if s_1 and s_2 are *distinct* (i.e., not equal) roots of the characteristic equation (9.14).

As an example, the homogeneous equation corresponding to (9.4) in the previous section is given by

$$\frac{d^2 i_2}{dt^2} + 10 \frac{di_2}{dt} + 16 i_2 = 0 \tag{9.19}$$

and thus the characteristic equation is

$$s^2 + 10s + 16 = 0$$

The roots are $s = -2$ and $s = -8$, so that the general solution is given by

$$i_2 = A_1 e^{-2t} + A_2 e^{-8t} \tag{9.20}$$

The reader may verify by direct substitution that (9.20) satisfies (9.19), regardless of the value of the arbitrary constants.

Because (9.18) is the natural response, the numbers s_1 and s_2 are sometimes called the *natural frequencies* of the circuit. Evidently they play the same role as the negative reciprocal of the time constants considered in Chapter 7. There are, of course, two time constants in the second-order case as compared to one in the first-order case. For example, the natural frequencies of the circuit of Fig. 9.1 are $s = -2, -8$, as displayed in (9.20); the time constants of the two terms are then $\frac{1}{2}$ and $\frac{1}{8}$.

The unit of natural frequency, which is the inverse of that of the time constant, is the reciprocal of seconds. That is, it is a dimensionless quantity divided by seconds. Therefore st is dimensionless, as it must be in e^{st}.

EXERCISES

9.3.1 Given the *linear* differential equation

$$(t - 1)\frac{d^2 x}{dt^2} + (t - 2)\frac{dx}{dt} = 0$$

show that $x_1 = te^{-t}$, $x_2 = 1$, and $x_1 + x_2$ are all solutions.

9.3.2 Given the *nonlinear* differential equation

$$x \frac{d^2x}{dt^2} - t \frac{dx}{dt} = 0$$

show that $x_1 = t^2$ and $x_2 = 1$ are both solutions but that $x_1 + x_2$ is *not* a solution.

9.3.3 Given

(a) $\dfrac{d^2x}{dt^2} + 4 \dfrac{dx}{dt} + 3x = 0$

(b) $\dfrac{d^2x}{dt^2} + 2 \dfrac{dx}{dt} + x = 0$

find the characteristic equation and the natural frequencies in each case.

Ans. (a) $-1, -3$, (b) $-1, -1$

9.4 TYPES OF NATURAL FREQUENCIES

Since the natural frequencies of a second-order circuit are the roots of a quadratic characteristic equation, they may be real, imaginary, or complex numbers. The nature of the roots is determined by the discriminant $a_1^2 - 4a_0$ of (9.15), which may be positive (corresponding to real, distinct roots), negative (complex roots), or zero (real, equal roots).

For example, consider the circuit of Fig. 9.3, where the response to be found is the mesh current i_2. The loop equations, written around the left mesh and the outer loop, are given by

$$(R + 4)i_1 + \frac{di_1}{dt} - Ri_2 - \frac{di_2}{dt} = v_g$$

$$4i_1 + 4 \int_0^t i_2 \, dt + v(0) = v_g$$

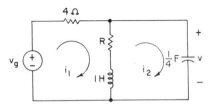

4 Ω

R

$\frac{1}{4}$ F

v_g

i_1

1 H

i_2

+ v −

FIGURE 9.3 *Second-order circuit*

Substituting for i_1 from the second equation into the first, we have

$$\left(\frac{R + 4}{4} \right) \left[v_g - 4 \int_0^t i_2 \, dt - v(0) \right] + \frac{1}{4} \left(\frac{dv_g}{dt} - 4i_2 \right) - Ri_2 - \frac{di_2}{dt} = v_g$$

Differentiating to eliminate the integral and rearranging, we have

$$\frac{d^2 i_2}{dt^2} + (R+1)\frac{di_2}{dt} + (R+4)i_2 = \frac{1}{4}\frac{d^2 v_g}{dt^2} + \frac{R}{4}\frac{dv_g}{dt}$$

The natural component $i_{2_n} = i_n$ satisfies the homogeneous equation

$$\frac{d^2 i_n}{dt^2} + (R+1)\frac{di_n}{dt} + (R+4)i_n = 0$$

from which the characteristic equation is

$$s^2 + (R+1)s + R + 4 = 0$$

Using the quadratic formula, we have the natural frequencies, given by

$$s_{1,2} = \frac{-(R+1) \pm \sqrt{R^2 - 2R - 15}}{2} \tag{9.21}$$

If $R = 6\,\Omega$ in (9.21), the natural frequencies are real and distinct, given by

$$s_{1,2} = -2, -5 \tag{9.22}$$

If $R = 5\,\Omega$, the natural frequencies are real and equal, given by

$$s_{1,2} = -3, -3 \tag{9.23}$$

Finally, if $R = 1\,\Omega$, the natural frequencies are complex numbers, given by

$$s_{1,2} = -1 \pm j2 \tag{9.24}$$

where $j = \sqrt{-1}$. (In electrical engineering we cannot use i, as the mathematicians do, for the imaginary number unit, since this would result in confusion with the current. Complex numbers are considered in Appendix C for the reader who needs to review the subject.)

Distinct Real Roots

If the natural frequencies $s_{1,2}$ are real and distinct, then the natural response is given by (9.18). For example, in the case of (9.22), we have

$$i_n = A_1 e^{-2t} + A_2 e^{-5t}$$

Complex Roots

If the natural frequencies are complex, then in general we have

$$s_{1,2} = \alpha \pm j\beta$$

where α and β are real numbers. For example, in the case of (9.24), $\alpha = -1$ and $\beta = 2$. By (9.18) the natural response in the general case is

$$x_n = A_1 e^{(\alpha + j\beta)t} + A_2 e^{(\alpha - j\beta)t} \tag{9.25}$$

This appears to be a complex quantity and not a suitable answer for a real current or voltage. However, because A_1 and A_2 are complex numbers, it is mathematically correct, although somewhat inconvenient.

To put the natural response (9.25) in a better form, let us consider *Euler's formula*, given by

$$e^{j\theta} = \cos\theta + j\sin\theta \tag{9.26}$$

and its alternative form, obtained by replacing j by $-j$,

$$e^{-j\theta} = \cos\theta - j\sin\theta \tag{9.27}$$

These results are derived in Appendix D. They are named for the great Swiss mathematician Leonhard Euler (pronounced "oiler"), who lived from 1707 to 1783. Euler's greatness is attested to by the fact that the number e, the base of the natural logarithmic system, was chosen in his honor.

Using (9.26) and (9.27), we may write (9.25) as

$$\begin{aligned}
x_n &= e^{\alpha t}(A_1 e^{j\beta t} + A_2 e^{-j\beta t}) \\
&= e^{\alpha t}[A_1(\cos\beta t + j\sin\beta t) + A_2(\cos\beta t - j\sin\beta t)] \\
&= e^{\alpha t}[(A_1 + A_2)\cos\beta t + (jA_1 - jA_2)\sin\beta t]
\end{aligned}$$

Since A_1 and A_2 are arbitrary, let us rename the constants as

$$A_1 + A_2 = B_1$$
$$jA_1 - jA_2 = B_2$$

so that

$$x_n = e^{\alpha t}(B_1\cos\beta t + B_2\sin\beta t) \tag{9.28}$$

As an example, in the case of (9.24) we have $\alpha = -1$ and $\beta = 2$, so that

$$i_n = e^{-t}(B_1\cos 2t + B_2\sin 2t)$$

where B_1 and B_2 are, of course, arbitrary.

Real Equal Roots

The last type of natural frequencies we may have are those that are real and equal, say

$$s_1 = s_2 = k \tag{9.29}$$

In this case (9.18) is not the general solution since both x_{n1} and x_{n2} are of the form Ae^{kt}, and thus there is only one independent arbitrary constant. For (9.29) to be the natural frequencies, the characteristic equation must be

$$(s - k)^2 = s^2 - 2ks + k^2 = 0$$

and therefore the homogeneous equation must be

$$\frac{d^2 x_n}{dt^2} - 2k \frac{dx_n}{dt} + k^2 x_n = 0 \tag{9.30}$$

Since we know that Ae^{kt} is a solution for A arbitrary, let us *try*

$$x_n = h(t)e^{kt}$$

Substituting this expression into (9.30) and simplifying, we have

$$\frac{d^2 h}{dt^2} e^{kt} = 0$$

Therefore $h(t)$ must be such that its second derivative is zero for all t. This is true if $h(t)$ is a polynomial of degree 1, or

$$h(t) = A_1 + A_2 t$$

where A_1 and A_2 are arbitrary constants. The general solution in the repeated-root case, $s_{1,2} = k$, is thus

$$x_n = (A_1 + A_2 t)e^{kt} \tag{9.31}$$

which may be verified by direct substitution into the homogeneous equation (9.30). As an example, in the case of (9.23) we have $s_{1,2} = -3, -3$, and thus

$$i_n = (A_1 + A_2 t)e^{-3t}$$

EXERCISES

9.4.1 Find the natural frequencies of a circuit described by

$$\frac{d^2 x}{dt^2} + a_1 \frac{dx}{dt} + a_0 x = 0$$

if (a) $a_1 = 5$, $a_0 = 6$; (b) $a_1 = 4$, $a_0 = 13$; and (c) $a_1 = 8$, $a_0 = 16$.
    Ans. (a) $-2, -3$; (b) $-2 \pm j3$; (c) $-4, -4$

9.4.2 Find x in Ex. 9.4.1 with the arbitrary constants determined so that $x(0) = 1$ and $dx(0)/dt = 4$.

Ans. (a) $7e^{-2t} - 6e^{-3t}$, (b) $e^{-2t}(\cos 3t + 2 \sin 3t)$, (c) $(1 + 8t)e^{-4t}$

9.4.3 Find x if

$$\frac{d^2x}{dt^2} + 9x = 0$$

Ans. $x = A_1 \cos 3t + A_2 \sin 3t$

9.5 THE FORCED RESPONSE

The forced response x_f of the general second-order circuit must satisfy (9.10) and contain no arbitrary constants. There are a number of methods for finding x_f, but for our purposes we shall use the procedure of guessing the solution, which has worked so well for us in the past. We know from our experience with first-order circuits that the forced response has the form of the driving function. A constant source results in a constant forced response, etc. However, the response must satisfy (9.10) identically, which means that first and second derivatives of x_f, as well as x_f itself, will appear in the left member of (9.10). Thus we are led to *try* as x_f a combination of the right member of (9.10) and its derivatives.

As an example, let us consider the case $v_g = 16$ V in Fig. 9.1. Then by (9.4), for $i_2 = x$, we have

$$\frac{d^2x}{dt^2} + 10\frac{dx}{dt} + 16x = 32 \qquad (9.32)$$

The natural response was given earlier in (9.20) by

$$x_n = A_1 e^{-2t} + A_2 e^{-8t} \qquad (9.33)$$

Since the right member of (9.32) is a constant and all its derivatives are constant (namely zero), let us try

$$x_f = A$$

where A is a constant to be determined. We note that A is not arbitrary but is a particular value that hopefully makes x_f a solution of (9.32). Substituting x_f into (9.32) yields

$$16A = 32$$

or

$$x_f = A = 2$$

Therefore the general solution of (9.32) is

$$x(t) = A_1 e^{-2t} + A_2 e^{-8t} + 2$$

A knowledge of the initial energy stored in the inductors can now be used to evaluate A_1 and A_2.

In the case of constant forcing functions we may often obtain x_f from the circuit itself. In the example just considered x_f is the steady-state value of i_2 in Fig. 9.1 when $v_g = 16$ V. At steady state the inductors look like short circuits, as shown in Fig. 9.4, so that, from the figure, we have

$$x_f = i_2 = 2\text{A}$$

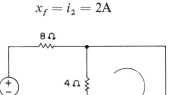

FIGURE 9.4 *Circuit of Fig. 9.1 in the steady state*

As another example, suppose in Fig. 9.1 we have

$$v_g = 20 \cos 4t \text{ V}$$

Then by (9.4), again for $i_2 = x$, we have

$$\frac{d^2x}{dt^2} + 10\frac{dx}{dt} + 16x = 40 \cos 4t \tag{9.34}$$

The natural response x_n is given by (9.33), as before. To find the forced response x_f we need to seek a solution which contains all the terms, and their possible derivatives, in the right member of (9.34). The coefficients of these terms will then be determined by requiring x_f to satisfy the differential equation. In the case under consideration, the only term is a cos 4t term and the trial,

$$x_f = A \cos 4t + B \sin 4t \tag{9.35}$$

contains this term and all its possible derivatives (which are cos 4t and sin 4t).

From (9.35) we have

$$\frac{dx_f}{dt} = -4A \sin 4t + 4B \cos 4t$$

$$\frac{d^2x_f}{dt^2} = -16A \cos 4t - 16B \sin 4t$$

Substituting these values and (9.35) into (9.34) and collecting terms, we have

$$40B \cos 4t - 40A \sin 4t = 40 \cos 4t$$

Since this must be an identity, the coefficients of like terms must be the same on both sides of the equation. In the case of the cos 4t terms we have

$$40B = 40$$

and for the sin 4t terms we have

$$-40A = 0$$

Thus $A = 0$ and $B = 1$, so that

$$x_f = \sin 4t \tag{9.36}$$

The general solution of (9.34), from (9.33) and (9.36), is given by

$$i_2 = x = A_1 e^{-2t} + A_2 e^{-8t} + \sin 4t \tag{9.37}$$

This may be readily verified by direct substitution.

Some of the more common forcing functions $f(t)$ which occur in (9.7) are listed in the first column of Table 9.1. The general form of the corresponding forced response is given in the second column, which may be useful for formulating the trial solution x_f.

TABLE 9.1 *Trial forced responses*

$f(t)$	x_f
k	A
t	$At + B$
t^2	$At^2 + Bt + C$
e^{at}	Ae^{at}
$\sin bt, \cos bt$	$A \sin bt + B \cos bt$
$e^{at} \sin bt, e^{at} \cos bt$	$e^{at}(A \sin bt + B \cos bt)$

EXERCISES

9.5.1 Find the forced response if

$$\frac{d^2x}{dt^2} + 4\frac{dx}{dt} + 3x = f(t)$$

where $f(t)$ is given by (a) 6, (b) $4e^{-2t}$, and (c) $9t$.

Ans. (a) 2, (b) $-4e^{-2t}$, (c) $3t - 4$

9.5.2 If $x(0) = 2$ and $dx(0)/dt = 4$, find the complete solution in Ex. 9.5.1.

Ans. (a) $2e^{-t} - 2e^{-3t} + 2$, (b) $7e^{-t} - 4e^{-2t} - e^{-3t}$, (c) $9.5e^{-t} - 3.5e^{-3t} + 3t - 4$

9.6 EXCITATION AT A NATURAL FREQUENCY

Suppose the circuit equation to be solved is given by

$$\frac{d^2x}{dt^2} - (a + b)\frac{dx}{dt} + abx = f(t) \tag{9.38}$$

where a and $b \neq a$ are known constants. In this case the characteristic equation is

$$s^2 - (a + b)s + ab = 0$$

from which the natural frequencies are

$$s_1 = a, \qquad s_2 = b$$

Therefore we have the natural response

$$x_n = A_1 e^{at} + A_2 e^{bt} \tag{9.39}$$

where A_1 and A_2 are arbitrary.

Let us suppose now that the excitation function contains a natural frequency, say

$$f(t) = e^{at} \tag{9.40}$$

The usual procedure is to seek a forced response,

$$x_f = A e^{at} \tag{9.41}$$

and determine A so that x_f satisfies (9.38), which in this case is

$$\frac{d^2x}{dt^2} - (a + b)\frac{dx}{dt} + abx = e^{at} \tag{9.42}$$

However, substituting x_f into (9.42) yields

$$0 = e^{at}$$

which is an impossible situation.

This difficulty could have been foreseen by observing that x_f in (9.41) has the form of one of the components of x_n in (9.39), and thus x_f will satisfy the homogeneous equation corresponding to (9.38). That is, x_f substituted into (9.38) makes its left member identically zero. There is no point then in trying such a forced response as (9.41).

Let us consider what happens if we multiply by t the part of x_f that is duplicated in x_n. That is, let us try

$$x_f = At e^{at} \tag{9.43}$$

instead of (9.41). We then have

$$\frac{dx_f}{dt} = A(at + 1)e^{at}$$

$$\frac{d^2x_f}{dt^2} = A(a^2t + 2a)e^{at}$$

Substituting these values, along with (9.43), into (9.42), we have

$$Ae^{at}[a^2t + 2a - (a + b)(at + 1) + abt] = e^{at}$$

which, upon simplification, becomes

$$A(a - b)e^{at} = e^{at}$$

Since this must be an identity for all t, we must have

$$A = \frac{1}{a - b}$$

The general solution of (9.38), using (9.39) and (9.43), is then

$$x = A_1 e^{at} + A_2 e^{bt} + \frac{te^{at}}{a - b}$$

As an example, suppose the excitation in Fig. 9.1 is given by

$$v_g = 6e^{-2t} + 32$$

Then if $i_2 = x$, we have, by (9.4),

$$\frac{d^2x}{dt^2} + 10\frac{dx}{dt} + 16x = 12e^{-2t} + 64 \tag{9.44}$$

The natural response, as before, is

$$x_n = A_1 e^{-2t} + A_2 e^{-8t}$$

Noting that the right member of the differential equation has the term e^{-2t} in common with x_n, we try

$$x_f = Ate^{-2t} + B$$

The factor t has been inserted into the natural trial solution of x_f to remove the duplication of the term e^{-2t}. Substituting x_f into (9.44) and simplifying, we have

$$6Ae^{-2t} + 16B = 12e^{-2t} + 64$$

Therefore we have $A = 2$ and $B = 4$, so that

$$x_f = 2te^{-2t} + 4$$

The general solution is now

$$i_2 = x = x_n + x_f$$

As a final example, let us consider the case of (9.38) where $b = a$ and $f(t)$ is given by (9.40). That is,

$$\frac{d^2x}{dt^2} - 2a\frac{dx}{dt} + a^2x = e^{at} \qquad (9.45)$$

The characteristic equation is

$$s^2 - 2as + a^2 = 0$$

and thus the natural frequencies are

$$s_1 = s_2 = a$$

The natural response is then

$$x_n = (A_1 + A_2t)e^{at}$$

We know it is fruitless to try as the forced response x_f given in (9.41) because it is duplicated in the natural response. In this case, (9.43) will not work either because it, too, is duplicated. The lowest power of t that is not duplicated is 2; thus we are led to try

$$x_f = At^2e^{at}$$

Substituting this expression into (9.45) we have

$$2Ae^{at} = e^{at}$$

so that $A = \frac{1}{2}$. The forced and complete responses follow as before.

EXERCISES

9.6.1 Find the forced response if

$$\frac{d^2x}{dt^2} + 4\frac{dx}{dt} + 3x = f(t)$$

where $f(t)$ is given by (a) $2e^{-3t} - e^{-2t}$ and (b) $4e^{-t} + 2e^{-3t}$.

$Ans.$ (a) $e^{-2t} - te^{-3t}$, (b) $t(2e^{-t} - e^{-3t})$

9.6.2 Find the forced response if

$$\frac{d^2x}{dt^2} + 4\frac{dx}{dt} + 4x = f(t)$$

where $f(t)$ is given by (a) $6e^{-2t}$ and (b) $6te^{-2t}$. [Suggestion: In (b), try $x_f = At^3e^{-2t}$.]   Ans. (a) $3t^2e^{-2t}$, (b) t^3e^{-2t}

9.6.3 Find the complete response if

$$\frac{d^2x}{dt^2} + 4x = 8\sin 2t$$

and $x(0) = dx(0)/dt = 0$. [Suggestion: Try $x_f = t(A\cos 2t + B\sin 2t)$.]
                                                      Ans. $\sin 2t - 2t\cos 2t$

9.7 THE COMPLETE RESPONSE

In the previous sections we have noted that the complete response of a circuit is the sum of a natural and a forced response and that the natural response, and thus the complete response, contains arbitrary constants. These constants, as in the first-order cases of Chapter 8, are determined so that the complete response satisfies specified initial energy-storage conditions.

To illustrate the procedure, let us find $x(t)$, for $t > 0$, which satisfies the system of equations

$$\frac{dx}{dt} + 2x + 5\int_0^t x\,dt = 16e^{-3t} \tag{9.46}$$

$$x(0) = 2$$

To begin, let us differentiate the first of these equations to eliminate the integral; this results in

$$\frac{d^2x}{dt^2} + 2\frac{dx}{dt} + 5x = -48e^{-3t}$$

The characteristic equation is

$$s^2 + 2s + 5 = 0$$

with roots

$$s_{1,2} = -1 \pm j2$$

Therefore the natural response is

$$x_n = e^{-t}(A_1\cos 2t + A_2\sin 2t)$$

Trying as the forced response

$$x_f = Ae^{-3t}$$

we see that

$$8Ae^{-3t} = -48e^{-3t}$$

so that $A = -6$. The complete response is therefore

$$x(t) = e^{-t}(A_1 \cos 2t + A_2 \sin 2t) - 6e^{-3t} \tag{9.47}$$

To determine the arbitrary constants we need two initial conditions. One, $x(0) = 2$, is given in (9.46). To obtain the other we may evaluate the first of (9.46) at $t = 0$, resulting in

$$\frac{dx(0)}{dt} + 2x(0) + 5 \int_0^0 x \, dt = 16$$

Noting the value of $x(0)$ and that the integral term is zero, we have

$$\frac{dx(0)}{dt} = 12 \tag{9.48}$$

Applying the second of (9.46) to (9.47), we have

$$x(0) = A_1 - 6 = 2$$

or $A_1 = 8$. To apply (9.48) we may differentiate (9.47), obtaining

$$\frac{dx}{dt} = e^{-t}(-2A_1 \sin 2t + 2A_2 \cos 2t)$$

$$-e^{-t}(A_1 \cos 2t + A_2 \sin 2t) + 18e^{-3t}$$

from which

$$\frac{dx(0)}{dt} = 2A_2 - A_1 + 18 = 12 \tag{9.49}$$

From this, knowing A_1, we find $A_2 = 1$.

At this point let us digress for a moment to note a very easy way to get (9.49). We may differentiate $x(t)$ and immediately replace t by 0 before we write down the result. That is, in (9.47), the derivative of x at $t = 0$ is e^{-t} at $t = 0$ (which is 1) times the derivative of $(A_1 \cos 2t + A_2 \sin 2t)$ at $t = 0$ (which is $2A_2$) plus $(A_1 \cos 2t + A_2 \sin 2t)$ at $t = 0$ (which is A_1) times the derivative of e^{-t} at $t = 0$ (which is -1) plus the derivative of $-6e^{-3t}$ at $t = 0$ (which is 18). These steps are written down in (9.49) and can be done mentally, avoiding the intermediate prior step.

Returning to our problem, we now have the arbitrary constants, so that by (9.47), the final answer is

$$x = e^{-t}(8 \cos 2t + \sin 2t) - 6e^{-3t}$$

As a last example, let us find v, $t > 0$, in the circuit of Fig. 9.5 if $v_1(0) = v(0) = 0$ and $v_g = 5 \cos 2000t$ V. The nodal equation at node v_1 is

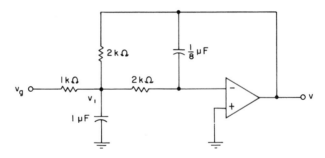

FIGURE 9.5 *Example*

$$2 \times 10^{-3} v_1 - 10^{-3} v_g - \tfrac{1}{2} \times 10^{-3} v + 10^{-6} \frac{dv_1}{dt} = 0$$

or

$$4v_1 - v + 2 \times 10^{-3} \frac{dv_1}{dt} = 2v_g = 10 \cos 2000t \qquad (9.50)$$

and the nodal equation at the inverting input of the op amp is

$$\tfrac{1}{2} \times 10^{-3} v_1 + \tfrac{1}{8} \times 10^{-6} \frac{dv}{dt} = 0$$

or

$$v_1 = -\tfrac{1}{4} \times 10^{-3} \frac{dv}{dt} \qquad (9.51)$$

Substituting (9.51) into (9.50) and simplifying, we have

$$\frac{d^2 v}{dt^2} + 2 \times 10^3 \frac{dv}{dt} + 2 \times 10^6 v = -2 \times 10^7 \cos 2000t$$

The characteristic equation is

$$s^2 + 2 \times 10^3 s + 2 \times 10^6 = 0$$

so that the natural frequencies are $s_{1,2} = 1000(-1 \pm j1)$. The natural response is therefore

$$v_n = e^{-1000t}(A_1 \cos 1000t + A_2 \sin 1000t)$$

For the forced response we shall try

$$v_f = A \cos 2000t + B \sin 2000t$$

which substituted into the differential equation yields

$$(-2A + 4B) \cos 2000t + (-4A - 2B) \sin 2000t = -20 \cos 2000t$$

Therefore equating coefficients of like terms, we have

$$-2A + 4B = -20$$
$$-4A - 2B = 0$$

from which $A = 2$ and $B = -4$. The complete response is then

$$v = e^{-1000t}(A_1 \cos 1000t + A_2 \sin 1000t) + 2 \cos 2000t - 4 \sin 2000t \quad (9.52)$$

From (9.51), for $t = 0^+$, we see that

$$v_1(0^+) = -\tfrac{1}{4} \times 10^{-3} \frac{dv(0^+)}{dt}$$

and since $v_1(0^+) = v_1(0^-) = 0$ we have

$$\frac{dv(0^+)}{dt} = 0 \qquad\qquad (9.53)$$

Like v_1, v is also a capacitor voltage (across the $\tfrac{1}{8}$-μF capacitor), so that

$$v(0^+) = v(0^-) = 0 \qquad\qquad (9.54)$$

From (9.52) and (9.54) we have

$$A_1 + 2 = 0$$

or $A_1 = -2$, and from (9.52) and (9.53) we have

$$1000A_2 - 1000A_1 - 8000 = 0$$

from which $A_2 = 6$. Thus the complete response is

$$v = e^{-1000t}(-2 \cos 1000t + 6 \sin 1000t) + 2 \cos 2000t - 4 \sin 2000t \text{ V}$$

EXERCISES

9.7.1 Find x, $t > 0$, where

$$\frac{dx}{dt} + 2x + \int_0^t x \, dt = f(t)$$

$$x(0) = -1$$

and (a) $f(t) = 1$ and (b) $f(t) = -2t$.

Ans. (a) $(-1 + 2t)e^{-t}$, (b) $(1 + 3t)e^{-t} - 2$

9.7.2 Find i, v, di/dt, and dv/dt at $t = 0^+$.   *Ans.* 0, 0, 2 A/s, 40 V/s

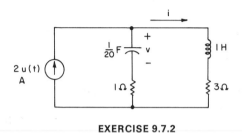

EXERCISE 9.7.2

9.7.3 For $t > 0$ in Ex. 9.7.2, find (a) v and (b) i. (Suggestion: Because of Kirchhoff's laws and the terminal relationships of the elements, i has the same natural frequencies as v. Thus v_n is easily obtained after i_n is found; its forced response is evident by inspection of the circuit.)
    Ans. (a) $e^{-2t}(-6 \cos 4t + 7 \sin 4t) + 6$, (b) $e^{-2t}(-2 \cos 4t - \frac{1}{2} \sin 4t) + 2$

9.8 THE PARALLEL *RLC* CIRCUIT

One of the most important second-order circuits is the parallel *RLC* circuit, the source-free version of which is shown in Fig. 9.6. We shall assume that at $t = 0$ there is an initial inductor current,

$$i(0) = I_0 \tag{9.55}$$

and an initial capacitor voltage,

$$v(0) = V_0 \tag{9.56}$$

and analyze the circuit by finding v for $t > 0$.

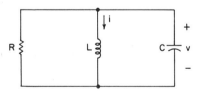

FIGURE 9.6 *Source-free parallel RLC circuit*

The single nodal equation that is necessary is given by

$$\frac{v}{R} + \frac{1}{L} \int_0^t v \, dt + I_0 + C \frac{dv}{dt} = 0 \tag{9.57}$$

which is an integrodifferential equation that becomes, upon differentiation,

$$C\frac{d^2v}{dt^2} + \frac{1}{R}\frac{dv}{dt} + \frac{1}{L}v = 0 \tag{9.58}$$

Since there is no forcing function, the natural response is the complete response. The characteristic equation is

$$Cs^2 + \frac{1}{R}s + \frac{1}{L} = 0$$

from which the natural frequencies are

$$s_{1,2} = -\frac{1}{2RC} \pm \sqrt{\left(\frac{1}{2RC}\right)^2 - \frac{1}{LC}} \tag{9.59}$$

As in the general second-order case already discussed, there are three types of responses, depending on the nature of the discriminant, $(1/2RC)^2 - (1/LC)$, in (9.59). We shall now look briefly at these three cases.

Overdamped Case

If the discriminant is positive, i.e.,

$$\left(\frac{1}{2RC}\right)^2 - \frac{1}{LC} > 0$$

or, equivalently,

$$L > 4R^2C \tag{9.60}$$

then the natural frequencies of (9.59) are real and distinct negative numbers. In this case, known as the *overdamped case*, the response has the form

$$v = A_1 e^{s_1 t} + A_2 e^{s_2 t} \tag{9.61}$$

From the initial conditions and (9.57) evaluated at $t = 0^+$, we obtain

$$\frac{dv(0^+)}{dt} = -\frac{V_0 + RI_0}{RC} \tag{9.62}$$

which together with (9.56) can be used to determine the arbitrary constants.

As an example, suppose $R = 1\,\Omega$, $L = \frac{4}{3}\,\text{H}$, $C = \frac{1}{4}\,\text{F}$, $V_0 = 2\,\text{V}$, and $I_0 = -3\,\text{A}$. Then by (9.59) we have $s_{1,2} = -1, -3$, and hence

$$v = A_1 e^{-t} + A_2 e^{-3t}$$

Also, by (9.56) and (9.62) we have

$$v(0) = 2 \text{ V}$$

$$\frac{dv(0^+)}{dt} = 4 \text{ V/s}$$

which may be used to obtain $A_1 = 5$ and $A_2 = -3$, and thus

$$v = 5e^{-t} - 3e^{-3t}$$

This overdamped case is easily sketched, as shown by the solid line of Fig. 9.7, by sketching the two components and adding them graphically.

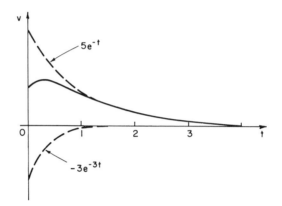

FIGURE 9.7 *Sketch of an overdamped response*

The reason for the term *overdamped* may be seen from the absence of oscillations (fluctuations in sign). The element values are such as to "damp out" any oscillatory tendencies. It is, of course, possible for the response to change signs *once*, depending on the initial conditions.

Underdamped Case

If the discriminant in (9.59) is negative, i.e.,

$$L < 4R^2 C \qquad (9.63)$$

then the circuit is said to be *underdamped*. In this case the natural frequencies are complex, and the response contains sines and cosines, which of course are oscillatory-type functions. In this case it is convenient to define a *resonant frequency*,

$$\omega_0 = \frac{1}{\sqrt{LC}} \qquad (9.64)$$

a *damping coefficient*,

$$\alpha = \frac{1}{2RC} \qquad (9.65)$$

and a *damped frequency*,

$$\omega_d = \sqrt{\omega_0^2 - \alpha^2} \qquad (9.66)$$

Each of these is a dimensionless quantity "per second." The resonant and damped frequencies are defined to be radians per second (rad/s) and the damping coefficient is nepers per second (Np/s).

Using these definitions, the natural frequencies, by (9.59), are

$$s_{1,2} = -\alpha \pm j\omega_d$$

and therefore the response is

$$v = e^{-\alpha t}(A_1 \cos \omega_d t + A_2 \sin \omega_d t) \qquad (9.67)$$

which is oscillatory in nature, as expected.

As an example, suppose $R = 5 \, \Omega$, $L = 1 \, \mathrm{H}$, $C = \frac{1}{10} \, \mathrm{F}$, $V_0 = 0$, and $I_0 = -\frac{3}{2} \, \mathrm{A}$. Then we have

$$v = e^{-t}(A_1 \cos 3t + A_2 \sin 3t)$$

From the initial conditions we have $v(0) = 0$ and $dv(0^+)/dt = 15 \, \mathrm{V/s}$, from which $A_1 = 0$ and $A_2 = 5$. Therefore the underdamped response is

$$v = 5e^{-t} \sin 3t$$

This response is readily sketched if it is observed that since $\sin 3t$ varies between $+1$, and -1, v must be a sinusoid that varies between $5e^{-t}$ and $-5e^{-t}$. The response is shown in Fig. 9.8, where it may be seen that it is oscillatory in nature. The response goes through zero at the points where the sinusoid is zero, which is determined, in general, by the damped frequency ω_d.

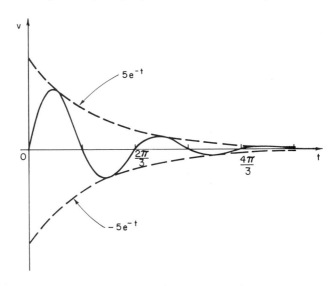

FIGURE 9.8 *Sketch of an underdamped response*

Critically Damped Case

The dividing line between the overdamped and the underdamped cases is the *critically damped* case, occurring when the discriminant of (9.59) is zero. That is,

$$L = 4R^2C \tag{9.68}$$

In this case the natural frequencies are real and equal, given by

$$s_{1,2} = -\alpha, \ -\alpha$$

where α is given by (9.65). The response is then

$$v = (A_1 + A_2t)e^{-\alpha t} \tag{9.69}$$

As an example, consider the case $R = 1\ \Omega$, $L = 1$ H, $C = \frac{1}{4}$ F, $V_0 = 0$, and $I_0 = -1$ A. In this case we have $\alpha = 2$, $A_1 = 0$, and $A_2 = 4$. Thus the response is

$$v = 4te^{-2t}$$

This is easily sketched by plotting $4t$ and e^{-2t} and multiplying the two together. The result is shown in Fig. 9.9.

For every case in the parallel *RLC* circuit, the steady-state value of the natural response is zero, because each term in the response contains a factor e^{at}, where $a < 0$.

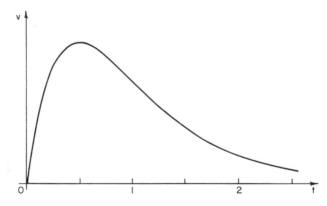

FIGURE 9.9 *Sketch of a critically damped response*

EXERCISES

9.8.1   In a source-free parallel *RLC* circuit, $R = 1$ kΩ and $C = 1$ μF. Find L so that the circuit is (a) overdamped with $s_{1,2} = -250, -750$ s^{-1}, (b) underdamped with $\omega_d = 250$ rad/s, and (c) critically damped.

Ans. (a) $\frac{16}{3}$ H, (b) $\frac{16}{5}$ H, (c) 4 H

9.8.2 (a) Find the differential equation satisfied by i in Fig. 9.6. (b) Use this result to find i, for $t > 0$, if $R = 5\,\Omega$, $L = 1$ H, $C = 0.1$ F, $v(0) = 0$, and $i(0) = 3$ A.

$$\text{Ans. (a) } \frac{d^2i}{dt^2} + \frac{1}{RC}\frac{di}{dt} + \frac{1}{LC}i = 0, \text{ (b) } e^{-t}(3\cos 3t + \sin 3t)$$

9.8.3 The larger the value of R, the less damping there is in the underdamped case of the parallel RLC circuit (because $\alpha = 1/2RC$). Let $R = \infty$ (open circuit) and show that

$$\frac{d^2v}{dt^2} + \omega_0^2 v = 0$$

For this case, find the general solution for v.

$$\text{Ans. } A_1 \cos \omega_0 t + A_2 \sin \omega_0 t$$

9.9 THE SERIES *RLC* CIRCUIT

The source-free series RLC circuit, shown in Fig. 9.10, is the dual of the parallel circuit, considered in the previous section. Therefore all the results for the parallel circuit have dual counterparts for the series circuit, which may be written down by inspection. In this section we shall simply list these results using duality and leave to the reader their verification by conventional means.

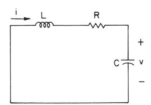

FIGURE 9.10 *Series RLC circuit*

Referring to Fig. 9.10, the initial conditions will be taken as

$$v(0) = V_0$$
$$i(0) = I_0$$

The single-loop equation necessary in the analysis is

$$L\frac{di}{dt} + Ri + \frac{1}{C}\int_0^t i\,dt + V_0 = 0 \tag{9.70}$$

which is valid for $t > 0$. The resulting characteristic equation is

$$Ls^2 + Rs + \frac{1}{C} = 0 \tag{9.71}$$

and the natural frequencies are

$$s_{1,2} = -\frac{R}{2L} \pm \sqrt{\left(\frac{R}{2L}\right)^2 - \frac{1}{LC}} \qquad (9.72)$$

The series *RLC* circuit is overdamped if

$$C > \frac{4L}{R^2} \qquad (9.73)$$

and the response is

$$i = A_1 e^{s_1 t} + A_2 e^{s_2 t} \qquad (9.74)$$

The circuit is critically damped if

$$C = \frac{4L}{R^2} \qquad (9.75)$$

in which case $s_1 = s_2 = -R/2L$, and the response is

$$i = (A_1 + A_2 t)e^{s_1 t} \qquad (9.76)$$

Finally, the circuit is underdamped if

$$C < \frac{4L}{R^2} \qquad (9.77)$$

in which case the resonant frequency is

$$\omega_0 = \frac{1}{\sqrt{LC}} \qquad (9.78)$$

the damping coefficient is

$$\alpha = \frac{R}{2L} \qquad (9.79)$$

and the damped frequency is

$$\omega_d = \sqrt{\omega_0^2 - \alpha^2} \qquad (9.80)$$

The underdamped response is

$$i = e^{-\alpha t}(A_1 \cos \omega_d t + A_2 \sin \omega_d t) \qquad (9.81)$$

Since the series and parallel *RLC* circuits are duals, we shall not consider any examples in this section. In the following section we shall illustrate the series circuit with a forcing function.

EXERCISES

9.9.1 Let $R = 4\,\Omega$, $L = 1\,\text{H}$, $v(0) = 4\,\text{V}$, and $i(0) = 2\,\text{A}$ in Fig. 9.10. Find i for $t > 0$ if C is (a) $\frac{1}{20}\,\text{F}$, (b) $\frac{1}{4}\,\text{F}$, and (c) $\frac{1}{3}\,\text{F}$.

Ans. (a) $2e^{-2t}(\cos 4t - \sin 4t)\,\text{A}$, (b) $2(1 - 4t)e^{-2t}\,\text{A}$, (c) $-3e^{-t} + 5e^{-3t}\,\text{A}$

9.9.2 Find v for $t > 0$ if $R = 40\,\Omega$, $L = 10\,\text{mH}$, and $C = 5\,\mu\text{F}$.

Ans. $e^{-2000t}(4 \cos 4000t - 3 \sin 4000t)\,\text{V}$

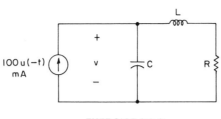

EXERCISE 9.9.2

9.10 THE COMPLETE RESPONSE OF THE *RLC* CIRCUIT

In this section we shall consider the series *RLC* circuit excited by a source, an example of which is shown in Fig. 9.11. The parallel *RLC* circuit is the dual circuit and we shall leave examples of it to the problems.

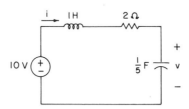

FIGURE 9.11 *Driven series RLC circuit*

Suppose it is required to find v, for $t > 0$, in Fig. 9.11, given that

$$v(0) = 6\,\text{V}, \qquad i(0) = 2\,\text{A}$$

We know that

$$v = v_n + v_f$$

where the natural response v_n contains the natural frequencies. The natural frequencies of the current i are the same as those of v because obtaining one from the other, in general, requires only Kirchhoff's laws and the operations of addition, subtraction, multiplication by constants, integration, and differentiation. None of these operations changes the natural frequencies. Therefore, since the natural frequencies of i are easier to get (only one loop equation is required), let us obtain them. Around the loop we have

$$\frac{di}{dt} + 2i + 5 \int_0^t i \, dt + 6 = 10 \tag{9.82}$$

The characteristic equation, following differentiation, is

$$s^2 + 2s + 5 = 0$$

with roots

$$s_{1,2} = -1 \pm j2$$

Thus we have

$$v_n = e^{-t}(A_1 \cos 2t + A_2 \sin 2t)$$

[We could have obtained $s_{1,2}$ from (9.72) directly, but it is probably easier to write the characteristic equation, since it can be done by inspection.]

The forced response v_f is a constant in this case and may be obtained by inspection of the circuit in the steady state. Since in the steady state the capacitor is an open circuit and the inductor is a short circuit, $i_f = 0$ and $v_f = 10$. Therefore the complete response is

$$v = e^{-t}(A_1 \cos 2t + A_2 \sin 2t) + 10$$

From the initial voltage, we have

$$v(0) = 6 = A_1 + 10$$

or $A_1 = -4$. Also, we have

$$\frac{1}{5} \frac{dv(0^+)}{dt} = i(0) = 2$$

or

$$\frac{dv(0^+)}{dt} = 10 = 2A_2 - A_1$$

Therefore $A_2 = 3$, and we have

$$v = e^{-t}(-4 \cos 2t + 3 \sin 2t) + 10$$

As a final example, let us find v in Fig. 9.11 if the source is

$$v_g = 4 \cos t \text{ V}$$

In this case we shall need the differential equation for v, which we may get from (9.82) and

$$i = \frac{1}{5}\frac{dv}{dt} \tag{9.83}$$

We may also obtain it directly from the figure since the voltages across the inductor and resistor are

$$\frac{di}{dt} = \frac{1}{5}\frac{d^2v}{dt^2}$$

and

$$2i = \frac{2}{5}\frac{dv}{dt}$$

and that across the capacitor is v, of course. In either case, the result is

$$\frac{d^2v}{dt^2} + 2\frac{dv}{dt} + 5v = 20 \cos t$$

The natural response is the same as in the previous example. To get the forced response we shall try

$$v_f = A \cos t + B \sin t$$

which substituted into the differential equation results in

$$(4A + 2B) \cos t + (4B - 2A) \sin t = 20 \cos t$$

Equating like coefficients and solving for A and B, we have

$$A = 4, \quad B = 2$$

Therefore the general solution is

$$v = e^{-t}(A_1 \cos 2t + A_2 \sin 2t) + 4 \cos t + 2 \sin t$$

From the initial voltage, we have

$$v(0) = 6 = A_1 + 4$$

or $A_1 = 2$. From the initial current and (9.83), we have

$$\frac{dv(0^+)}{dt} = 10 = 2A_2 - A_1 + 2$$

or $A_2 = 5$.

The complete response is therefore

$$v = e^{-t}(2 \cos 2t + 5 \sin 2t) + 4 \cos t + 2 \sin t$$

In the case of a driven, parallel RLC circuit, the complete response is obtained in the same manner as that of the series case. Examples are given in Ex. 9.10.1 and in the problems at the end of the chapter.

EXERCISES

9.10.1   Find i for $t > 0$ if (a) $L = \frac{8}{3}$ H and (b) $L = 2$ H.

$Ans.$ (a) $-3e^{-t} + e^{-3t} + 2$, (b) $(-2 - 4t)e^{-2t} + 2$

EXERCISE 9.10.1

9.10.2   Find v for $t > 0$ if (a) $C = \frac{1}{5}$ F and (b) $C = \frac{1}{10}$ F.

$Ans.$ (a) $-25e^{-t} + e^{-5t} + 24$, (b) $e^{-3t}(-24\cos t - 32\sin t) + 24$

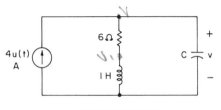

EXERCISE 9.10.2

9.11   ALTERNATIVE METHODS FOR OBTAINING THE DESCRIBING EQUATIONS*

We shall close this chapter by considering two methods of expediting the process of obtaining the describing equation of the circuit. In the case of the parallel and series *RLC* circuits, a single equation is required, which, after differentiation, is the describing equation. However, in many second-order circuits there are two simultaneous circuit equations from which the describing equation is obtained after a tedious elimination process.

As an example, let us consider the circuit of Fig. 9.12 for $t > 0$. Taking node b as reference and writing nodal equations at nodes a and v_1 we have

$$\frac{v - v_g}{4} + \frac{v - v_1}{6} + \frac{1}{4}\frac{dv}{dt} = 0$$

$$\frac{v_1 - v}{6} + \int_0^t v_1 \, dt + i(0) = 0$$

(9.84)

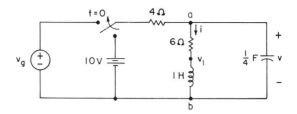

FIGURE 9.12 *Circuit with two storage elements*

If we are interested in finding v we must eliminate v_1 and obtain the describing equation in terms of v. The result, as the reader may verify, is

$$\frac{d^2v}{dt^2} + 7\frac{dv}{dt} + 10v = \frac{dv_g}{dt} + 6v_g \tag{9.85}$$

In this case the process is not overly complicated but it can be shortened by the methods we shall consider in this chapter.

The first method we shall discuss is a systematic way of obtaining a describing equation, such as (9.85), from the circuit equations, such as (9.84). To develop the method, let us first introduce the differentiation *operator D*, which is defined by

$$D = \frac{d}{dt}$$

That is, $Dx = dx/dt$, $D(Dx) = D^2x = d^2x/dt^2$, etc. Also, we have, for example,

$$a\frac{dx}{dt} + bx = aDx + bx = (aD + b)x$$

It is important here to note that x is factored out of the middle member and placed *after* the operator expression, indicating that the operation is to be performed on x. Otherwise the meaning is changed radically.

With these ideas in mind, let us rewrite (9.84) in operator form. This result, after first differentiating the second equation, is

$$\left(\frac{1}{4}D + \frac{5}{12}\right)v - \frac{1}{6}v_1 = \frac{1}{4}v_g$$

$$-\frac{1}{6}Dv + \left(\frac{1}{6}D + 1\right)v_1 = 0$$

which when cleared of fractions becomes

$$(3D + 5)v - 2v_1 = 3v_g$$
$$-Dv + (D + 6)v_1 = 0 \tag{9.86}$$

To eliminate v_1, let us "operate" on the first equation with $D + 6$, by which we mean multiply it through by $D + 6$ on the left of each term. Then let us multiply the second equation by 2, resulting in

$$(D + 6)(3D + 5)v - 2(D + 6)v_1 = 3(D + 6)v_g$$
$$-2Dv + 2(D + 6)v_1 = 0$$

[We note that constants such as 2 commute with operators, i.e., $(D + 6)2 = 2(D + 6)$, but variables do not.] Adding these last two equations eliminates v_1 and results in

$$[(D + 6)(3D + 5) - 2D]v = 3(D + 6)v_g \tag{9.87}$$

Multiplying the operators as if they were polynomials, collecting terms, and dividing out the common factor 3, we have

$$(D^2 + 7D + 10)v = (D + 6)v_g$$

which is the same as (9.85).

The procedure may be carried out in a more direct manner by using determinants. For example, we may use Cramer's rule to obtain the expression for v from (9.86), given by

$$v = \frac{\Delta_1}{\Delta} \tag{9.88}$$

where Δ is the coefficient determinant

$$\Delta = \begin{vmatrix} 3D + 5 & -2 \\ -D & D + 6 \end{vmatrix} \tag{9.89}$$

and Δ_1 is given by

$$\Delta_1 = \begin{vmatrix} 3v_g & -2 \\ 0 & D + 6 \end{vmatrix} = 3(D + 6)v_g \tag{9.90}$$

We note that in this last expression we must be careful to write v_g *after* the operator. Writing (9.88) as

$$\Delta v = \Delta_1$$

we see from (9.89) and (9.90) that we have the describing equation (9.87).

The second method we shall consider is a mixture of the loop and nodal methods in which we select the inductor currents and the capacitor voltages as the unknowns, rather than the loop currents or the node voltages. We then write *KVL* around loops which contain only a single inductor and *KCL* at nodes, or generalized nodes, to which only a single capacitor is connected. In this manner each equation contains only one derivative, that of an inductor current or a capacitor voltage, and no integrals. The equations are then relatively easy to manipulate to find the describing equation.

As an example using Fig. 9.12 again, let i, the inductor current, and v, the capacitor voltage, be the unknowns. (These unknowns are sometimes called the *state variables* of the circuit.) Then the nodal equation at node a is

$$\frac{v - v_g}{4} + i + \frac{1}{4}\frac{dv}{dt} = 0 \tag{9.91}$$

and the loop equation around the right mesh is

$$v = 6i + \frac{di}{dt} \tag{9.92}$$

It is a relatively simple matter to solve for i in (9.91), substitute its value into (9.92), and simplify the result to obtain (9.85).

Some advantages of this method are that no integrals appear (thus no second derivatives occur as a result of differentiation), one unknown is easily found in terms of the others, and the initial conditions on the first derivatives are easily obtained for use in determining the arbitrary constants in the general solution. For example, it may be seen from Fig. 9.12 that $i(0) = 1$ A and $v(0) = 6$ V. Thus from (9.91) and the value of $v_g(0^+)$, we have

$$\frac{dv(0^+)}{dt} = v_g(0^+) - 10$$

This last method can be greatly facilitated, particularly in the case of complex circuits, by using graph theory. Since we are looking for inductor currents and capacitor voltages (the state variables), we put the inductors in the links, whose currents constitute an independent set, and the capacitors in the tree, whose branch voltages constitute an independent set, as we recall from Chapter 6. (The tree should also contain voltage sources, and the links should contain current sources, etc., if possible.)

Each inductor L is then a link with current i, which forms a loop whose only other elements are tree branches. Therefore KVL around this loop will contain only one derivative term, $L(di/dt)$, and no integrals. This loop can easily be found since it is the only loop in the graph if the only link added to the tree is L. For example, the graph of Fig. 9.12 is shown in Fig. 9.13 with tree branches shown as solid lines and links as dashed lines. The loop containing the 1-H inductor is a,v_1,b,a through the branch labeled v, and KVL around it is (9.92).

Each capacitor is a tree branch whose current, together with link currents, consti-

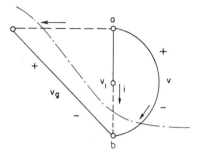

FIGURE 9.13 *Graph of Fig. 9.12*

tutes a set of currents flowing out of a node or a generalized node, because if the capacitor is cut from the circuit, the tree is separated into two parts connected only by links. In the example of Fig. 9.13 the three currents, shown crossing the dotted line through the capacitor, labeled v, and two links, add to zero by *KCL*. This is stated by (9.91).

EXERCISES

9.11.1 Find the describing equation for v, $t > 0$, in Ex. 9.10.2 by (a) the first method of this section using nodal equations and (b) the second method of this section.
Ans. $C(d^2v/dt^2) + 6C(dv/dt) + v = 24$

9.11.2 Solve Ex. 9.1.1 using (a) the first method of this section applied to the mesh equations and (b) the second method of this section.

PROBLEMS

9.1 Find i_1 and i_2, for $t > 0$, in Ex. 9.1.1 if $v_g = 0$ (a short circuit), $i_1(0) = 2$ A, and $i_2(0) = 3$ A.

9.2 Find i_2, for $t > 0$, in Ex. 9.1.1 if $v_g = 12$ V, $i_1(0) = -1$ A, and $i_2(0) = 2$ A.

9.3 Find i, for $t > 0$, if $v(0) = 4$ V, $i(0) = 2$ A, and (a) $L = 1$ H and $R = 1\ \Omega$, (b) $L = 1$ H and $R = 3\ \Omega$, and (c) $L = 2$ H and $R = 5\ \Omega$.

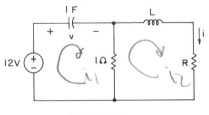

PROBLEM 9.3

9.4 The circuit is in steady state when the switch is opened at $t = 0$. Find v and i for $t > 0$.

9.5 Insert a 1-Ω resistor in series with v_g in Fig. 9.2, thereby making the source a practical rather than an ideal one. Show that in this case v_2 satisfies the second-order equation

$$5\frac{d^2v_2}{dt^2} + 11\frac{dv_2}{dt} + 4v_2 = 4\frac{dv_g}{dt} + 4v_g$$

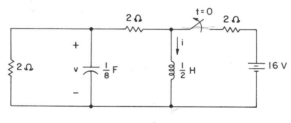

PROBLEM 9.4

9.6 Find i_2, for $t > 0$, if $i_1(0) = 3$ A, $i_2(0) = -1$ A, and $v_g = 0$.

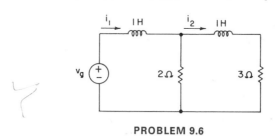

PROBLEM 9.6

9.7 Find v, for $t > 0$, if the circuit is in steady state when the switch is moved from position 1 to position 2 at $t = 0$ and L is (a) 2 H, (b) 4 H, and (c) $\frac{100}{21}$ H.

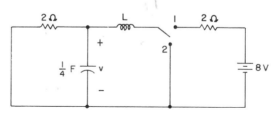

PROBLEM 9.7

9.8 Find i, for $t > 0$, if the circuit is in steady state when the switch is opened at $t = 0$.

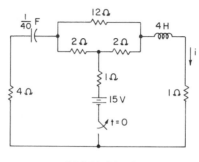

PROBLEM 9.8

9.9 Find i_2 in Prob. 9.6 if (a) $v_g = 15$ V, (b) $v_g = 10e^{-2t}$ V, and (c) $v_g = 5e^{-t}$ V.

9.10 Find v, for $t > 0$, if (a) $v_g = 12u(t)$ V and (b) $v_g = 12tu(t)$ V.

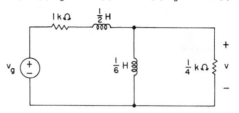

PROBLEM 9.10

9.11 Find v, for $t > 0$, if the circuit is in steady state when the switch is moved from position 1 to position 2 at $t = 0$.

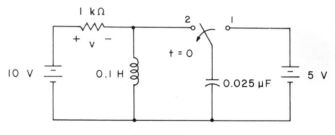

PROBLEM 9.11

9.12 Find v, $t > 0$, if there is no initial stored energy and (a) $i_g = 2$ mA and (b) $i_g = 4e^{-1000t}$ mA.

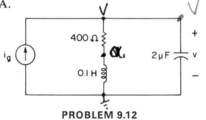

PROBLEM 9.12

9.13 Find i, $t > 0$, if the circuit is in steady state when the switch is opened at $t = 0$.

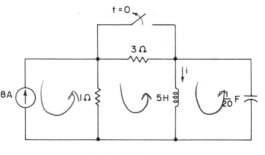

PROBLEM 9.13

9.14 Find v, $t > 0$.

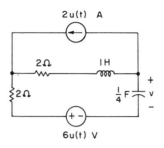

2u(t) A

2Ω

1H

2Ω

$\frac{1}{4}$F   + V −

6u(t) V

PROBLEM 9.14

9.15 Find v, $t > 0$, in the figure for Prob. 9.7 if $L = 1$ H and the switch is moved from position 2 to position 1 at $t = 0$.

9.16 Find v, $t > 0$, if $v(0) = 4$ V and $i(0) = 2$ A.

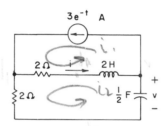

$3e^{-t}$ A

2Ω

2H

2Ω

$\frac{1}{2}$F   + V −

PROBLEM 9.16

9.17 Find the maximum value of the overdamped response of Prob. 9.7(c) and the time at which it occurs.

9.18 Find i, $t > 0$, if there is no initial stored energy and (a) $R = \frac{1}{2}\,\Omega$, $\mu = 2$; (b) $R = \frac{1}{2}\,\Omega$, $\mu = 1$; and (c) $R = \frac{1}{4}\,\Omega$, $\mu = 2$.

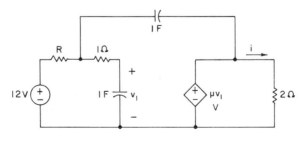

1F

R   1Ω

+

12V   1F V_1 μV_1 2Ω

−

PROBLEM 9.18

9.19 Find i, $t > 0$, if there is no initial stored energy and (a) $C = \frac{1}{6}$ F, (b) $C = \frac{1}{8}$ F, and (c) $C = \frac{1}{16}$ F.

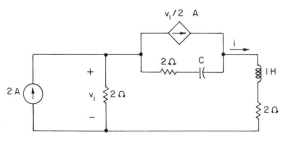

PROBLEM 9.19

9.20 Find v, $t > 0$, in the circuit of Fig. 9.12 if $v_g = 29 \cos 2t$ V and the circuit is in steady state when the switch is moved at $t = 0$.

9.21 Find v, $t > 0$ if there is no initial stored energy and $v_g = 4 \cos 2t$ V.

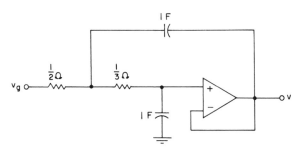

PROBLEM 9.21

9.22 Find v, $t > 0$, if there is no initial stored energy and $v_g = 5 \cos 4t$ V.

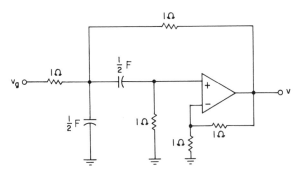

PROBLEM 9.22

9.23 Find v, $t > 0$, if there is no initial stored energy and $v_g = 2 \cos t$ V.

9.24 Find v, $t > 0$, if (a) $v_1(0) = 4$ V, $v(0) = 0$; (b) $v_1(0) = 0$, $v(0) = 2$ V; and (c) $v_1(0) = 4$ V, $v(0) = 2$ V. (Note that the response is an unforced sinusoidal response. Such a circuit is called a *harmonic oscillator*.)

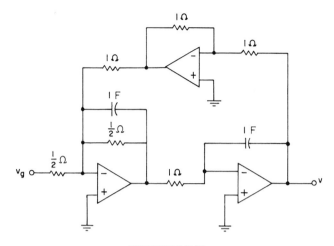

PROBLEM 9.23

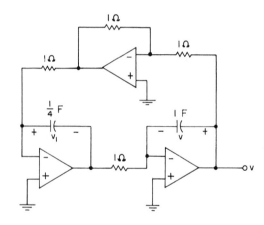

PROBLEM 9.24

9.25 Higher-order differential equations may be solved in the same manner as second-order equations. There are more natural frequencies and thus more terms in the natural response. For example, if

$$\frac{d^3x}{dt^3} + 6\frac{d^2x}{dt^2} + 11\frac{dx}{dt} + 6x = 12$$

show that the characteristic equation

$$s^3 + 6s^2 + 11s + 6 = 0$$

has the natural frequencies

$$s = -1, \ -2, \ -3$$

as its roots. Thus show that the natural response is

$$x_n = A_1e^{-t} + A_2e^{-2t} + A_3e^{-3t}$$

Show also that the forced response is

$$x_f = 2$$

and that the general solution is

$$x = x_n + x_f$$

9.26 Using the results of Prob. 9.25, find i, $t > 0$, if there is no initial stored energy.

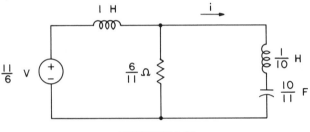

PROBLEM 9.26

10

SINUSOIDAL EXCITATION
AND PHASORS

In the two previous chapters we have analyzed circuits containing dynamic elements and have seen that the complete response is the sum of a natural and a forced response. The natural response for a given circuit is obtained from the dead circuit and therefore is independent of the sources, or excitations. The forced response, on the other hand, depends directly on the type of excitation applied to the circuit. In the case of a dc source, the forced response is a dc steady-state response, an exponential input results in an exponential forced response, and so forth.

One of the most important excitations is the sinusoidal forcing function. Sinusoids abound in nature, as, for example, in the motion of a pendulum, in the bouncing of a ball, and in the vibrations of strings and membranes. Also, as we have seen, the natural response of an underdamped second-order circuit is a damped sinusoid and in the absence of damping is a pure sinusoid.

In electrical engineering, sinusoidal functions are extremely important for a number of reasons. The carrier signals generated for communication purposes are sinusoids, and, of course, the sinusoid is the dominant signal in the electric power industry, to name two very important examples. Indeed, as we shall see later in the study of Fourier series, almost every useful signal in electrical engineering can be resolved into sinusoidal components.

Because of their importance, circuits with a sinusoidal forcing function will be considered in detail in this chapter. Since the natural response is independent of the sources and can be found by the methods of the previous chapters, we shall concentrate on finding only the forced response. This response is important in itself since it is the ac steady-state response that is left after the short time required for the transitory natural response to die.

Since we are interested only in the ac steady-state response, we shall not limit ourselves, as we did in Chapters 8 and 9, to first- and second-order circuits. As we shall see, higher-order *RLC* circuits may be handled, insofar as the ac steady-state response is concerned, in the same way as resistive circuits.

10.1 PROPERTIES OF SINUSOIDS

Because of the importance of sinusoidal functions, we shall devote this section to a review of some of their properties. Let us begin with the sine wave,

$$v(t) = V_m \sin \omega t \tag{10.1}$$

which is sketched in Fig. 10.1. The *amplitude* of the sinusoid is V_m, which is the maximum value that the function attains. The *radian frequency*, or *angular frequency*, is ω, measured in radians per second (rad/s).

The sinusoid is a *periodic* function, defined generally by the property

$$v(t + T) = v(t) \tag{10.2}$$

where *T* is the *period*. That is, the function goes through a complete cycle, or period, which is then repeated, every *T* seconds. In the case of the sinusoid, the period is

$$T = \frac{2\pi}{\omega} \tag{10.3}$$

as may be seen from (10.1) and (10.2). Thus in 1 s the function goes through $1/T$ cycles, or periods. Its *frequency f* is then

$$f = \frac{1}{T} = \frac{\omega}{2\pi} \tag{10.4}$$

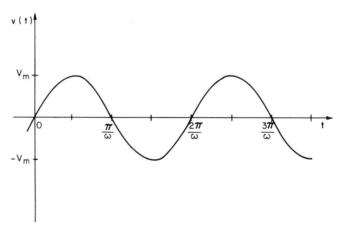

FIGURE 10.1 *Sinusoidal function*

cycles per second, or *hertz* (abbreviated Hz). The latter term, named for the German physicist Heinrich R. Hertz (1857–1894), is now the standard unit for frequency. Some older books use the former term, but it is being discontinued. The relation between frequency and radian frequency is seen by (10.4) to be

$$\omega = 2\pi f \tag{10.5}$$

A more general sinusoidal expression is given by

$$v(t) = V_m \sin(\omega t + \phi) \tag{10.6}$$

where ϕ is the *phase angle*, or simply the *phase*. To be consistent, since ωt is in radians, ϕ should be expressed in radians. However, in electrical engineering it is often convenient to specify ϕ in degrees. For example, we may write

$$v = V_m \sin\left(2t + \frac{\pi}{4}\right)$$

or

$$v = V_m \sin(2t + 45°)$$

interchangeably, even though the latter expression contains a mathematical inconsistency.

A sketch of (10.6) is shown in Fig. 10.2 by the solid line, along with a sketch of (10.1), shown dashed. The solid curve is simply the dashed curve displaced ϕ/ω seconds, or ϕ radians to the left. Therefore points on the solid curve, such as its peaks, occur ϕ rad, or ϕ/ω s, earlier than corresponding points on the dashed curve. Accordingly, we shall say that $V_m \sin(\omega t + \phi)$ *leads* $V_m \sin \omega t$ by ϕ rad (or degrees). In general, the sinusoid

$$v_1 = V_{m1} \sin(\omega t + \alpha)$$

leads the sinusoid

$$v_2 = V_{m2} \sin(\omega t + \beta)$$

by $\alpha - \beta$. An equivalent expression is that v_2 *lags* v_1 by $\alpha - \beta$.

As an example, consider

$$v_1 = 4 \sin(2t + 30°)$$

and

$$v_2 = 6 \sin(2t - 12°)$$

Then v_1 leads v_2 (or v_2 lags v_1) by $30 - (-12) = 42°$.

Thus far we have considered sine functions rather than cosine functions in defining sinusoids. It does not matter which form we use since

$$\cos\left(\omega t - \frac{\pi}{2}\right) = \sin \omega t \tag{10.7}$$

or

$$\sin\left(\omega t + \frac{\pi}{2}\right) = \cos \omega t \tag{10.8}$$

The only difference between sines and cosines is thus the phase angle. For example, we may write (10.6) as

$$v(t) = V_m \cos\left(\omega t + \phi - \frac{\pi}{2}\right)$$

To determine how much one sinusoid leads or lags another of the same frequency, we must first express both as sine waves or as cosine waves with positive amplitudes. For example, let

$$v_1 = 4 \cos (2t + 30°)$$

and

$$v_2 = -2 \sin (2t + 18°)$$

Then, since

$$-\sin \omega t = \sin (\omega t + 180°)$$

we have

$$v_2 = 2 \sin (2t + 18° + 180°)$$
$$= 2 \cos (2t + 18° + 180° - 90°)$$
$$= 2 \cos (2t + 108°)$$

Comparing this last expression with v_1, we see that v_1 leads v_2 by $30° - 108° = -78°$, which is the same as saying that v_1 lags v_2 by $78°$.

The sum of a sine wave and a cosine wave of the same frequency is another sinusoid of that frequency. To show this, consider

$$A \cos \omega t + B \sin \omega t = \sqrt{A^2 + B^2}\left[\frac{A}{\sqrt{A^2 + B^2}} \cos \omega t + \frac{B}{\sqrt{A^2 + B^2}} \sin \omega t\right]$$

which by Fig. 10.3 may be written

$$A \cos \omega t + B \sin \omega t = \sqrt{A^2 + B^2} (\cos \omega t \cos \theta + \sin \omega t \sin \theta)$$

By a formula from trigonometry, this is

$$A \cos \omega t + B \sin \omega t = \sqrt{A^2 + B^2} \cos (\omega t - \theta) \tag{10.9}$$

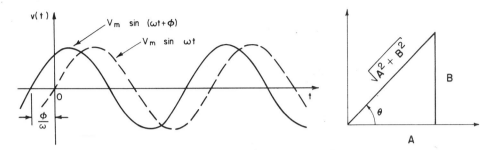

FIGURE 10.2 *Two sinusoids with different phases*   **FIGURE 10.3** *Triangle useful in adding two sinusoids*

where, by Fig. 10.3,

$$\theta = \tan^{-1}\frac{B}{A} \tag{10.10}$$

A similar result may be established if the sine and cosine terms have phase angles other than zero, indicating that, in general, the sum of two sinusoids of a given frequency is another sinusoid of the same frequency.

As an example, we have

$$-5\cos 3t + 12\sin 3t = \sqrt{5^2 + 12^2}\cos\left[3t - \tan^{-1}\left(\frac{12}{-5}\right)\right]$$

$$= 13\cos(3t - 112.6°)$$

since $\tan^{-1}(12/{-5})$ is in the second quadrant, because $A = -5 < 0$ and $B = 12 > 0$.

EXERCISES

10.1.1 Find the period of the following sinusoids:
(a) $6\sin(5t + 17°)$
(b) $\cos\left(2t + \frac{\pi}{4}\right) + 2\sin\left(2t - \frac{\pi}{6}\right)$.
(c) $4\cos 2\pi t$.   *Ans.* (a) $2\pi/5$, (b) π, (c) 1

10.1.2 Find the amplitude and phase of the following sinusoids:
(a) $3\cos 2t + 4\sin 2t$.
(b) $(4\sqrt{3} - 3)\cos(2t + 30°) + (3\sqrt{3} - 4)\cos(2t + 60°)$.
[*Suggestion*: In (b), expand both functions and use (10.9).]
Ans. (a) 5, $-53.1°$; (b) 5, 36.9°

10.1.3 Find the frequency of the following sinusoids:
(a) $3\cos(4\pi t - 10°)$.
(b) $4\sin 377t$.   *Ans.* (a) 2, (b) 60 Hz

10.2 AN RL CIRCUIT EXAMPLE

As an example of a circuit with a sinusoidal excitation, let us find the forced component i_f of the current i in Fig. 10.4. The describing equation is

$$L\frac{di}{dt} + Ri = V_m \cos \omega t \tag{10.11}$$

and, following the method of the previous chapter, let us assume the trial solution

$$i_f = A \cos \omega t + B \sin \omega t$$

Substituting the trial solution into (10.11), we have

$$L(-\omega A \sin \omega t + \omega B \cos \omega t) + R(A \cos \omega t + B \sin \omega t) = V_m \cos \omega t$$

Therefore equating coefficients of like terms, we must have

$$RA + \omega LB = V_m$$
$$-\omega LA + RB = 0$$

from which

$$A = \frac{RV_m}{R^2 + \omega^2 L^2}$$

$$B = \frac{\omega L V_m}{R^2 + \omega^2 L^2}$$

The forced response is then

$$i_f = \frac{RV_m}{R^2 + \omega^2 L^2} \cos \omega t + \frac{\omega L V_m}{R^2 + \omega^2 L^2} \sin \omega t$$

which by (10.9) and (10.10) may be written as

$$i_f = \frac{V_m}{\sqrt{R^2 + \omega^2 L^2}} \cos \left(\omega t - \tan^{-1}\frac{\omega L}{R}\right) \tag{10.12}$$

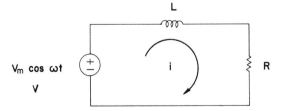

FIGURE 10.4 *RL circuit*

The forced response is, therefore, a sinusoid like the excitation, as we predicted when we chose the trial solution. We may write it in the form

$$i_f = I_m \cos(\omega t + \phi) \tag{10.13}$$

where

$$I_m = \frac{V_m}{\sqrt{R^2 + \omega^2 L^2}}$$

and

$$\phi = -\tan^{-1}\frac{\omega L}{R} \tag{10.14}$$

Since the natural response is

$$i_n = A_1 e^{-Rt/L}$$

it is clear that after a short time $i_n \to 0$, and the current settles down to its ac steady-state value, given by (10.12).

The method we have used is straightforward and conventional but, as the reader might agree, is rather laborious for such a simple problem. For a second-order circuit, the method is more tedious, as was illustrated by the example of (9.34) in the previous chapter. For very high-order circuits the procedure is, of course, even more complicated. Evidently we need a better method. One such method is developed in the remainder of this chapter, and its use allows us to treat circuits with storage elements in the same way that we treated resistive circuits in Chapters 2, 4, and 5.

EXERCISES

10.2.1 Find the forced response i_f in Fig. 10.4 if $L = 4$ mH, $R = 3$ kΩ, $V_m = 5$ V, and $\omega = 10^6$ rad/s.   *Ans.* $\cos(10^6 t - 53.1°)$ mA

10.2.2 Find the forced component of v.

Ans. $(RI_m/\sqrt{1 + \omega^2 R^2 C^2}) \cos(\omega t - \tan^{-1} \omega RC)$ V

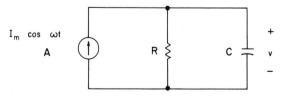

EXERCISE 10.2.2

10.3 AN ALTERNATIVE METHOD USING COMPLEX NUMBERS

The alternative method of analyzing circuits with sinusoidal excitations, which we shall consider in the remainder of the chapter, relies heavily on the concept of complex numbers. The reader who is unfamiliar with complex numbers, or who needs to review the subject, should consult Appendices C and D, where complex numbers and their properties are discussed in some detail. For convenience, we shall list a few of these properties in this section before considering the alternative method of analysis.

The complex number A is written in rectangular form as

$$A = a + jb \tag{10.15}$$

where $j = \sqrt{-1}$ and the real numbers a and b are the real and imaginary parts of A, respectively. Equivalently we may say that

$$a = \text{Re } A, \qquad b = \text{Im } A$$

where Re and Im denote "the real part of" and "the imaginary part of."

The number A may be written also in the polar form,

$$A = |A|e^{j\alpha} = |A|\underline{/\alpha} \tag{10.16}$$

where $|A|$ is the magnitude, given by

$$|A| = \sqrt{a^2 + b^2}$$

and α is the angle or argument, given by

$$\alpha = \tan^{-1}\frac{b}{a}$$

These relations between rectangular and polar forms are illustrated in Fig. 10.5.

As an example, suppose we have $A = 4 + j3$. Then $|A| = \sqrt{4^2 + 3^2} = 5$ and $\alpha = \tan^{-1}\frac{3}{4} = 36.9°$. Therefore the polar form is

$$A = 5\underline{/36.9°}$$

As another example, consider $A = -5 - j12$. Since both a and b are negative, the line segment representing A lies in the third quadrant, as shown in Fig. 10.6, from which we see that

$$|A| = \sqrt{5^2 + 12^2} = 13$$

and

$$\alpha = 180° + \tan^{-1}\frac{12}{5} = 247.4°$$

Thus we have $A = 13\underline{/247.4°}$.

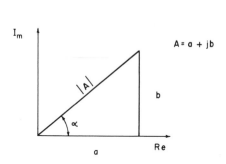

FIGURE 10.5 *Geometrical representation of a complex number A*

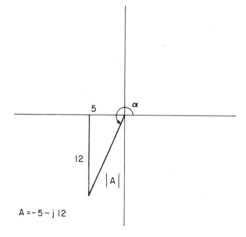

FIGURE 10.6 *Complex number with negative real and imaginary parts*

Other useful results are

$$j = 1\underline{/90°}$$

$$j^2 = -1 = 1\underline{/180°}$$

and so forth.

By Euler's formula, which is discussed in Appendix D and which we have used previously in Chapter 9, we know that

$$V_m \cos \omega t + jV_m \sin \omega t = V_m e^{j\omega t}$$

Therefore we may say that

$$V_m \cos \omega t = \text{Re}\,(V_m e^{j\omega t}) \qquad (10.17)$$

and

$$V_m \sin \omega t = \text{Im}\,(V_m e^{j\omega t})$$

Returning to the *RL* circuit example of Fig. 10.4, we know that exponentials are mathematically easier to handle as excitations than sinusoids. Therefore let us see what happens if we apply the complex excitation

$$v_1 = V_m e^{j\omega t} \qquad (10.18)$$

instead of the real excitation

$$v_g = V_m \cos \omega t = \text{Re}\, v_1 \qquad (10.19)$$

We cannot duplicate such a complex excitation in the laboratory, but there is no reason we cannot consider it abstractly. In this case the forced component of the current, which we shall call i_1, satisfies

$$L\frac{di_1}{dt} + Ri_1 = v_1 = V_m e^{j\omega t} \tag{10.20}$$

To solve this equation, we try

$$i_f = Ae^{j\omega t}$$

which, substituted into (10.20), yields

$$(j\omega L + R)Ae^{j\omega t} = V_m e^{j\omega t}$$

from which

$$A = \frac{V_m}{R + j\omega L}$$

$$= \frac{V_m}{\sqrt{R^2 + \omega^2 L^2}} e^{-j\,\tan^{-1}\omega L/R}$$

using complex number division. Therefore we have

$$i_1 = \frac{V_m}{\sqrt{R^2 + \omega^2 L^2}} e^{j(\omega t - \tan^{-1}\omega L/R)}$$

Let us now observe that

$$\text{Re}\,i_1 = \text{Re}\left[\frac{V_m}{\sqrt{R^2 + \omega^2 L^2}} e^{j(\omega t - \tan^{-1}\omega L/R)}\right]$$

$$= \frac{V_m}{\sqrt{R^2 + \omega^2 L^2}} \cos\left(\omega t - \tan^{-1}\frac{\omega L}{R}\right)$$

which, by (10.12), is the correct forced response of Fig. 10.4. That is,

$$i_f = \text{Re}\,i_1 \tag{10.21}$$

We have established for this example the interesting result that if i_1 is the complex response to the complex forcing function v_1, then $i_f = \text{Re}\,i_1$ is the response to $v_g = \text{Re}\,v_1$. That is, v_1 yields i_1 and $\text{Re}\,v_1 = v_g$ yields $\text{Re}\,i_1 = i_f$. The reason for this is that the describing equation (10.20) contains only real coefficients. Thus, from (10.20), we have

$$\text{Re}\left(L\frac{di_1}{dt} + Ri_1\right) = \text{Re}\,v_1$$

or

$$L\frac{d}{dt}(\text{Re}\,i_1) + R(\text{Re}\,i_1) = V_m \cos \omega t$$

and therefore, by (10.11),

$$i = i_f = \text{Re } i_1 \qquad (10.22)$$

Thus we see that it is easier to use the complex forcing function v_1 to find the complex response i_1. Then since the real forcing function is Re v_1, the real response is Re i_1. This principle holds for all our circuit analyses, since the describing equations are linear with real coefficients, as may be seen in a development analogous to that leading to (10.22).

EXERCISES

10.3.1   Replace the real forcing function $I_m \cos \omega t$ in Ex. 10.2.2 by the complex forcing function $I_m e^{j\omega t}$, find the resulting complex response v_1, and show that the real response is $v = \text{Re } v_1$.

10.3.2   Show that, for a real,

$$\text{Re}\left(a\frac{dx}{dt}\right) = a\frac{d}{dt}(\text{Re } x)$$

and use this result to establish (10.22). (*Suggestion*: Let $x = f + jg$, where f and g are real.)

10.4   COMPLEX EXCITATIONS

Let us now generalize the results using complex excitation functions in the previous section. The excitation, as well as the forced response, may be a sinusoidal voltage or current. However, to be specific, let us consider the input to be a voltage source and the output to be a current through some element. The other cases may be considered in an analogous way.

In general, we know that if

$$v_g = V_m \cos (\omega t + \theta) \qquad (10.23)$$

then the forced response is of the form

$$i = I_m \cos (\omega t + \phi) \qquad (10.24)$$

as indicated in the general circuit of Fig. 10.7. Therefore if by some means we can find I_m and ϕ, we have our answer, since ω, θ, and V_m are known.

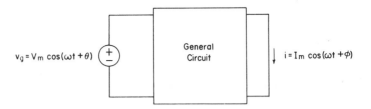

FIGURE 10.7 *General circuit with input and output*

To solve for i in Fig. 10.7, let us apply the complex excitation

$$v_1 = V_m e^{j(\omega t + \theta)} \qquad (10.25)$$

and find the complex response i_1, as shown in Fig. 10.8. Then we know from the results of the previous section that the real response of Fig. 10.7 is

$$i = \text{Re } i_1 \qquad (10.26)$$

This is a consequence of the fact that the coefficients in the describing equation are real, as pointed out previously.

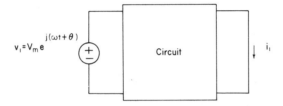

FIGURE 10.8 *General circuit with a complex excitation*

The describing equation may be solved for the forced response by the method of Chapter 9. That is, since we may write the excitation as

$$v_1 = V_m e^{j\theta} e^{j\omega t} \qquad (10.27)$$

which is a constant times $e^{j\omega t}$, then the trial solution is

$$i_1 = A e^{j\omega t}$$

Comparing (10.24) and (10.26), we must have

$$I_m \cos(\omega t + \phi) = \text{Re } [A e^{j\omega t}]$$

which requires that

$$A = I_m e^{j\phi}$$

and hence

$$i_1 = I_m e^{j\phi} e^{j\omega t} \tag{10.28}$$

Taking the real part, we have the solution (10.24), of course.

As an example, let us find the forced response i_f of

$$\frac{d^2 i}{dt^2} + 2\frac{di}{dt} + 8i = 12\sqrt{2}\,\cos{(2t + 15°)}$$

First we replace the real excitation by the complex excitation,

$$v_1 = 12\sqrt{2}\,e^{j(2t+15°)}$$

where for convenience the phase is written in degrees. (This is, of course, an inconsistent mathematical expression, but as long as we interpret it correctly it should present no difficulty.) The complex response i_1 satisfies

$$\frac{d^2 i_1}{dt^2} + 2\frac{di_1}{dt} + 8i_1 = 12\sqrt{2}\,e^{j(2t+15°)}$$

and it must have the general form

$$i_1 = Ae^{j2t}$$

Therefore we must have

$$(-4 + j4 + 8)Ae^{j2t} = 12\sqrt{2}\,e^{j2t}e^{j15°}$$

or

$$A = \frac{12\sqrt{2}\,e^{j15°}}{4 + j4} = \frac{12\sqrt{2}\,\underline{/15°}}{4\sqrt{2}\,\underline{/45°}} = 3\underline{/-30°}$$

which gives

$$i_1 = (3\underline{/-30°})e^{j2t}$$
$$= 3e^{j(2t-30°)}$$

Thus the real answer is

$$i_f = \text{Re}\,i_1 = 3\cos{(2t - 30°)}$$

EXERCISES

10.4.1 Find the forced response v using the method of this section.

Ans. $5\cos{(16t - 28.1°)}$ V

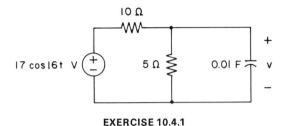

EXERCISE 10.4.1

10.4.2 Find the forced response i using the method of this section if $v_g = 6 \sin 4t$.
[*Suggestion*: $\sin 4t = \text{Im } e^{j4t}$]   *Ans.* $3 \sin 4t$ A

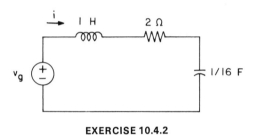

EXERCISE 10.4.2

10.4.3 Repeat Ex. 10.4.2 if $v_g = 6 \cos 4t$.   *Ans.* $3 \cos 4t$ A

10.5 PHASORS

The results obtained in the previous section may be put in much more compact form by the use of quantities called *phasors*, which we shall introduce in this section. The phasor method of analyzing circuits is credited generally to Charles Proteus Steinmetz (1865–1923), a famous electrical engineer with the General Electric Company in the early part of this century.

To begin, let us recall the general sinusoidal voltage,

$$v = V_m \cos (\omega t + \theta) \tag{10.29}$$

which, of course, is the source voltage v_g of the previous section. If the frequency ω is known, then v is completely specified by its amplitude V_m and its phase θ. These quantities are displayed in a related complex number,

$$\mathbf{V} = V_m e^{j\theta} = V_m \underline{/\theta} \tag{10.30}$$

which is defined as a *phasor*, or a phasor *representation*. To distinguish them from other complex numbers, phasors are printed in boldface type, as indicated.

The motivation for the phasor definition may be seen from the equivalence, by Euler's formula, of

$$V_m \cos (\omega t + \theta) = \text{Re } (V_m e^{j\theta} e^{j\omega t}) \tag{10.31}$$

Therefore, in view of (10.29) and (10.30), we have

$$v = \text{Re} (\mathbf{V}e^{j\omega t}) \tag{10.32}$$

As an example, suppose we have

$$v = 10 \cos (4t + 30°) \text{ V}$$

The phasor representation is then

$$\mathbf{V} = 10\underline{/30°} \text{ V}$$

since $V_m = 10$ and $\theta = 30°$. Conversely, since $\omega = 4$ rad/s is assumed to be known, v is readily obtained from $\mathbf{V}$.

In an identical fashion we define the phasor representation of the time-domain current

$$i = I_m \cos (\omega t + \phi) \tag{10.33}$$

to be

$$\mathbf{I} = I_m e^{j\phi} = I_m\underline{/\phi} \tag{10.34}$$

Thus if we know, for example, that $\omega = 6$ rad/s and that $\mathbf{I} = 2\underline{/15°}$ A, then we have

$$i = 2 \cos (6t + 15°) \text{ A}$$

We have chosen to represent sinusoids and their related phasors on the basis of cosine functions, though we could have chosen sine functions just as easily. Therefore if a function such as

$$v = 8 \sin (3t + 30°)$$

is given, we may change it to

$$v = 8 \cos (3t + 30° - 90°)$$
$$= 8 \cos (3t - 60°)$$

Then the phasor representation is

$$\mathbf{V} = 8\underline{/-60°}$$

To see how the use of phasors can greatly shorten the work, let us reconsider Fig. 10.4 and its describing equation (10.11), rewritten as

$$L\frac{di}{dt} + Ri = V_m \cos \omega t \tag{10.35}$$

Following our method, we replace the excitation $V_m \cos \omega t$ by the complex forcing function

$$v_1 = V_m e^{j\omega t}$$

which may be written

$$v_1 = \mathbf{V}e^{j\omega t}$$

since $\theta = 0$, and therefore $\mathbf{V} = V_m \underline{/0} = V_m$. Substituting this value and $i = i_1$ into (10.35), we have

$$L\frac{di_1}{dt} + Ri_1 = \mathbf{V}e^{j\omega t}$$

whose solution i_1 is related to the real solution i by

$$i = \text{Re}\, i_1$$

Next, trying

$$i_1 = \mathbf{I}e^{j\omega t}$$

as a solution, we have

$$j\omega L\mathbf{I}e^{j\omega t} + R\mathbf{I}e^{j\omega t} = \mathbf{V}e^{j\omega t}$$

Dividing out the factor $e^{j\omega t}$, we have the phasor equation

$$j\omega L\mathbf{I} + R\mathbf{I} = \mathbf{V} \tag{10.36}$$

Therefore

$$\mathbf{I} = \frac{\mathbf{V}}{R + j\omega L} = \frac{V_m}{\sqrt{R^2 + \omega^2 L^2}}\underline{\left/-\tan^{-1}\frac{\omega L}{R}\right.}$$

Substituting this value into the expression for i_1, we have

$$i_1 = \frac{V_m}{\sqrt{R^2 + \omega^2 L^2}}e^{j(\omega t - \tan^{-1}\omega L/R)}$$

Taking the real part, we have $i = i_f$, obtained earlier in (10.12).

It is important to note that if we can go directly from (10.35) to (10.36), there is a vast saving of time and effort. Also, in the process we have converted the differential equation into an algebraic equation, somewhat like those encountered in resistive circuits. Indeed, the only difference is that the numbers here are complex, whereas in resistive circuits they were real. With the hand calculator as commonly available as it is today, even the complexity of the numbers presents little difficulty.

In the remainder of the chapter we shall see how to bypass all the steps between (10.35) and (10.36) by studying the phasor relationships of the circuit elements and considering Kirchhoff's laws as they pertain to phasors. Indeed, as we shall see, we may go directly from the circuit to (10.36), bypassing even the step of writing down the differential equation.

In general, the real solutions are time-domain functions, and their phasors are *frequency-domain* functions; i.e., they are functions of the frequency ω. This is illustrated by the phasor $\mathbf{I}$ of the last example. Thus to solve the time-domain problems, we may convert to phasors and solve the corresponding frequency-domain problems, which are generally much easier. Finally, we convert back to the time domain by finding the time function from its phasor representation.

EXERCISES

10.5.1 Find the phasor representation of (a) $10 \cos (2t + 45°)$, (b) $4 \cos 2t + 3 \sin 2t$, and (c) $-6 \sin (5t - 75°)$.

Ans. (a) $10/\underline{45°}$, (b) $5/\underline{-36.9°}$, (c) $6/\underline{15°}$

10.5.2 Find the time-domain function represented by the phasors (a) $10/\underline{-12°}$, (b) $6 + j8$, and (c) $-j4$. In all cases $\omega = 3$.

Ans. (a) $10 \cos (3t - 12°)$, (b) $10 \cos (3t + 53.1°)$, (c) $4 \cos (3t - 90°)$

10.6 VOLTAGE-CURRENT RELATIONSHIPS FOR PHASORS

In this section we shall show that relations between phasor voltage and phasor current for resistors, inductors, and capacitors are very similar to Ohm's law for resistors. In fact, the phasor voltage is proportional to the phasor current, as in Ohm's law, with the proportionality factor being a constant or a function of the frequency ω.

We begin by considering the voltage-current relation for the resistor,

$$v = Ri \tag{10.37}$$

where

$$\begin{aligned} v &= V_m \cos (\omega t + \theta) \\ i &= I_m \cos (\omega t + \phi) \end{aligned} \tag{10.38}$$

If we apply the complex voltage $V_m e^{j(\omega t + \theta)}$, the complex current which results is $I_m e^{j(\omega t + \phi)}$, which substituted into (10.37) yields

$$V_m e^{j(\omega t + \theta)} = RI_m e^{j(\omega t + \phi)}$$

Dividing out the factor $e^{j\omega t}$ results in

$$V_m e^{j\theta} = RI_m e^{j\phi} \tag{10.39}$$

which, since $V_m e^{j\theta}$ and $I_m e^{j\phi}$ are the phasors **V** and **I**, respectively, reduces to

$$\mathbf{V} = R\mathbf{I} \tag{10.40}$$

Thus the phasor or frequency-domain relation for the resistor is exactly like the time-domain relation. The voltage-current relations for the resistor are illustrated in Fig. 10.9.

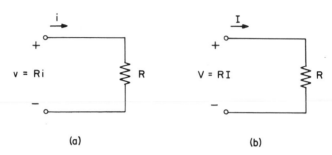

(a)   (b)

FIGURE 10.9 *Voltage-current relations for a resistor R in the (a) time and (b) frequency domains*

From (10.39) we have $V_m = RI_m$ and $\theta = \phi$. Thus the sinusoidal voltage and current for a resistor have the same phase angle, in which case they are said to be *in phase*. This phase relationship is shown in Fig. 10.10, where the voltage is represented by the solid line and the current by the dashed line.

As an illustration, suppose that the voltage

$$v = 10 \cos (100t + 30°) \text{ V} \tag{10.41}$$

is applied across a 5-Ω resistor, with the polarity indicated in Fig. 10.9(a). Then the phasor voltage is

$$\mathbf{V} = 10/\underline{30°} \text{ V}$$

and the phasor current is

$$\mathbf{I} = \frac{\mathbf{V}}{R} = \frac{10/\underline{30°}}{5} = 2/\underline{30°} \text{ A}$$

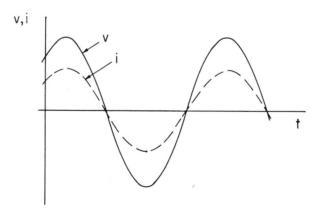

FIGURE 10.10 *Voltage and current waveforms for a resistor*

Therefore in the time domain we have

$$i = 2 \cos (100t + 30°) \text{ A} \qquad (10.42)$$

This is, of course, simply the result we would have obtained using Ohm's law.

In the case of the inductor, substituting the complex current and voltage into the time-domain relation,

$$v = L\frac{di}{dt}$$

gives the complex relation

$$V_m e^{j(\omega t + \theta)} = L\frac{d}{dt}[I_m e^{j(\omega t + \phi)}]$$

$$= j\omega L I_m e^{j(\omega t + \phi)}$$

Again, dividing out the factor $e^{j\omega t}$ and identifying the phasors, we obtain the phasor relation

$$\mathbf{V} = j\omega L\mathbf{I} \qquad (10.43)$$

Thus the phasor voltage $\mathbf{V}$, as in Ohm's law, is proportional to the phasor current $\mathbf{I}$, with the proportionality factor $j\omega L$. The voltage-current relations for the inductor are shown in Fig. 10.11.

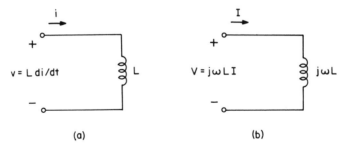

(a)   (b)

FIGURE 10.11 *Voltage-current relations for an inductor L in the (a) time and (b) frequency domains*

If the current in the inductor is given by the second of (10.38), then by (10.43), the phasor voltage is

$$\mathbf{V} = (j\omega L)(I_m \underline{/\phi})$$

$$= \omega L I_m \underline{/\phi + 90°}$$

since $j = 1\underline{/90°}$. Therefore in the time domain we have

$$v = \omega L I_m \cos (\omega t + \phi + 90°)$$

Comparing this result with the second of (10.38), we see that in the case of an inductor the current *lags* the voltage by 90°. Another expression that is used is that the current and voltage are 90° out of phase. This is shown graphically in Fig. 10.12.

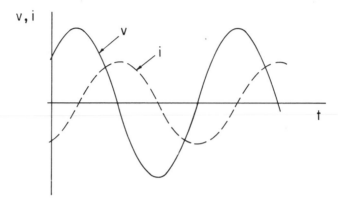

FIGURE 10.12 *Voltage and current waveforms for an inductor*

Finally, let us consider the capacitor. Substituting the complex current and voltage into the time-domain relation,

$$i = C\frac{dv}{dt}$$

gives the complex relation

$$I_m e^{j(\omega t + \phi)} = C\frac{d}{dt}[V_m e^{j(\omega t + \theta)}]$$
$$= j\omega C V_m e^{j(\omega t + \theta)}$$

Again dividing by $e^{j\omega t}$ and identifying the phasors, we obtain the phasor relation

$$\mathbf{I} = j\omega C\mathbf{V} \tag{10.44}$$

or

$$\mathbf{V} = \frac{\mathbf{I}}{j\omega C} \tag{10.45}$$

Thus the phasor voltage $\mathbf{V}$ is proportional to the phasor current $\mathbf{I}$, with the proportionality factor given by $1/j\omega C$. The voltage-current relations for a capacitor in the time and frequency domains are shown in Fig. 10.13.

In the general case, if the capacitor voltage is given by the first equation of (10.38), then by (10.44), the phasor current is

$$I = (j\omega C)(V_m\underline{/\theta})$$
$$= \omega C V_m\underline{/\theta + 90°}$$

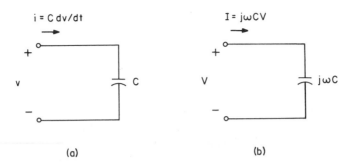

FIGURE 10.13 *Voltage-current relations for a capacitor in the (a) time and (b) frequency domains*

Therefore in the time domain we have

$$i = \omega C V_m \cos(\omega t + \theta + 90°)$$

which, by comparison with the first equation of (10.38), indicates that in the case of a capacitor the current and voltage are out of phase with the current *leading* the voltage by 90°. This is shown graphically in Fig. 10.14.

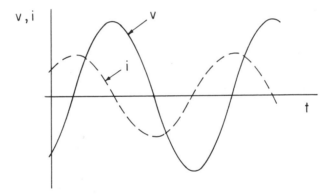

FIGURE 10.14 *Voltage and current waveforms for a capacitor*

As an example, if the voltage of (10.41) is applied across a 1-μF capacitor, then by (10.44) the phasor current is

$$\mathbf{I} = j(100)(10^{-6})(10\underline{/30°}) \text{ A}$$
$$= 1\underline{/120°} \text{ mA}$$

The time-domain current is then

$$i = \cos(100t + 120°) \text{ mA}$$

and therefore the current leads the voltage by 90°.

EXERCISES

10.6.1 Using phasors, find the ac steady-state current i if $v = 12 \cos (1000t + 30°)$ V in (a) Fig. 10.9(a) for $R = 4$ kΩ, (b) Fig. 10.11(a) for $L = 20$ mH, and (c) Fig. 10.13(a) for $C = 1$ μF.

Ans. (a) 3 cos $(1000t + 30°)$ mA, (b) 0.6 cos $(1000t - 60°)$ A, (c) 12 cos $(1000t + 120°)$ mA

10.6.2 In Ex. 10.6.1, find i in each case at $t = 1$ ms.

Ans. (a) 0.142 mA, (b) 0.599 A, (c) −11.987 mA

10.7 IMPEDANCE AND ADMITTANCE

Let us now consider a general phasor circuit with two accessible terminals, as shown in Fig. 10.15. If the time-domain voltage and current at the terminals are given by (10.38), then the phasor quantities at the terminals are

$$\mathbf{V} = V_m \underline{/\theta}$$
$$\mathbf{I} = I_m \underline{/\phi} \qquad (10.46)$$

We define the ratio of the phasor voltage to the phasor current as the *impedance* of the circuit, which we denote by **Z**. That is,

$$\mathbf{Z} = \frac{\mathbf{V}}{\mathbf{I}} \qquad (10.47)$$

which by (10.46) is

$$\mathbf{Z} = |\mathbf{Z}| \underline{/\theta_z} = \frac{V_m}{I_m} \underline{/\theta - \phi} \qquad (10.48)$$

where $|\mathbf{Z}|$ is the magnitude and θ_z the angle of **Z**. Evidently,

$$|\mathbf{Z}| = \frac{V_m}{I_m}, \qquad \theta_z = \theta - \phi$$

Impedance, as is seen from (10.47), plays the role, in a general circuit, of resistance in resistive circuits. Indeed, (10.47) looks very much like Ohm's law; also like resistance, impedance is measured in ohms, being a ratio of volts to amperes.

It is important to stress that impedance is a complex number, being the ratio of two complex numbers, but it is *not* a phasor. That is, it has no corresponding sinusoidal time-domain function of any physical meaning, as current and voltage phasors have.

The impedance **Z** is written in polar form in (10.48); in rectangular form it is generally denoted by

$$\mathbf{Z} = R + jX \qquad (10.49)$$

where $R = \text{Re } \mathbf{Z}$ is the *resistive component*, or simply *resistance*, and $X = \text{Im } \mathbf{Z}$ is the *reactive component*, or *reactance*. In general, $\mathbf{Z} = \mathbf{Z}(j\omega)$ is a complex function of $j\omega$, but $R = R(\omega)$ and $X = X(\omega)$ are real functions of ω. Both R and X, like $\mathbf{Z}$, are measured in ohms. Evidently, comparing (10.48) and (10.49) we may write

$$|\mathbf{Z}| = \sqrt{R^2 + X^2}$$

$$\theta_z = \tan^{-1} \frac{X}{R}$$

and

$$R = |\mathbf{Z}| \cos \theta_z$$

$$X = |\mathbf{Z}| \sin \theta_z$$

These relations are shown graphically in Fig. 10.16.

As an example, suppose in Fig. 10.15 that $\mathbf{V} = 10\underline{/56.1°}$ V and $\mathbf{I} = 2\underline{/20°}$ A. Then we have

$$\mathbf{Z} = \frac{10\underline{/56.1°}}{2\underline{/20°}} = 5\underline{/36.1°}\ \Omega$$

In rectangular form this is

$$\mathbf{Z} = 5(\cos 36.1° + j \sin 36.1°)$$
$$= 4 + j3\ \Omega$$

The impedances of resistors, inductors, and capacitors are readily found from their **V-I** relations of (10.40), (10.43), and (10.45). Distinguishing their impedances with subscripts R, L, and C, respectively, we have, from these equations and (10.47),

$$\mathbf{Z}_R = R$$
$$\mathbf{Z}_L = j\omega L = \omega L \underline{/90°}$$
$$\mathbf{Z}_C = \frac{1}{j\omega C} = -j\frac{1}{\omega C} = \frac{1}{\omega C}\underline{/-90°} \qquad (10.50)$$

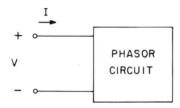

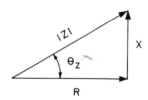

FIGURE 10.15 *General phasor circuit*   **FIGURE 10.16** *Graphical representation of impedance*

In the case of a resistor, the impedance is purely resistive, its reactance being zero. Impedances of inductors and capacitors are purely reactive, having zero resistive components. The *inductive reactance* is denoted by

$$X_L = \omega L \tag{10.51}$$

so that

$$\mathbf{Z}_L = jX_L$$

and the *capacitive reactance* is denoted by

$$X_C = -\frac{1}{\omega C} \tag{10.52}$$

and thus

$$\mathbf{Z}_C = jX_C \tag{10.53}$$

Since ω, L, and C are positive, we see that inductive reactance is positive and that capacitive reactance is negative. In the general case of (10.49), we may have $X = 0$, in which case the circuit appears to be resistive; $X > 0$, in which case its reactance appears to be inductive; and $X < 0$, in which case its reactance appears to be capacitive. These cases are possible when resistance, inductance, and capacitance are all present in the circuit, as we shall see. As an example, the circuit with impedance given by $\mathbf{Z} = 4 + j3$, which we have just considered, has reactance $X = 3$, which is of the inductive type. In all cases of passive circuits, as we shall see in Chapter 12, the resistance R is nonnegative.

The reciprocal of impedance, denoted

$$\mathbf{Y} = \frac{1}{\mathbf{Z}} \tag{10.54}$$

is called *admittance* and is analogous to conductance (the reciprocal of resistance) in resistive circuits. Evidently, since $\mathbf{Z}$ is a complex number, then so is $\mathbf{Y}$, the standard representation being

$$\mathbf{Y} = G + jB \tag{10.55}$$

The quantities $G = \text{Re}\,\mathbf{Y}$ and $B = \text{Im}\,\mathbf{Y}$ are, respectively, called *conductance* and *susceptance* and are related to the impedance components by

$$\mathbf{Y} = G + jB = \frac{1}{\mathbf{Z}} = \frac{1}{R + jX} \tag{10.56}$$

The units of $\mathbf{Y}$, G, and B are all mhos, since in general $\mathbf{Y}$ is the ratio of a current to a voltage phasor.

To obtain the relation between components of **Y** and **Z** we may rationalize the last member of (10.56), which results in

$$G + jB = \frac{1}{R + jX} \cdot \frac{R - jX}{R - jX}$$

$$= \frac{R - jX}{R^2 + X^2}$$

Equating real and imaginary parts results in

$$G = \frac{R}{R^2 + X^2}$$

(10.57)

$$B = -\frac{X}{R^2 + X^2}$$

Therefore we note that R and G are *not* reciprocals except in the purely resistive case $(X = 0)$. Similarly, X and B are never reciprocals, but in the purely reactive case $(R = 0)$ they are negative reciprocals.

As an example, if we have

$$\mathbf{Z} = 4 + j3$$

then

$$\mathbf{Y} = \frac{1}{4 + j3} = \frac{4 - j3}{4^2 + 3^2} = \frac{4}{25} - j\frac{3}{25} \quad .$$

Therefore $G = \frac{4}{25}$ and $B = -\frac{3}{25}$.

Further examples are

$$\mathbf{Y}_R = G$$

$$\mathbf{Y}_L = \frac{1}{j\omega L}$$

$$\mathbf{Y}_C = j\omega C$$

which are the admittances of a resistor, with $R = 1/G$, an inductor, and a capacitor.

EXERCISES

10.7.1 Find the impedance seen at the terminals of the source in Fig. 10.4 in both rectangular and polar form.

Ans. $R + j\omega L, \sqrt{R^2 + \omega^2 L^2} \underline{/\tan^{-1} \omega L/R}$

10.7.2 Find the admittance seen at the terminals of the source in Fig. 10.4 in both rectangular and polar form.

$$Ans. \; \frac{R}{R^2 + \omega^2 L^2} - j\frac{\omega L}{R^2 + \omega^2 L^2}, \; \frac{1}{\sqrt{R^2 + \omega^2 L^2}} \Big/ -\tan^{-1}\frac{\omega L}{R}$$

10.7.3 Find the conductance and susceptance if $\mathbf{Z}$ is (a) $5 + j12$, (b) $0.8 - j0.6$, and (c) $\sqrt{2}\,/45°$. Ans (a) $\frac{5}{169}$, $-\frac{12}{169}$, (b) 0.8, 0.6, (c) $\frac{1}{2}$, $-\frac{1}{2}$

10.8 KIRCHHOFF'S LAWS AND IMPEDANCE COMBINATIONS

Kirchhoff's laws hold for phasors as well as for their corresponding time-domain voltages or currents. We may see this by observing that if a complex excitation, say $V_m e^{j(\omega t + \theta)}$, is applied to a circuit, then complex voltages, such as $V_1 e^{j(\omega t + \theta_1)}$, $V_2 e^{j(\omega t + \theta_2)}$, etc., appear across the elements in the circuit. Since Kirchhoff's laws hold in the time domain, KVL applied around a typical loop results in an equation such as

$$V_1 e^{j(\omega t + \theta_1)} + V_2 e^{j(\omega t + \theta_2)} + \ldots + V_N e^{j(\omega t + \theta_N)} = 0$$

Dividing out the common factor $e^{j\omega t}$, we have

$$\mathbf{V}_1 + \mathbf{V}_2 + \ldots + \mathbf{V}_N = 0$$

where

$$\mathbf{V}_n = V_n \underline{/\theta_n}, \qquad n = 1, 2, \ldots, N$$

are the phasor voltages around the loop. Thus KVL holds for phasors. A similar development will also establish KCL.

In circuits having sinusoidal excitations with a common frequency ω, if we are interested only in the forced, or ac steady-state, response, we may find the phasor voltages or currents of every element and use Kirchhoff's laws to complete the analysis. The ac steady-state analysis is therefore identical to the resistive circuit analysis of Chapters 2, 4, and 5, with impedances replacing resistances and phasors replacing time-domain quantities. Once we have found the phasors, we can convert immediately to the time-domain sinusoidal answers.

As an example, consider the circuit of Fig. 10.17, which consists of N impedances connected in series. By KCL for phasors, the single phasor current $\mathbf{I}$ flows in each element. Therefore the voltages shown across each element are

$$\mathbf{V}_1 = \mathbf{Z}_1 \mathbf{I}$$
$$\mathbf{V}_2 = \mathbf{Z}_2 \mathbf{I}$$
$$\vdots$$
$$\mathbf{V}_N = \mathbf{Z}_N \mathbf{I}$$

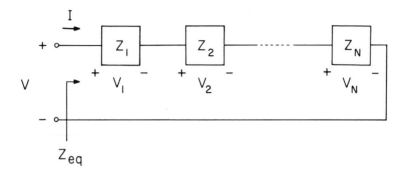

FIGURE 10.17 *Impedances connected in series*

and by KVL around the circuit,

$$\mathbf{V} = \mathbf{V}_1 + \mathbf{V}_2 + \ldots + \mathbf{V}_N$$
$$= (\mathbf{Z}_1 + \mathbf{Z}_2 + \ldots + \mathbf{Z}_N)\mathbf{I}$$

Since we must also have, from Fig. 10.17,

$$\mathbf{V} = \mathbf{Z}_{eq}\mathbf{I}$$

where $\mathbf{Z}_{eq}$ is the *equivalent* impedance seen at the terminals, it follows that

$$\mathbf{Z}_{eq} = \mathbf{Z}_1 + \mathbf{Z}_2 + \ldots + \mathbf{Z}_N \qquad (10.58)$$

as in the case of series resistors.

Similarly, as was the case for parallel conductances in Chapter 2, the equivalent admittance $\mathbf{Y}_{eq}$ of N parallel admittances is

$$\mathbf{Y}_{eq} = \mathbf{Y}_1 + \mathbf{Y}_2 + \ldots + \mathbf{Y}_N \qquad (10.59)$$

In the case of two parallel elements ($N = 2$), we have

$$\mathbf{Z}_{eq} = \frac{1}{\mathbf{Y}_{eq}} = \frac{1}{\mathbf{Y}_1 + \mathbf{Y}_2} = \frac{\mathbf{Z}_1\mathbf{Z}_2}{\mathbf{Z}_1 + \mathbf{Z}_2} \qquad (10.60)$$

In like manner, voltage and current division rules hold for phasor circuits, with impedances and frequency-domain quantities, in exactly the same way that they held for resistive circuits, with resistances and time-domain quantities. The reader is asked to establish these rules in Ex. 10.8.2.

For example, let us to return to the *RL* circuit considered in Sec. 10.2. The circuit and its phasor counterpart are shown in Figs. 10.18(a) and (b), respectively. By KVL in the phasor circuit we have

$$\mathbf{Z}_L\mathbf{I} + R\mathbf{I} = V_m\underline{/0}$$

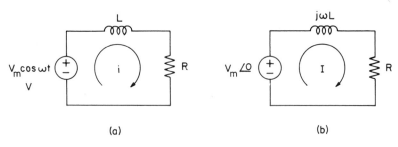

FIGURE 10.18 *(a) Time-domain circuit; (b) equivalent phasor circuit*

or

$$(j\omega L + R)\mathbf{I} = V_m\underline{/0}$$

from which the phasor current is

$$\mathbf{I} = \frac{V_m\underline{/0}}{R + j\omega L}$$

$$= \frac{V_m}{\sqrt{R^2 + \omega^2 L^2}}\underline{\bigg/ -\tan^{-1}\frac{\omega L}{R}}$$

Therefore in the time domain we have, as before,

$$i = \frac{V_m}{\sqrt{R^2 + \omega^2 L^2}} \cos\left(\omega t - \tan^{-1}\frac{\omega L}{R}\right)$$

An alternative method of solution is to observe that the impedance **Z** seen at the source terminals is the impedance of the inductor, $j\omega L$, and the resistor, R, connected in series. Therefore

$$\mathbf{Z} = j\omega L + R$$

and

$$\mathbf{I} = \frac{\mathbf{V}}{\mathbf{Z}} = \frac{V_m\underline{/0}}{R + j\omega L}$$

as obtained earlier.

EXERCISES

10.8.1 Derive (10.59).

10.8.2 Show in (a) that the voltage division rule,

$$\mathbf{V} = \frac{\mathbf{Z}_2}{\mathbf{Z}_1 + \mathbf{Z}_2}\mathbf{V}_g$$

and in (b) that the current division rule,

$$\mathbf{I} = \frac{\mathbf{Y}_2}{\mathbf{Y}_1 + \mathbf{Y}_2}\mathbf{I}_g = \frac{\mathbf{Z}_1}{\mathbf{Z}_1 + \mathbf{Z}_2}\mathbf{I}_g$$

are valid, where $\mathbf{Z}_1 = 1/\mathbf{Y}_1$ and $\mathbf{Z}_2 = 1/\mathbf{Y}_2$.

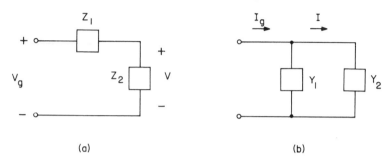

(a)   (b)

EXERCISE 10.8.2

10.8.3   Find the steady-state current *i* using phasors.

Ans. 2 cos (4*t* − 45°) A

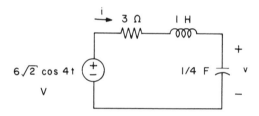

EXERCISE 10.8.3

10.8.4   Find the steady-state value of *v* in Ex. 10.8.3 using voltage division. Check by
using the fact that the current leads the voltage by 90° in a capacitor.

Ans. 2 cos (4*t* − 135°) V

10.9 PHASOR CIRCUITS

As the discussion in the previous section suggests, we may omit the steps of finding the
describing equation in the time domain, replacing the excitations and responses by
their complex forcing functions and then dividing the equation through by $e^{j\omega t}$ to
obtain the phasor equation. We may simply start with the phasor circuit, which is the
time-domain circuit with the voltages and currents replaced by their phasors and the
elements identified by their impedances, as illustrated previously in Fig. 10.18(b). The
describing equation obtained from this circuit is then the phasor equation. Solving
this equation yields the phasor of the answer, which then may be converted to the time-
domain answer.

The procedure from starting with the phasor circuit to obtaining the phasor answer is identical to that used earlier in resistive circuits. The only difference is that the numbers are complex.

As an example, let us find the steady-state current i in Fig. 10.19(a). The phasor circuit, shown in Fig. 10.19(b), is obtained by replacing the voltage source and the currents by their phasors and labeling the elements with their impedances. In the phasor circuit the impedance seen from the source terminals is

$$\mathbf{Z} = 1 + \frac{(3 + j3)(-j3)}{3 + j3 - j3}$$

$$= 4 - j3 \; \Omega$$

Therefore we have

$$\mathbf{I}_1 = \frac{5/0}{4 - j3} = \frac{5/0}{5/-36.9°} = 1/36.9°$$

and by current division,

$$\mathbf{I} = \left(\frac{3 + j3}{3 + j3 - j3}\right)\mathbf{I}_1 = \sqrt{2}/81.9° \; \text{A}$$

In the time domain, the answer is

$$i = \sqrt{2} \cos (3t + 81.9°) \; \text{A}$$

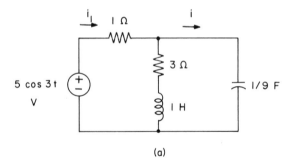

(a)

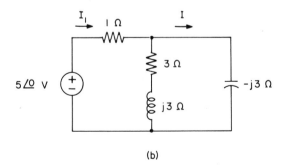

(b)

FIGURE 10.19 *RLC time-domain and phasor circuit*

In the case of a dependent source, such as a source kv_x volts controlled by a voltage v_x, it will appear in the phasor circuit as a source $k\mathbf{V}_x$, where $\mathbf{V}_x$ is the phasor representation of v_x, because $v_x = V_m \cos(\omega t + \phi)$ in the time domain will become $V_m e^{j(\omega t + \phi)}$ when a complex excitation is applied. Then dividing $e^{j\omega t}$ out of the equations leaves v_x represented by its phasor $V_m e^{j\phi}$. In the same way, $kv_x = kV_m \cos(\omega t + \phi)$ is represented by its phasor $kV_m e^{j\phi}$, which is k times the phasor of v_x.

As an example of a circuit containing a dependent source, let us consider Fig. 10.20(a), in which it is required to find the steady-state value of i. The corresponding phasor circuit is shown in Fig. 10.20(b). Since phasor circuits are analyzed exactly like resistive circuits, we may write a nodal equation at node a in Fig. 10.20(b), resulting in

$$\mathbf{I} + \frac{\mathbf{V}_1 - \frac{1}{2}\mathbf{V}_1}{-j2} = 3\underline{/0} \tag{10.61}$$

By Ohm's law we have $\mathbf{V}_1 = 4\mathbf{I}$, which substituted into (10.61) yields

$$-j2\mathbf{I} + \tfrac{1}{2}(4\mathbf{I}) = -j6$$

or

$$\mathbf{I} = \frac{-j6}{2 - j2} = \frac{6\underline{/-90}}{2\sqrt{2}\underline{/-45}} = \frac{3}{\sqrt{2}}\underline{/-45°}\ \text{A}$$

Therefore we have

$$i = \frac{3}{\sqrt{2}} \cos(4t - 45°)\ \text{A}$$

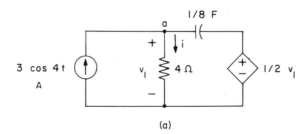

(a)

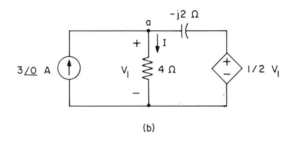

(b)

FIGURE 10.20 *(a) Circuit containing a dependent source; (b) corresponding phasor circuit*

In the case of an op amp, the phasor circuit is the same as the time-domain circuit. That is, an ideal op amp in the time-domain circuit appears as an ideal op amp in the phasor circuit, because the time-domain equations

$$i = 0, \qquad v = 0$$

which characterize the current into and the voltage across the input terminals retain the identical form,

$$\mathbf{I} = 0, \qquad \mathbf{V} = 0$$

in the phasor equations.

EXERCISES

10.9.1 Solve Ex. 10.2.2 by means of the phasor circuit.

10.9.2 Find the forced response v in the figure for Ex. 10.4.1 by means of the phasor circuit.

10.9.3 Find the forced response i in the figure for Ex. 10.4.2 by means of the phasor circuit.

PROBLEMS

10.1 Given

$$v_1 = 10 \cos 4t$$
$$v_2 = 20(\cos 4t + \sqrt{3} \sin 4t)$$

determine if v_1 leads or lags v_2 and by what amount.

10.2 Find i_4, using only the properties of sinusoids, if $i_1 = 2 \cos 3t$, $i_2 = 4 \cos (3t - 30°)$, and $i_3 = 6 \cos (3t - 60°)$.

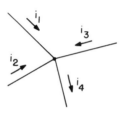

PROBLEM 10.2

10.3 In the figure of Ex. 10.2.2, if the source is $2 \cos 5000t$ mA and the output is $v = 4 \cos (5000t - 45°)$ V, find R and C.

10.4 A voltage $V_m \sin \omega t$ V, a resistor R, and an inductor L are all in series. Find the forced response i in the following two ways. (a) Noting that $V_m \sin \omega t = V_m \cos (\omega t - 90°)$, use the complex excitation $V_m e^{j(\omega t - 90°)}$ and take the real part of the current response. (b) Noting that $V_m \sin \omega t = \mathrm{Im}\,(V_m e^{j\omega t})$, use the complex excitation $V_m e^{j\omega t}$ and take the imaginary part of the current response.

10.5 A complex voltage source $4e^{j(3t+30°)}$ V produces a current response $0.5e^{j(3t-10°)}$ A. Find the current response to a voltage source of (a) $16e^{j(3t+45°)}$ V, (b) $8 \cos (3t + 35°)$ V, and (c) $2 \sin (3t - 60°)$ V.

10.6 If $\omega = 4$ rad/s, find the time-domain functions represented by the phasors (a) $-6 - j8$, (b) $-6 + j8$, (c) $-j10$, and (d) -10.

10.7 Find the impedance of the circuit of Fig. 10.15 if the time-domain functions represented by the phasors $\mathbf{V}$ and $\mathbf{I}$ are
(a) $v = -5 \cos 2t + 12 \sin 2t$ V, $i = 1.3 \cos (2t + 40°)$ A.
(b) $v = \mathrm{Re}\,[je^{j2t}]$ V, $i = \mathrm{Re}\,[(1 + j)e^{j(2t+30°)}]$ mA.
(c) $v = aV_m \cos (\omega t + \theta)$ V, $i = V_m \cos (\omega t + \theta - \alpha)$ A.

10.8 Three admittances, $5 + j6$, $7 - j1$, and $17/\tan^{-1} \frac{15}{8}$, are connected in parallel. Find the equivalent impedance of the combination.

10.9 Show that if $\mathrm{Re}\,\mathbf{Z} = R$ is positive, then $\mathrm{Re}\,\mathbf{Y} = \mathrm{Re}\,[1/\mathbf{Z}] = G$ is also positive.

10.10 A circuit has an impedance

$$\mathbf{Z} = \frac{16(2 + j\omega)(8 - \omega^2 - j2\omega)}{\omega^4 - 15\omega^2 + 64}\,\Omega$$

Find the reactance at $\omega = 1$ rad/s, 2 rad/s, and 3 rad/s. If the time-domain voltage applied to the circuit is $64 \cos \omega t$ V, find the steady-state current i produced in each case.

10.11 Find the steady-state response v.

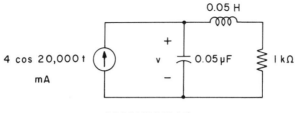

4 cos 20,000 t

mA

0.05 H

+

v   0.05 µF   1 kΩ

−

PROBLEM 10.11

10.12 Find the power being delivered by the source in Prob. 10.11 at $t = \pi/20$ ms.

10.13 Show that if the capacitance in the figure for Prob. 10.11 is changed to 0.025 μF, then the impedance seen at the terminals of the source is real. In this case, find the power being delivered by the source at $t = \pi/20$ ms.

10.14 Find the steady-state value of v using phasors.

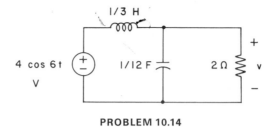

PROBLEM 10.14

10.15 Find the steady-state value of i using phasors.

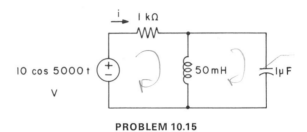

PROBLEM 10.15

10.16 Find the steady-state value of v using phasors and voltage division.

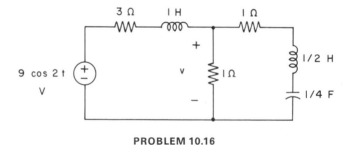

PROBLEM 10.16

10.17 Find the steady-state value of i if (a) $\omega = 4$ rad/s and (b) $\omega = 2$ rad/s. Note that in the latter case the impedance seen at the terminals of the source is purely resistive.

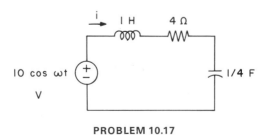

PROBLEM 10.17

10.18 Find the steady-state current i.

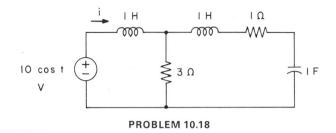

PROBLEM 10.18

10.19 Find the steady-state values of i and v.

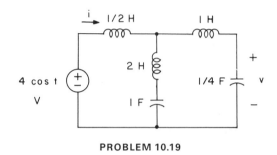

PROBLEM 10.19

10.20 Find the steady-state value of i.

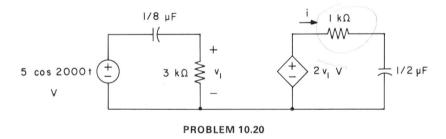

PROBLEM 10.20

10.21 Find the steady-state value of v.

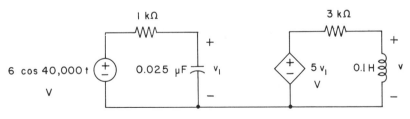

PROBLEM 10.21

10.22 Find the steady-state value of v if $v_g = 5 \cos 10{,}000t$ V.

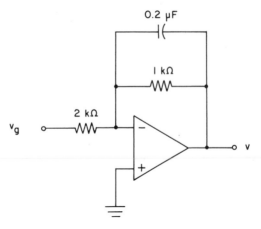

PROBLEM 10.22

10.23 Find the steady-state current i if $v_g = 2 \cos 4000t$ V.

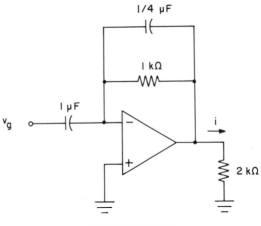

PROBLEM 10.23

10.24 Find the forced response i. [*Suggestion*: Note that $\mathbf{I} = \mathbf{V}/\mathbf{Z}$ fails to work since $\mathbf{Z}(j1) = 0$. Solve the describing equation by the method of Sec. 9.6]

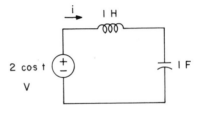

PROBLEM 10.24

10.25 Find the *complete* response i if $i(0) = 2$ A and $v(0) = 6$ V.

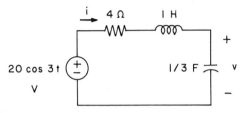

PROBLEM 10.25

10.26 Determine $i(0)$ and $v(0)$ in Prob. 10.25 so that the natural component vanishes and i is simply the forced component.

11

AC STEADY-STATE ANALYSIS

In the previous chapter we have seen that in the case of circuits with sinusoidal inputs we may obtain the ac steady-state response by analyzing the corresponding phasor circuits. The circuits encountered were relatively simple ones which could be analyzed by the use of voltage-current relations and current and voltage division rules.

It should be clear, because of the close kinship between phasor circuits and resistive circuits, that we may extend the methods of Chapter 10 to more general circuits using nodal analysis, loop analysis, Thevenin's and Norton's theorems, superposition, etc. In this chapter we shall formally consider these more general analysis procedures, limiting ourselves to obtaining the forced, or ac steady-state, response.

11.1 NODAL ANALYSIS

As we have seen, the voltage-current relation

$$\mathbf{V} = \mathbf{ZI}$$

for passive elements is identical in form to Ohm's law, and KVL and KCL hold in phasor circuits exactly as they did in resistive circuits. Therefore the only difference in analyzing phasor circuits and resistive circuits is that the excitations and responses are complex quantities in the former case and real quantities in the latter case. Thus we may analyze phasor circuits in exactly the same manner in which we analyzed resistive circuits. Specifically, nodal and mesh, or loop, analysis methods apply. We shall illustrate nodal analysis in this section and loop analysis in the following section.

To illustrate the nodal method, let us find the ac steady-state voltages v_1 and v_2 of Fig. 11.1. First we shall obtain the phasor circuit by replacing the element values by their impedances for $\omega = 2$ rad/s and the sources and node voltages by their phasors.

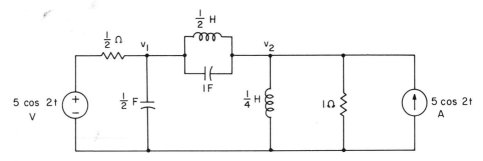

FIGURE 11.1 *Circuit to be analyzed by the phasor method*

This results in the circuit of Fig. 11.2(a). Since we are interested in finding V_1 and V_2, the node voltage phasors, we may replace the two sets of parallel impedances by their equivalent impedances, resulting in the simpler, equivalent circuit of Fig. 11.2(b).

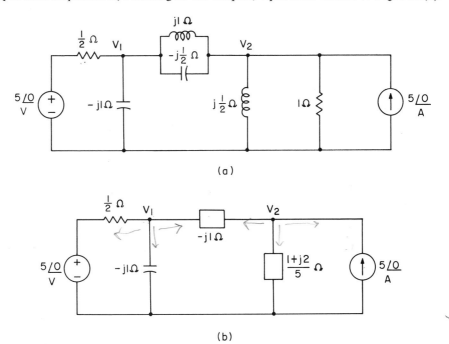

(a)

(b)

FIGURE 11.2 *Two versions of the phasor circuit corresponding to Fig. 11.1*

The nodal equations, from Fig. 11.2(b), are

$$2(V_1 - 5\underline{/0}) + \frac{V_1}{-j1} + \frac{V_1 - V_2}{-j1} = 0$$

$$\frac{V_2 - V_1}{-j1} + \frac{V_2}{(1 + j2)/5} = 5\underline{/0}$$

which in simplified form are

$$(2 + j2)\mathbf{V}_1 - j1\mathbf{V}_2 = 10$$
$$-j1\mathbf{V}_1 + (1 - j1)\mathbf{V}_2 = 5$$

Solving these equations by determinants, we have

$$\mathbf{V}_1 = \frac{\begin{vmatrix} 10 & -j1 \\ 5 & 1 - j1 \end{vmatrix}}{\begin{vmatrix} 2 + j2 & -j1 \\ -j1 & 1 - j1 \end{vmatrix}} = \frac{10 - j5}{5} = 2 - j1 \text{ V}$$

$$\mathbf{V}_2 = \frac{\begin{vmatrix} 2 + j2 & 10 \\ -j1 & 5 \end{vmatrix}}{5} = \frac{10 + j20}{5} = 2 + j4 \text{ V}$$

In polar form these quantities are

$$\mathbf{V}_1 = \sqrt{5}\ \underline{/-26.6°}\ \text{V}$$
$$\mathbf{V}_2 = 2\sqrt{5}\ \underline{/63.4°}\ \text{V}$$

Therefore the time-domain solutions are

$$v_1 = \sqrt{5}\ \cos(2t - 26.6°)\ \text{V}$$
$$v_2 = 2\sqrt{5}\ \cos(2t + 63.4°)\ \text{V}$$

As an example involving a dependent source, let us consider Fig. 11.3, in which it is required to find the forced response i. Taking the ground node as shown, we have the two unknown node voltages v and $v + 3000i$, as indicated. The phasor circuit in its simplest form is shown in Fig. 11.4, from which we may observe that only one nodal equation is needed. Writing KCL at the generalized node, shown dashed, we have

$$\frac{V - 4}{\frac{1}{2}(10^3)} + \frac{V}{\frac{2}{5}(1 - j2)(10^3)} + \frac{V + 3000I}{(2 - j1)(10^3)} = 0$$

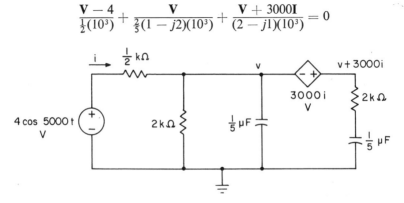

FIGURE 11.3 *Circuit containing a dependent source*

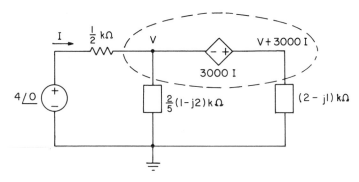

FIGURE 11.4 *Phasor circuit of Fig. 11.3*

Also, from the phasor circuit we have

$$\mathbf{I} = \frac{4 - \mathbf{V}}{\frac{1}{2}(10^3)}$$

Eliminating **V** between these two equations and solving for **I**, we have

$$\mathbf{I} = 24 \times 10^{-3}\underline{/53.1°}\ \text{A}$$
$$= 24\underline{/53.1°}\ \text{mA}$$

Therefore in the time domain, we have

$$i = 24 \cos (5000t + 53.1°)\ \text{mA}$$

As a final example illustrating nodal analysis, let us find the forced response v in Fig. 11.5 if

$$v_g = V_m \cos \omega t\ \text{V}$$

We note first that the op amp and the two 2-kΩ resistors constitute a VCVS with

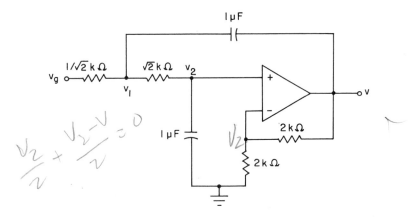

FIGURE 11.5 *Circuit containing an op amp*

gain $1 + \frac{2000}{2000} = 2$ (see Sec. 3.4). Therefore $v = 2v_2$, or $v_2 = v/2$, as indicated in the phasor circuit of Fig. 11.6.

Writing nodal equations at the nodes labeled $\mathbf{V}_1$ and $\mathbf{V}/2$, we have

$$\frac{\mathbf{V}_1 - V_m\underline{/0}}{(1/\sqrt{2})(10^3)} + \frac{\mathbf{V}_1 - (\mathbf{V}/2)}{\sqrt{2}(10^3)} + \frac{\mathbf{V}_1 - \mathbf{V}}{-j10^6/\omega} = 0$$

$$\frac{(\mathbf{V}/2) - \mathbf{V}_1}{\sqrt{2}(10^3)} + \frac{\mathbf{V}/2}{-j10^6/\omega} = 0$$

Eliminating $\mathbf{V}_1$ and solving for $\mathbf{V}$ results in

$$\mathbf{V} = \frac{2V_m}{[1 - (\omega^2/10^6)] + j(\sqrt{2}\,\omega/10^3)}$$

In polar form this is

$$\mathbf{V} = \frac{2V_m\underline{/\theta}}{\sqrt{1 + (\omega/1000)^4}} \tag{11.1}$$

where

$$\theta = -\tan^{-1}\left[\frac{\sqrt{2}\,\omega/1000}{1 - (\omega/1000)^2}\right] \tag{11.2}$$

In the time domain we have

$$v = \frac{2V_m}{\sqrt{1 + (\omega/1000)^4}}\cos{(\omega t + \theta)} \tag{11.3}$$

We might note in this example that for low frequencies, say $0 < \omega < 1000$, the amplitude of the output voltage v is relatively large, and for higher frequencies, its amplitude is relatively small. Thus the circuit of Fig. 11.5 *filters* out higher frequencies and allows lower frequencies to "pass." Such a circuit is called a *filter* and will be considered in more detail in Chapter 15.

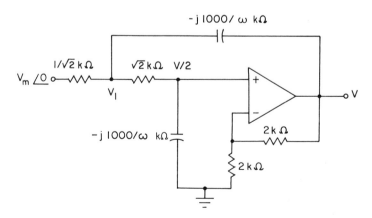

FIGURE 11.6 *Phasor circuit of Fig. 11.5*

EXERCISES

11.1.1 Find the forced response v using nodal analysis.   *Ans.* 2 sin 2*t* V

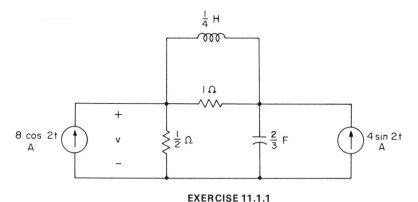

EXERCISE 11.1.1

11.1.2 Find the forced response v.   *Ans.* 4.8 cos $(2t - 53.1°)$

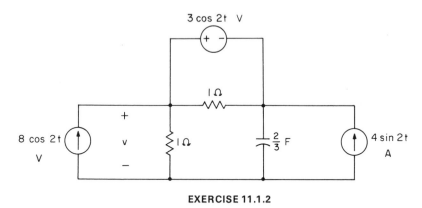

EXERCISE 11.1.2

11.1.3 Find the amplitude of v in (11.3) if $V_m = 10$ V for (a) $\omega = 0$, (b) $\omega = 1000$ rad/s, (c) $\omega = 10,000$ rad/s, and (d) $\omega = 100,000$ rad/s.

Ans. (a) 20, (b) 14.14, (c) 0.2, (d) 0.002 V

11.2 MESH ANALYSIS

To illustrate mesh analysis of an ac steady-state circuit, let us find v_1 in Fig. 11.1, which was obtained, using nodal analysis, in the previous section. We shall use the phasor circuit of Fig. 11.2(b), which is redrawn in Fig. 11.7, with mesh currents $\mathbf{I}_1$ and

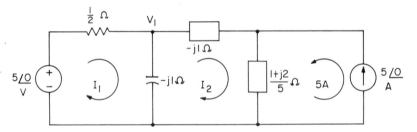

FIGURE 11.7 *Circuit of Fig. 11.2 redrawn for mesh analysis*

I_2, as indicated. Evidently, the phasor voltage V_1 may be obtained from

$$I_1 = 2(5 - V_1)$$

or

$$V_1 = 5 - \frac{I_1}{2} \tag{11.4}$$

The two mesh equations are

$$\tfrac{1}{2}I_1 - j1(I_1 - I_2) = 5$$
$$-j1(I_2 - I_1) - j1I_2 + \left(\frac{1 + j2}{5}\right)(I_2 + 5) = 0 \tag{11.5}$$

Solving these equations for I_1, we have

$$I_1 = 6 + j2 \text{ A}$$

which substituted into (11.4) yields

$$V_1 = 2 - j1 \text{ V}$$

This is the same result that was obtained in the previous section and may be used to obtain the time-domain voltage v_1.

The same shortcut procedures for writing loop and nodal equations, discussed in Secs. 4.1 and 4.5 for resistive circuits, apply to phasor circuits. For example, in Fig. 11.7 if $I_3 = -5$ is the mesh current in the right mesh in the clockwise direction, the two mesh equations are written down by inspection as

$$(\tfrac{1}{2} - j1)I_1 - (-j1)I_2 = 5$$
$$-(-j1)I_1 + \left(-j1 - j1 + \frac{1 + j2}{5}\right)I_2 - \left(\frac{1 + j2}{5}\right)I_3 = 0$$

These are equivalent to (11.5) and are formed as in the resistive circuit case. That is, in the first equation the coefficient of the first variable is the sum of the impedances around the first mesh. The other coefficients are the negatives of the impedances common to the first mesh and the meshes whose numbers correspond to the currents. The right member is the sum of the voltage sources in the mesh with polarities consistent with the direction of the mesh current. Replacing "first" by "second" applies to the next equation, and so forth. The dual development, as described in Sec. 4.1, holds for nodal equations.

As a final example, let us consider the circuit of Fig. 11.8(a), where the response is the steady-state value of v_1. The phasor circuit is shown in Fig. 11.8(b), with the loop currents as indicated.

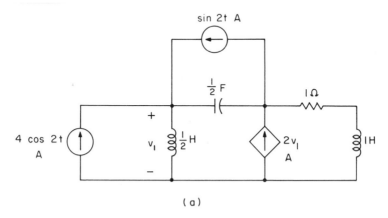

(a)

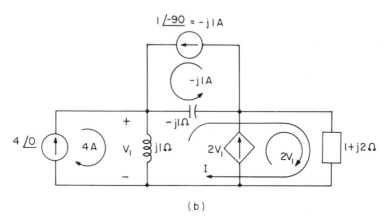

(b)

FIGURE 11.8 *(a) Time-domain circuit; (b) phasor counterpart*

Applying KVL around the loop labeled I, we have

$$-\mathbf{V}_1 - j1(-j1 + \mathbf{I}) + (1 + j2)(\mathbf{I} + 2\mathbf{V}_1) = 0$$

Also, from the figure we see that

$$\mathbf{V}_1 = j1(4 - \mathbf{I})$$

Eliminating **I** from these equations and solving for $\mathbf{V}_1$, we have

$$\mathbf{V}_1 = \frac{-4 + j3}{5} = 1\underline{/143.1^\circ}\ \text{V}$$

Therefore, in the time domain, the voltage is

$$v_1 = \cos\,(2t + 143.1^\circ)\ \text{V}$$

EXERCISES

11.2.1 Find the forced response i in Fig. 11.3 using mesh analysis.

11.2.2 Solve Ex. 11.1.1 using loop analysis.

11.2.3 Find the steady-state current i using loop analysis.   *Ans.* $\cos\,(t - 36.9^\circ)$ A

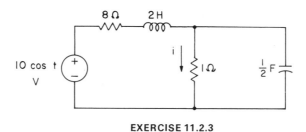

EXERCISE 11.2.3

11.3 NETWORK THEOREMS

Because the phasor circuits are exactly like the resistive circuits except for the nature of the currents, voltages, and impedances, all the network theorems discussed in Chapter 5 for resistive circuits apply to phasor circuits. In this section we shall illustrate superposition, Thevenin's theorem, Norton's theorem, and the proportionality principle, as applied to phasor circuits.

   In the case of superposition, if a phasor circuit has two or more inputs, we may find the phasor currents or voltages due to each input acting alone (i.e., with the others dead) and add the individual corresponding time-domain responses to obtain the total. In the case of a circuit like Fig. 11.1, we may solve the corresponding phasor circuit of Fig. 11.2 by mesh or nodal analysis or by superposition because both sources are operating at the same frequency, namely $\omega = 2$ rad/s. If the sources have *different* frequencies, we *must* use superposition, because the definition of $\mathbf{Z}(j\omega)$ allows us to use only one frequency at a time, and thus we cannot even construct a phasor circuit.

To illustrate superposition, let us find the forced response i in Fig. 11.9. There are two sources, one an ac source with $\omega = 2$ rad/s and one a dc source with $\omega = 0$. Therefore

$$i = i_1 + i_2$$

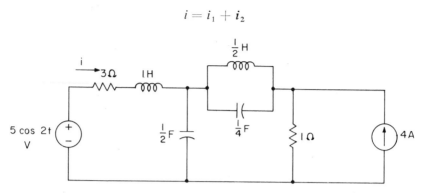

FIGURE 11.9 *Circuit with an ac and a dc source*

where i_1 is due to the voltage source acting alone and i_2 is due to the current source acting alone. Using phasors, we may find i_1 and i_2 by finding their respective phasor representations $\mathbf{I}_1$ and $\mathbf{I}_2$, which are the phasor currents shown in Figs. 11.10(a) and (b), respectively. Figure 11.10(a) is a phasor circuit representing the time-domain circuit with the current source killed and $\omega = 2$ rad/s. Figure 11.10(b) is also a phasor circuit, with the voltage source killed and $\omega = 0$.

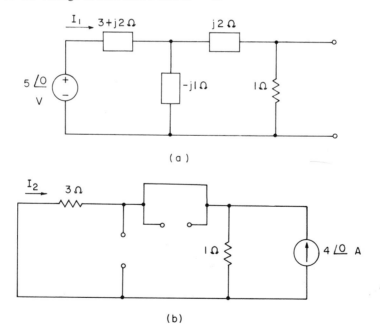

FIGURE 11.10 *Phasor circuits representing Fig. 11.9*

In the latter case, $\omega = 0$, the inductances are short circuits ($\mathbf{Z}_L = j0 = 0$) and the capacitance is an open circuit ($\mathbf{Z}_C = 1/j\omega C$, which becomes infinite as $\omega \to 0$). Also, since the current source is

$$i_g = 4 \cos (0t + 0) \text{ A}$$

its phasor representation is

$$\mathbf{I}_g = 4\underline{/0^\circ} \text{ A}$$

In short, this is just the dc case considered earlier, as is evident from Fig. 11.10(b).
From Fig. 11.10(a) we have

$$\mathbf{I}_1 = \frac{5\underline{/0^\circ}}{3 + j2 + [(1 + j2)(-j1)/(1 + j2 - j1)]}$$
$$= \sqrt{2}\underline{/-8.1^\circ}$$

from which

$$i_1 = \sqrt{2} \cos (2t - 8.1^\circ) \text{ A}$$

From Fig. 11.10(b) we have, by current division,

$$\mathbf{I}_2 = -\left(\frac{1}{1 + 3}\right)(4) = -1\underline{/0^\circ} \text{ A}$$

from which, since $\omega = 0$,

$$i_2 = -1 \text{ A}$$

Therefore the total forced response is

$$i = i_1 + i_2$$
$$= \sqrt{2} \cos (2t - 8.1^\circ) - 1 \text{ A}$$

For an example of a circuit with two sinusoidal sources with nonzero frequencies, the reader is referred to Ex. 11.3.1. The procedure is, of course, exactly the same as in the preceding example.

In the case of Thevenin's and Norton's theorems the procedure is identical to that for resistive circuits. The only change is that v_{oc} and i_{sc}, the time-domain open-circuit voltage and short-circuit current, are replaced by their phasor representations, $\mathbf{V}_{oc}$ and $\mathbf{I}_{sc}$, and R_{th}, the Thevenin resistance, is replaced by $\mathbf{Z}_{th}$, the Thevenin impedance (of the dead circuit). There must be only a single frequency present, of course; otherwise we must use superposition to break the problem up into single-frequency problems, in each of which Thevenin's or Norton's theorem may be applied.

In general, the Thevenin and Norton equivalent circuits in the frequency domain are as shown in Fig. 11.11. There is, of course, a close similarity with the resistive case.

As an example, let us use Thevenin's theorem to find the forced response v of Fig. 11.12. We shall find its phasor representation $\mathbf{V}$ using the Thevenin equivalent of the phasor circuit to the left of terminals a-b. The open-circuit phasor voltage $\mathbf{V}_{oc}$ is found from Fig. 11.13(a), and the short-circuit phasor current is found from Fig. 11.13(b).

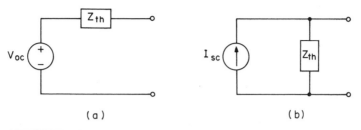

(a)   (b)

FIGURE 11.11 *(a) Thevenin and (b) Norton equivalent phasor circuits*

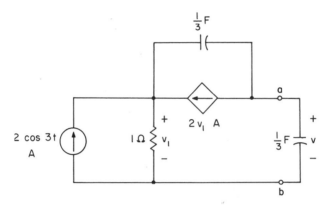

FIGURE 11.12 *Circuit with a dependent current source*

Then the Thevenin impedance, as for resistive circuits, is

$$\mathbf{Z}_{th} = \frac{\mathbf{V}_{oc}}{\mathbf{I}_{sc}} \tag{11.6}$$

In Fig. 11.13(a), since terminals *a-b* are open, the current $2/\underline{0^\circ}$ flows in the resistor and the current $2\mathbf{V}_1$ flows in the capacitor. Therefore by KVL we have

$$\mathbf{V}_{oc} = \mathbf{V}_1 - (-j1)(2\mathbf{V}_1)$$

where

$$\mathbf{V}_1 = 2(1) = 2 \text{ V}$$

Thus we have

$$\mathbf{V}_{oc} = 2 + j4 \text{ V}$$

In Fig. 11.13(b), the two nodal equations needed are

$$\frac{\mathbf{V}_1}{1} + \frac{\mathbf{V}_1}{-j1} - 2\mathbf{V}_1 = 2$$

and

$$\mathbf{I}_{sc} = -\mathbf{V}_1 + 2$$

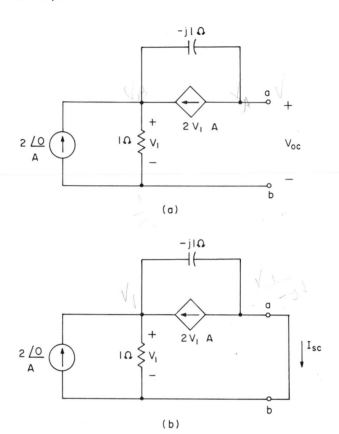

FIGURE 11.13 *Phasor circuits for use in Thevenin's theorem*

From these we find

$$\mathbf{I}_{sc} = 3 + j1 \text{ A}$$

The Thevenin impedance, by (11.6), is therefore

$$\mathbf{Z}_{th} = \frac{2 + j4}{3 + j1} = 1 + j1 \ \Omega$$

The Thevenin equivalent circuit is shown in Fig. 11.14, with the $\frac{1}{3}$-F capacitor, corresponding to $-j1 \ \Omega$, connected to terminals *a-b*.

It is a simple matter now to see, by voltage division, that

$$\mathbf{V} = \left[\frac{-j1}{(1 + j1) + (-j1)} \right] (2 + j4)$$
$$= 4 - j2$$
$$= 2\sqrt{5} \underline{/-26.6°} \text{ V}$$

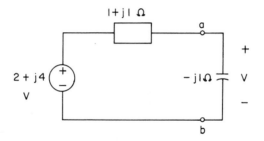

FIGURE 11.14 *Thevenin equivalent phasor circuit of Fig. 11.12*

Therefore, in the time-domain, we have

$$v = 2\sqrt{5}\ \cos{(3t - 26.6°)}\ V$$

As a final topic in this section, let us consider the ladder network of Fig. 11.15 and use the proportionality principle to obtain the steady-state response $\mathbf{V}$. This requires, as in the resistive case, that we assume $\mathbf{V}$ to have some convenient value like $\mathbf{V} = 1$ and work backward to find the corresponding $\mathbf{V}_g$. Then the correct value of $\mathbf{V}$ is found by multiplying the assumed value by an appropriate constant.

Let us begin by assuming

$$\mathbf{V} = 1\ V$$

Then from the circuit we have

$$\mathbf{I}_1 = \frac{\mathbf{V}}{1} + \frac{\mathbf{V}}{-j1} = 1 + j1\ A$$

Continuing, we have

$$\mathbf{V}_1 = j1\mathbf{I}_1 + \mathbf{V} = j1(1 + j1) + 1 = j1\ V$$

$$\mathbf{I}_2 = \frac{\mathbf{V}_1}{-j1} + \mathbf{I}_1 = -1 + (1 + j1) = j1\ A$$

$$\mathbf{V}_g = 1\mathbf{I}_2 + \mathbf{V}_1 = j1 + j1 = j2\ V$$

Therefore $\mathbf{V} = 1$ is the response to $\mathbf{V}_g = j2$. If we multiply this value of $\mathbf{V}_g$ by $6/j2$ to get the correct value of $\mathbf{V}_g$, then by the proportionality principle, we must multiply

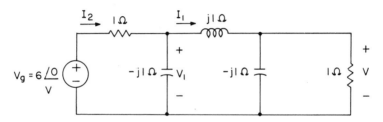

FIGURE 11.15 *Phasor ladder network*

the assumed response of 1 by the same factor, $6/j2$, to get the correct value of **V**. Therefore we have

$$\mathbf{V} = \left(\frac{6}{j2}\right)(1) = -j3 \text{ V}$$

EXERCISES

11.3.1 Find the forced response i.

Ans. $2.88 \cos(2t + 19.4°) + 1.8 \cos(3t + 111.9°)$

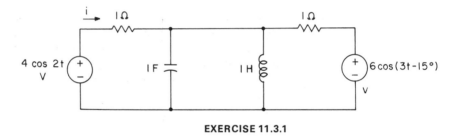

EXERCISE 11.3.1

11.3.2 Solve for v in Fig. 11.12 using Norton's theorem.

11.3.3 Find $\mathbf{V}_1$, $\mathbf{I}_1$, and $\mathbf{I}_2$ in Fig. 11.15.   *Ans.* $3 \text{ V}, 3 - j3 \text{ A}, 3 \text{ A}$

11.4 PHASOR DIAGRAMS

Since phasors are complex numbers, they may be represented by vectors in a plane, where operations, such as addition of phasors, may be carried out geometrically. Such a sketch is called a *phasor diagram* and may be quite helpful in analyzing ac steady-state circuits.

To illustrate, let us consider the phasor circuit of Fig. 11.16, for which we shall draw all the voltages and currents on a phasor diagram. To begin with, let us observe

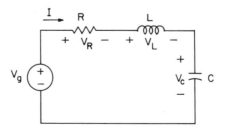

FIGURE 11.16 *RLC series phasor circuit*

that the current **I** is common to all elements and take it as our *reference* phasor, denoting it by

$$\mathbf{I} = |\mathbf{I}|\,\underline{/0^\circ}$$

We have taken the angle of **I** arbitrarily to be zero, since we want **I** to be our reference. We may always adjust this assumed value to the true value by the proportionality principle discussed in the previous section.

The voltage phasors of the circuit are

$$\mathbf{V}_R = R\mathbf{I} = R|\mathbf{I}|$$
$$\mathbf{V}_L = j\omega L\mathbf{I} = \omega L|\mathbf{I}|\underline{/90^\circ}$$
$$\mathbf{V}_C = -j\frac{1}{\omega C}\mathbf{I} = \frac{1}{\omega C}|\mathbf{I}|\underline{/-90^\circ}$$

and

$$\mathbf{V}_g = \mathbf{V}_R + \mathbf{V}_L + \mathbf{V}_C$$

These are shown in the phasor diagram of Fig. 11.17(a), where it is assumed that $|\mathbf{V}_L| > |\mathbf{V}_C|$. The cases, $|\mathbf{V}_L| < |\mathbf{V}_C|$ and $|\mathbf{V}_L| = |\mathbf{V}_C|$, are shown in Figs. 11.17(b) and (c), respectively. In all cases the lengths representing the units of current and voltage are not necessarily the same, so that for clarity **I** is shown longer than $\mathbf{V}_R$.

In case (a) the net reactance is inductive, and the current lags the source voltage by the angle θ that can be measured. In (b) the circuit has a net capacitive reactance, and the current leads the voltage. Finally, in (c) the current and voltage are in phase, since the inductive and capacitive reactance components exactly cancel each other. These conclusions follow also from the equation

$$\mathbf{I} = \frac{\mathbf{V}_g}{\mathbf{Z}} = \frac{\mathbf{V}_g}{R + j[\omega L - (1/\omega C)]} \tag{11.7}$$

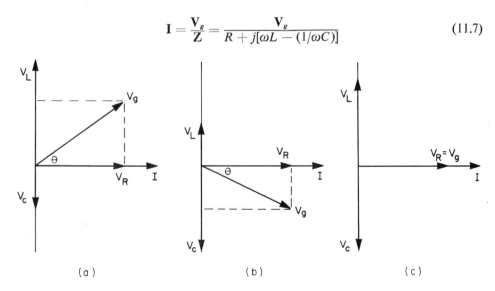

(a)   (b)   (c)

FIGURE 11.17 *Phasor diagrams for Fig. 11.16*

Case (c) is characterized by

$$\omega L - \frac{1}{\omega C} = 0$$

or

$$\omega = \frac{1}{\sqrt{LC}} \tag{11.8}$$

If the current in Fig. 11.16 is fixed, then the real component of the voltage $\mathbf{V}_g$ is fixed, since it is $R|\mathbf{I}|$. In this case the *locus* of the phasor $\mathbf{V}_g$ (its possible location on the phasor diagram) is the dashed line of Fig. 11.18. The voltage phasor varies up and down this line as ω varies between zero and infinity. The minimum amplitude of the voltage occurs when $\omega = 1/\sqrt{LC}$, as seen from the figure. For any other frequency, a larger amplitude of voltage is required for the same current.

As a last example illustrating the use of phasor diagrams, let us find the locus of $\mathbf{I}$ as R varies in Fig. 11.19. The current is given by

$$\mathbf{I} = \frac{V_m}{R + j\omega L} = \frac{V_m(R - j\omega L)}{R^2 + \omega^2 L^2}$$

Therefore if

$$\mathbf{I} = x + jy \tag{11.9}$$

we have

$$x = \text{Re } \mathbf{I} = \frac{RV_m}{R^2 + \omega^2 L^2} \tag{11.10}$$

$$y = \text{Im } \mathbf{I} = \frac{-\omega L V_m}{R^2 + \omega^2 L^2} \tag{11.11}$$

The equation of the locus is the equation satisfied by x and y as R varies; thus we need to eliminate R between these last two equations.

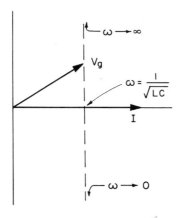

FIGURE 11.18 *Locus of the voltage phasor for a fixed current*

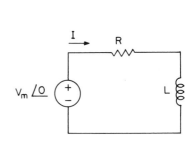

FIGURE 11.19 *RL phasor circuit*

If we divide the first of these two equations by the second, we have

$$\frac{x}{y} = -\frac{R}{\omega L}$$

from which

$$R = -\frac{\omega L x}{y}$$

Substituting this value of R into (11.11) we have, after some simplification,

$$x^2 + y^2 = -\frac{V_m y}{\omega L}$$

This result may be rewritten as

$$x^2 + \left(y + \frac{V_m}{2\omega L}\right)^2 = \left(\frac{V_m}{2\omega L}\right)^2 \tag{11.12}$$

which is the equation of a circle with center at $[0, -(V_m/2\omega L)]$ and radius $V_m/2\omega L$.

The circle (11.12) appears to be the locus, as R varies, of the phasor $\mathbf{I} = x + jy$. However, by (11.10), $x \geq 0$; thus the locus is actually the semicircle shown dashed on the phasor diagram of Fig. 11.20. The voltage $V_m\underline{/0}$, taken as reference, is also shown, along with the phasor $\mathbf{I}$. If $R = 0$, we have, from (11.10) and (11.11), $x = 0$ and $y = -V_m/\omega L$. If $R \to \infty$, then $x \to 0$ and $y \to 0$. Thus as R varies from 0 to ∞, the current phasor moves counterclockwise along the circle.

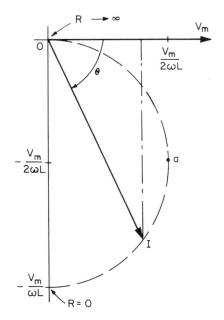

FIGURE 11.20 *Locus of the phasor* **I**

If **I** is as shown in Fig. 11.20, the current phasor may be resolved into two components, one having amplitude $I_m \cos \theta$ in phase with the voltage and one with amplitude $I_m \sin \theta$, which is 90° out of phase with the voltage. This construction is indicated by the dashed and dotted vertical line. As we shall see in the next chapter, the in-phase component of the current is important in calculating the average power delivered by the source. Thus the phasor diagram gives us a method of seeing at a glance the maximum in-phase component of current. Evidently this occurs at point a, which corresponds to $\theta = 45°$. This is the case $x = -y$, or $R = \omega L$.

EXERCISES

11.4.1   Eliminate ωL in (11.10) and (11.11) and show that, as ωL varies, the locus of the phasor $\mathbf{I} = x + jy$ is a semicircle.

$$Ans. \ \left(x - \frac{V_m}{2R}\right)^2 + y^2 = \left(\frac{V_m}{2R}\right)^2, y \leq 0$$

11.4.2   Find ωL in Ex. 11.4.1 so that Im **I** has its largest negative value. Also find **I** for this case.   *Ans.* $R, \ (V_m/\sqrt{2}\,R)/\underline{-45°}$

PROBLEMS

11.1   Solve for the steady-state response i in Fig. 10.19(a) using nodal analysis.

11.2   Solve for the steady-state response i in Fig. 10.20(a) using nodal analysis.

11.3   Solve Prob. 10.18 using nodal analysis.

11.4   Solve Prob. 10.19 using nodal analysis.

11.5   Find the steady-state value of v if $v_g = V_m \cos \omega t$ and find the amplitude of v for the cases $\omega =$ (a) 0, (b) 1000 and (c) 10^6 rad/s. Compare the results for this circuit with those for Fig. 11.6.

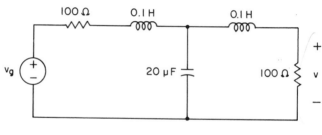

PROBLEM 11.5

11.6 Repeat Prob. 11.1 using mesh analysis.

11.7 Repeat Prob. 11.3 using mesh analysis.

11.8 Find the forced response v.

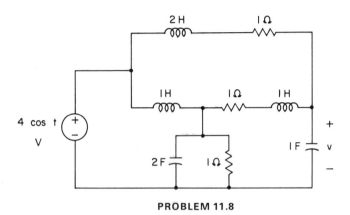

PROBLEM 11.8

11.9 Show that if $\mathbf{Z}_1\mathbf{Z}_4 = \mathbf{Z}_2\mathbf{Z}_3$ in the "bridge" circuit shown, then $\mathbf{I} = \mathbf{V} = 0$ and therefore all the other currents and voltages remain unchanged for any value of $\mathbf{Z}_5$. Thus it may be replaced by an open circuit, a short circuit, etc. In this case, the circuit is said to be a *balanced bridge*.

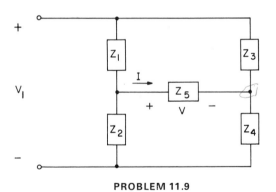

PROBLEM 11.9

11.10 Show that the circuit of Prob. 11.8 is a balanced bridge, with the series combination of 1 Ω and 1 H constituting $\mathbf{Z}_5$ in Prob. 11.9. Replace $\mathbf{Z}_5$ by a short, and show that the same v is obtained as before.

11.11 Find the steady-state response v.

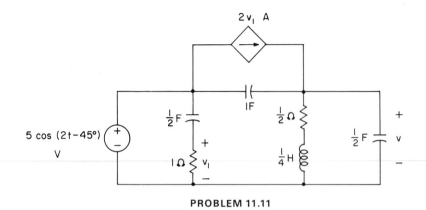

PROBLEM 11.11

11.12 Find the steady-state response i.

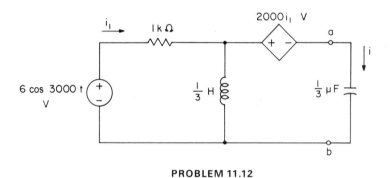

PROBLEM 11.12

11.13 Find the steady-state response v if $i_{g1} = 6 \cos 2t$ A and $i_{g2} = 2 \cos 2t$ A.

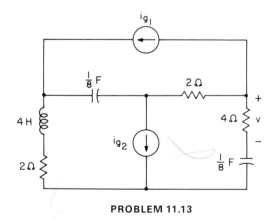

PROBLEM 11.13

11.14 Find the steady-state response v.

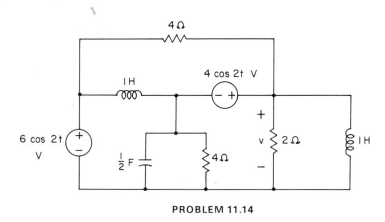

PROBLEM 11.14

11.15 Find the steady-state response v if $v_g = 2 \cos 2000t$ V.

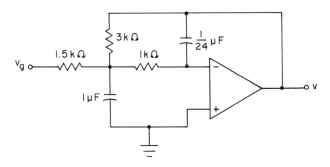

PROBLEM 11.15

11.16 Find the steady-state response v if $v_g = 5 \cos t$ V.

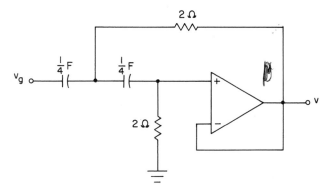

PROBLEM 11.16

11.17 Find the forced response v if $v_g = 2 \cos t$ V.

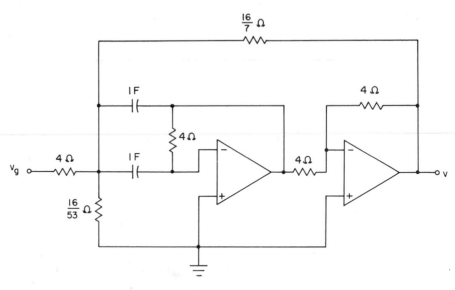

PROBLEM 11.17

11.18   Find the forced response v if (a) $v_g = 2 \cos 2t$ V and (b) $v_g = 2 \cos t$ V.

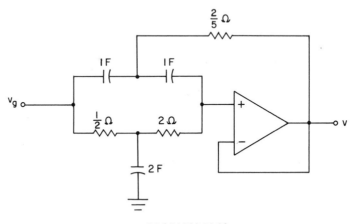

PROBLEM 11.18

11.19   Find the forced response i if $v_g = 4 \cos 1000t$ V.

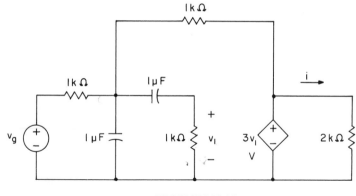

PROBLEM 11.19

11.20 Find the steady-state value of v if

$$v_g = 9 + 10 \cos (t - 30°) + 18 \cos (3t + 15°) \text{ V}$$

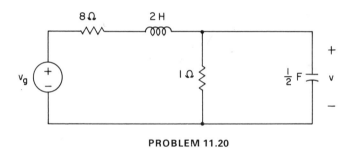

PROBLEM 11.20

11.21 Find the steady-state value of v.

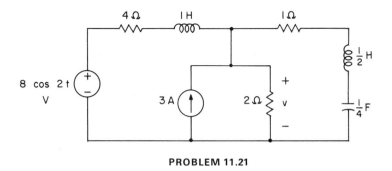

PROBLEM 11.21

11.22 Find the steady-state response v.

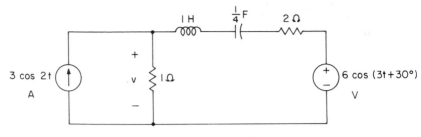

PROBLEM 11.22

11.23 For the phasor circuit corresponding to Prob. 11.12, replace the part to the left of terminals *a-b* by its Thevenin equivalent and find the steady-state value of *i*.

11.24 Replace the phasor circuit of the figure for Prob. 11.13, except for the 4-Ω resistor, by its Thevenin equivalent, and find the forced response *v*.

11.25 Replace the circuit to the left of terminals *a-b* by its Thevenin equivalent, and find **V**.

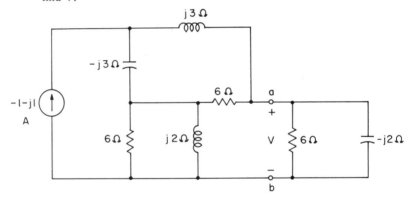

PROBLEM 11.25

11.26 Use the principle of proportionality on the corresponding phasor circuit to find the steady-state value of *v*.

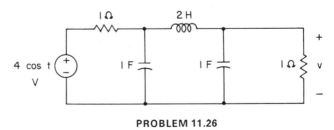

PROBLEM 11.26

11.27 Solve Prob. 11.16 by the proportionality principle.

11.28 Solve Prob. 11.19 by the proportionality principle.

11.29 Find **I**, the phasor representation of i, using a phasor diagram. Show the phasors of i_R, i_C, and i_L, with the phasor of the source voltage as reference.

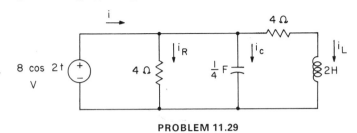

PROBLEM 11.29

11.30 Show that the locus of $\mathbf{V}_C$ in Fig. 11.16 is a circle if $R = 2\,\Omega$, $C = \frac{1}{4}\,F$, $\mathbf{V}_g = 1/\underline{0°}\,V$, $\omega = 2$ rad/s, and L varies from 0 to ∞. Select the value of L that gives the maximum amplitude of the time-domain voltage v_C, and find v_C in this case.

11.31 Find the locus of $\mathbf{V}_C$ in Prob. 11.30 if $L = 1$ H, $C = \frac{1}{2}$ F, $\mathbf{V}_g = 2/\underline{0°}$ V, $\omega = 2$ rad/s, and R varies from 0 to ∞. Select the value of R for which Im $\mathbf{V}_C$ has its largest negative value, and find v_C for this case.

12

AC STEADY-STATE POWER

In this chapter we shall consider power relationships for networks which are excited by periodic currents and voltages. We shall concern ourselves primarily with sinusoidal currents and voltages since nearly all electrical power is generated in this form. Instantaneous power, as we now well know, is the rate at which energy is absorbed by an element, and it varies as a function of time. The instantaneous power is an important quantity in engineering applications because its maximum value must be limited for all physical devices. For this reason, the maximum instantaneous power, or *peak power*, is a commonly used specification for characterizing electrical devices. In an electronic amplifier, for instance, if the specified peak power at the input is exceeded, the output signal will be distorted. Greatly exceeding this input rating may even permanently damage the amplifier.

A more important measure of power, particularly for periodic currents and voltages, is that of *average power*. The average power is equal to the average rate at which energy is absorbed by an element, and it is independent of time. This power, for example, is what is monitored by the electric company in determining monthly electricity bills. Average powers are encountered which range from a few picowatts, in applications such as satellite communications, to millions of watts, in applications such as supplying the electrical needs of a large city.

Our discussion will begin with a study of average power. We shall then consider superposition once again and introduce a mathematical measure for characterizing periodic currents or voltages, known as effective or rms values. We shall then consider the power factor associated with a load and present a complex power. Finally, we shall describe the measurement of power.

12.1 AVERAGE POWER

In linear networks which have inputs that are periodic functions of time, the steady-state currents and voltages produced are periodic, each having identical periods. Consider an instantaneous power

$$p = vi \tag{12.1}$$

where v and i are periodic of period T. That is, $v(t + T) = v(t)$, and $i(t + T) = i(t)$. In this case,

$$
\begin{aligned}
p(t + T) &= v(t + T)i(t + T) \\
&= v(t)i(t) \\
&= p(t)
\end{aligned}
\tag{12.2}
$$

Therefore the instantaneous power is also periodic of period T. That is, p repeats itself every T seconds.

The *fundamental* period T_1 of p (the *minimum* time in which p repeats itself) is not necessarily equal to T, however; but T must contain an integral number of periods T_1. In other words,

$$T = nT_1 \tag{12.3}$$

where n is a positive integer.

As an example, suppose that a resistor R carries a current $i = I_m \cos \omega t$ with period $T = 2\pi/\omega$. Then

$$
\begin{aligned}
p &= Ri^2 \\
&= RI_m^2 \cos^2 \omega t \\
&= \frac{RI_m^2}{2}(1 + \cos 2\omega t)
\end{aligned}
$$

Evidently, $T_1 = \pi/\omega$, and, therefore, $T = 2T_1$. Thus, for this case, $n = 2$ in (12.3). This is illustrated by the graph of p and i shown in Fig. 12.1(a).

If we now take $i = I_m(1 + \cos \omega t)$, then

$$p = RI_m^2(1 + \cos \omega t)^2$$

In this case, $T_1 = T = 2\pi/\omega$, and $n = 1$ in (12.3). This may be seen also from the graph of the function in Fig. 12.1(b).

Mathematically, the average value of a periodic function is defined as the time integral of the function over a complete period, divided by the period. Therefore the average power P for a periodic instantaneous power p is given by

$$P = \frac{1}{T_1} \int_{t_1}^{t_1+T_1} p \, dt \ \text{W} \tag{12.4}$$

where t_1 is arbitrary.

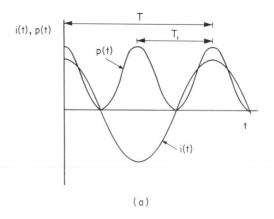

(a)

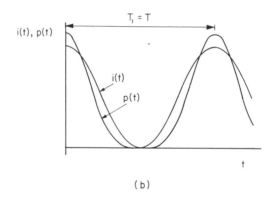

(b)

FIGURE 12.1 *Graphs of p and i*

A periodic instantaneous power p is shown in Fig. 12.2. It is clear that if we integrate over an integral number of periods, say mT_1 (where m is a positive integer), then the total area is simply m times that of the integral in (12.4). Thus we may write

$$P = \frac{1}{mT_1} \int_{t_1}^{t_1 + mT_1} p \, dt \tag{12.5}$$

If we select m such that $T = mT_1$ (the period of v or i), then

$$P = \frac{1}{T} \int_{t_1}^{t_1 + T} p \, dt \tag{12.6}$$

Therefore we may obtain the average power by integrating over the period of v or i.

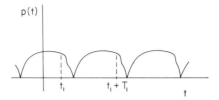

FIGURE 12.2 *Periodic instantaneous power*

Let us now consider several examples of the average power associated with sinusoidal currents and voltages. A number of very important integrals which often occur are tabulated in Table 12.1. Verification of these integrals is left as an exercise (see Ex. 12.1.1).

TABLE 12.1 *Integrals for Sinusoidal Functions and Their Products*

$f(t)$	$\int_0^{2\pi/\omega} f(t)\, dt, \; \omega \neq 0$
1. $\sin(\omega t + \alpha)$, $\cos(\omega t + \alpha)$	0
2. $\sin(n\omega t + \alpha)$, $\cos(n\omega t + \alpha)$*	0
3. $\sin^2(\omega t + \alpha)$, $\cos^2(\omega t + \alpha)$	π/ω
4. $\sin(m\omega t + \alpha)\cos(n\omega t + \alpha)$*	0
5. $\cos(m\omega t + \alpha)\cos(n\omega t + \beta)$*	$0, \; m \neq n$
	$\pi \cos(\alpha - \beta)/\omega, \; m = n$

* *m* and *n* are integers.

First, let us consider the general two-terminal device of Fig. 12.3, which is assumed to be in ac steady state. If, in the frequency domain,

$$\mathbf{Z} = |\mathbf{Z}| \, \underline{/\theta}$$

is the input impedance of the device, then for

$$v = V_m \cos(\omega t + \phi) \tag{12.7}$$

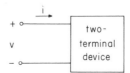

FIGURE 12.3 *General two-terminal device*

we have

$$i = I_m \cos(\omega t + \phi - \theta) \tag{12.8}$$

where

$$I_m = \frac{V_m}{|\mathbf{Z}|}$$

The average power delivered to the device, taking $t_1 = 0$ in (12.6), is

$$P = \frac{\omega V_m I_m}{2\pi} \int_0^{2\pi/\omega} \cos(\omega t + \phi) \cos(\omega + \phi - \theta)\, dt$$

Referring to Table 12.1, entry 5, we find, for $m = n = 1$, $\alpha = \phi$, and $\beta = \phi - \theta$,

$$P = \frac{V_m I_m}{2} \cos\theta \tag{12.9}$$

Thus the average power absorbed by a two-terminal device is determined by the amplitudes V_m and I_m and the angle θ by which the voltage v leads the current i.

In terms of the phasors of v and i,

$$\mathbf{V} = V_m\underline{/\phi} = |\mathbf{V}|\ \underline{/\phi}$$
$$\mathbf{I} = I_m\underline{/\phi - \theta} = |\mathbf{I}|\ \underline{/\phi - \theta}$$

we have, from (12.9),

$$P = \tfrac{1}{2}|\mathbf{V}||\mathbf{I}| \cos(\text{ang }\mathbf{V} - \text{ang }\mathbf{I}) \tag{12.10}$$

where

$$\text{ang }\mathbf{V} = \phi, \qquad \text{ang }\mathbf{I} = \phi - \theta$$

are the angles of the phasors $\mathbf{V}$ and $\mathbf{I}$.

If the two-terminal device is a resistor R, then $\theta = 0$ and $V_m = RI_m$, so that (12.9) becomes

$$P_R = \tfrac{1}{2}RI_m^2$$

It is worth noting at this point that if $i = I_{dc}$, a constant (dc) current, then $\omega = \phi = \theta = 0$, and $I_m = I_{dc}$ in (12.8). In this special case, we have, by (12.6),

$$P_R = RI_{dc}^2$$

In the case of an inductor, $\theta = 90°$, and in the case of a capacitor, $\theta = -90°$, and thus for either one, by (12.9), $P = 0$. Therefore an inductor or a capacitor, or for that matter any network composed entirely of ideal inductors and capacitors, in any combination, dissipates zero average power. For this reason, ideal inductors and capacitors are sometimes called *lossless* elements. Physically, lossless elements store energy during part of the period and release it during the other part, so that the average delivered power is zero.

A very useful alternative form of (12.9) may be obtained by recalling that

$$\mathbf{Z} = \text{Re } \mathbf{Z} + j \text{ Im } \mathbf{Z} = |\mathbf{Z}| \underline{/\theta}$$

and therefore

$$\cos \theta = \frac{\text{Re } \mathbf{Z}}{|\mathbf{Z}|}$$

Substituting this value into (12.9) and noting that $V_m = |\mathbf{Z}| I_m$, we have

$$P = \tfrac{1}{2} I_m^2 \text{ Re } \mathbf{Z} \tag{12.11}$$

Let us now consider this result if the device is a passive load. We know from the definition of passivity in (1.7) that the net energy delivered to a passive load is nonnegative. Since the average power is the average rate at which energy is delivered to a load, it follows that the average power is nonnegative. That is, $P \geq 0$. This requires, by (12.11), that

$$\text{Re } \mathbf{Z}(j\omega) \geq 0$$

or, equivalently,

$$-\frac{\pi}{2} \leq \theta \leq \frac{\pi}{2}$$

If $\theta = 0$, the device is equivalent to a resistance, and if $\theta = \pi/2$ (or $-\pi/2$), the device is equivalent to an inductance (or capacitance). For $-\pi/2 < \theta < 0$, the device is equivalent to an RC combination, whereas for $0 < \theta < \pi/2$, it is equivalent to an RL combination.

Finally, if $|\theta| > \pi/2$, then $P < 0$, and the device is active rather than passive. In this case the device is delivering power from its terminals and, of course, acts like a source.

As an example, let us find the power delivered to each element of Fig. 12.4. The impedance across the source is

$$\mathbf{Z} = 100 + j100$$
$$= 100\sqrt{2} \ \underline{/45°} \ \Omega$$

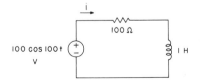

FIGURE 12.4 *RL circuit in the ac steady state*

The maximum current is

$$I_m = \frac{V_m}{|\mathbf{Z}|} = \frac{1}{\sqrt{2}} \text{ A}$$

Therefore, from (12.9), the power delivered to $\mathbf{Z}$ is

$$P = \frac{100}{2\sqrt{2}} \cos 45° = 25 \text{ W}$$

Alternatively, from (12.11),

$$P = \frac{1}{2}\left(\frac{1}{\sqrt{2}}\right)^2 (100) = 25 \text{ W}$$

The power absorbed by the 100-Ω resistor is

$$P_R = \frac{RI_m^2}{2} = \frac{(100)(1/\sqrt{2})^2}{2} = 25 \text{ W}$$

This power, of course, is equal to that delivered to $\mathbf{Z}$ since the inductor absorbs no power.

The power absorbed by the source is

$$P_g = -\frac{V_m I_m}{2} \cos \theta = -25 \text{ W}$$

where the minus sign is used because the current is flowing out of the positive terminal of the source. The source therefore is delivering 25 W to $\mathbf{Z}$. We note that the power flowing from the source is equal to that absorbed by the load, which illustrates the principle of conservation of power.

EXERCISES

12.1.1 Verify the integrals of Table 12.1.

12.1.2 For a capacitor of C farads carrying a current $i = I_m \cos \omega t$, verify that the average power is zero from (12.6). Repeat this for an inductor of L henrys.

12.1.3 Find the average power delivered to a 100-Ω resistor carrying a current given by (a) $10|\cos 10t|$ mA, (b) a square wave for which one cycle consists of 10 mA for 1 s followed by -10 mA for 1 s, and (c) a triangular wave for which one cycle consists of a current that increases linearly from 0 to 10 mA in 1 s.
    Ans. (a) 5 mW, (b) 10 mW, (c) 3.33 mW

12.1.4 Find the average power absorbed by each resistor, the capacitor, and the source.   *Ans.* 20 mW, 10 mW, 0, -30 mW

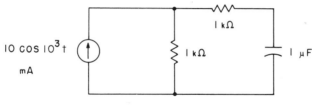

10 cos 10^3 t mA

1 kΩ

1 kΩ

1 μF

EXERCISE 12.1.4

12.1.5 If $f_1(t)$ is periodic of period T_1 and $f_2(t)$ is periodic of period T_2, show that $f_1(t) + f_2(t)$ is periodic of period T if positive integers m and n exist such that

$$T = mT_1 = nT_2$$

Extend this result to the function $(1 + \cos \omega t)^2$, considered in this section, to find its period, $T = 2\pi/\omega$.

12.2 SUPERPOSITION AND POWER

In this section we shall consider the power in networks containing two or more sources, such as the simple circuit of Fig. 12.5. By superposition, we know that

$$i = i_1 + i_2$$

where i_1 and i_2 are the currents in R due to v_{g1} and v_{g2}, respectively. The instantaneous power is

$$
\begin{aligned}
p &= R(i_1 + i_2)^2 \\
&= Ri_1^2 + Ri_2^2 + 2Ri_1i_2 \\
&= p_1 + p_2 + 2Ri_1i_2
\end{aligned}
$$

where p_1 and p_2 are the instantaneous powers, respectively, due to v_{g1} acting alone and to v_{g2} acting alone. In general, $2Ri_1i_2 \neq 0$; thus $p \neq p_1 + p_2$, and superposition does *not* apply for instantaneous power.

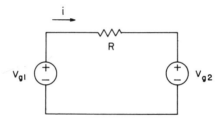

FIGURE 12.5 *Simple circuit with two sources*

In the case of p periodic with period T, the average power is

$$P = \frac{1}{T} \int_0^T p \, dt$$

$$= \frac{1}{T} \int_0^T (p_1 + p_2 + 2Ri_1i_2) \, dt$$

$$= P_1 + P_2 + \frac{2R}{T} \int_0^T i_1i_2 \, dt$$

where P_1 and P_2 are the average powers from v_{g1} and v_{g2}, respectively, acting alone. (We are assuming, of course, that each component of p is periodic of period T.) Superposition for average power applies when

$$P = P_1 + P_2 \tag{12.12}$$

Clearly, this condition holds if

$$\int_0^T i_1i_2 \, dt = 0 \tag{12.13}$$

The most important case in which this equation is satisfied is when $i(t)$ is composed of sinusoidal components of different frequencies. Suppose, for instance, that

$$i_1 = I_{m1} \cos(\omega_1 t + \phi_1)$$

and

$$i_2 = I_{m2} \cos(\omega_2 t + \phi_2)$$

Since we are assuming that $i = i_1 + i_2$ is periodic of period T, we must have

$$I_{m1} \cos[\omega_1(t + T) + \phi_1] + I_{m2} \cos[\omega_2(t + T) + \phi_2]$$
$$= I_{m1} \cos(\omega_1 t + \phi_1) + I_{m2} \cos(\omega_2 t + \phi_2)$$

which requires that

$$\omega_1 T = 2\pi m, \qquad \omega_2 T = 2\pi n$$

where m and n are positive integers. Therefore if ω is a number such that $T = 2\pi/\omega$, then $\omega_1 = m\omega$ and $\omega_2 = n\omega$. In this case the integral in (12.13) becomes, using Table 12.1,

$$\int_0^T i_1i_2 \, dt = I_{m1}I_{m2} \int_0^{2\pi/\omega} \cos(m\omega t + \phi_1) \cos(n\omega t + \phi_2) \, dt$$

$$= \frac{I_{m1}I_{m2}\pi \cos(\phi_1 - \phi_2)}{\omega}, \qquad m = n$$

$$= 0, \qquad\qquad\qquad\qquad m \neq n$$

Thus if $m = n$ ($\omega_1 = \omega_2$), superposition *does not* apply. However, if $m \neq n$, superposition *does* apply. We may generalize this result to the case of a periodic sinusoid with any number of sinusoidal components of different frequencies. *The average power due to the sum of the components is the sum of the average powers due to each component acting alone.*

It can be shown that superposition of average power holds for sinusoids whose frequencies are not integral multiples of some frequency ω, provided we generalize the definition of average power to

$$P = \lim_{\tau \to \infty} \frac{1}{\tau} \int_0^\tau p \, dt$$

This generalization applies to the periodic case just considered as well as to the case $i = i_1 + i_2$, where

$$i_1 = \cos t$$

$$i_2 = \cos \pi t$$

In this case i is not even periodic (the ratio $\omega_1/\omega_2 = 1/\pi$ is not a rational number m/n), but

$$\lim_{\tau \to \infty} \frac{1}{\tau} \int_0^\tau i_1 i_2 \, dt = 0$$

As an example, suppose in Fig. 12.5 that $v_{g1} = 100 \cos (377t + 60°)$ V, $v_{g2} = 50 \cos 377t$ V, and $R = 100 \, \Omega$. Since $\omega_1 = \omega_2$, we cannot apply superposition to power. However, we can use superposition to find the current and subsequently find the power. The phasor currents due to the respective sources are

$$\mathbf{I}_1 = 1 \underline{/60°} \text{ A}$$
$$\mathbf{I}_2 = -0.5 \text{ A}$$

Therefore

$$\mathbf{I} = \mathbf{I}_1 + \mathbf{I}_2 = j0.866 \text{ A}$$

so that $I_m = 0.866$ A. From (12.11), we find

$$P = \tfrac{1}{2}(100)(0.866)^2 = 37.5 \text{ W}$$

Let us now repeat the above example with $v_{g2} = 50$ V. Since v_{g1} and v_{g2} are sinusoids with $\omega_1 = 377$ and $\omega_2 = 0$ rad/s, respectively, superposition for the average power is applicable. Proceeding as before, we find

$$\mathbf{I}_1 = 1 \underline{/60°} \quad \text{for } \omega = 377$$
$$\mathbf{I}_2 = -0.5 \quad \text{for } \omega = 0$$

where I_2 is now a dc current. Therefore

$$P_1 = \frac{RI_{m1}^2}{2} = 50 \text{ W}$$

$$P_2 = RI_{m2}^2 = 25 \text{ W}$$

and the average power is

$$P = P_1 + P_2 = 75 \text{ W}$$

This example illustrates a very important case for electronic amplifiers with sinusoidal inputs. These amplifiers contain dc power supplies that produce dc currents which provide the energy for the amplified ac signals. Thus superposition is very useful in finding the average power associated with each frequency, including $\omega = 0$.

Extending the above procedure to a periodic current which is the sum of $N + 1$ sinusoids of *different* frequencies,

$$i = I_{dc} + I_{m1} \cos (\omega_1 t + \phi_1) + I_{m2} \cos (\omega_2 t + \phi_2)$$
$$+ \ldots + I_{mN} \cos (\omega_N t + \phi_N) \tag{12.14}$$

we find the average power delivered to a resistance R is

$$P = RI_{dc}^2 + \frac{R}{2}(I_{m1}^2 + I_{m2}^2 + \ldots + I_{mN}^2) \tag{12.15}$$

The first term, RI_{dc}^2, in which the factor $\frac{1}{2}$ is missing, is the special case of zero frequency and must be considered separately, as in Sec. 12.1. That is, by (12.6),

$$\frac{1}{T} \int_0^T RI_{dc}I_{mi} \cos (\omega_i t + \phi_i) \, dt = 0, \qquad i = 1, 2, 3, \ldots, N$$

and

$$\frac{1}{T} \int_0^T RI_{dc}^2 \, dt = RI_{dc}^2$$

EXERCISES

12.2.1   Find the average power delivered to the resistor in Fig. 12.5 if $R = 10 \, \Omega$ and
(a) $v_{g1} = 10 \cos 100t$ and $v_{g2} = 20 \cos (100t + 30°)$ V.
(b) $v_{g1} = 100 \cos (t + 60°)$ and $v_{g2} = 50 \cos (2t - 30°)$ V.
(c) $v_{g1} = 50 \cos (t + 45°)$ and $v_{g2} = 100 \sin (t + 30°)$ V.
(d) $v_{g1} = 20 \cos (t + 25°)$ and $v_{g2} = 30 \sin (5t - 50°)$ V.
                    Ans. (a) 7.68 W, (b) 625 W, (c) 754.4 W, (d) 65 W

12.2.2 Find the average power absorbed by each resistor and each source.

Ans. 3.5 W, 3.5 W, −5 W, −2 W

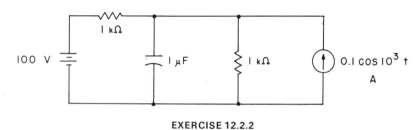

EXERCISE 12.2.2

12.3 RMS VALUES

We have seen in the previous sections that periodic currents and voltages deliver an average power to resistive loads. The amount of power that is delivered depends on the characteristics of the particular waveform. A method of comparing the power delivered by different waveforms is therefore very useful. One such method is the use of *rms* or *effective* values for periodic currents or voltages.

The *rms* value of a periodic current (voltage) is a constant that is equal to the dc current (voltage) which would deliver the same average power to a resistance R. Thus, if I_{rms} is the *rms* value of i, we may write

$$P = RI_{rms}^2 = \frac{1}{T} \int_0^T Ri^2 \, dt$$

from which the rms current is

$$I_{rms} = \sqrt{\frac{1}{T} \int_0^T i^2 \, dt} \tag{12.16}$$

In a similar manner, it is easily shown that the rms voltage is

$$V_{rms} = \sqrt{\frac{1}{T} \int_0^T v^2 \, dt}$$

The term rms is an abbreviation for *root-mean-square*. Inspecting (12.16), we see that we are indeed taking the square *root* of the average, or *mean*, value of the *square* of the current.

From our definition, the rms value of a constant (dc) is simply the constant itself. The dc case is a special case ($\omega = 0$) of the most important type of waveform, the sinusoidal current or voltage.

Suppose we now consider a sinusoidal current $i = I_m \cos(\omega t + \phi)$. Then, from

(12.16) and Table 12.1, we find

$$I_{rms} = \sqrt{\frac{\omega I_m^2}{2\pi} \int_0^{2\pi/\omega} \cos^2(\omega t + \phi)\, dt}$$

$$= \frac{I_m}{\sqrt{2}}$$

Thus a sinusoidal current having an amplitude I_m delivers the same average power to a resistance R as does a dc current which is equal to $I_m/\sqrt{2}$. We also see that the rms current is independent of the frequency ω or the phase ϕ of the current i. Similarly, in the case of a sinusoidal voltage, we find that

$$V_{rms} = \frac{V_m}{\sqrt{2}}$$

Substituting these values into the important power relations of (12.9) and (12.11) for the two-terminal network, we have

$$P = V_{rms} I_{rms} \cos \theta \qquad (12.17)$$

and

$$P = I_{rms}^2 \operatorname{Re} \mathbf{Z} \qquad (12.18)$$

In practice, rms values are usually used in the fields of power generation and distribution. For instance, the nominal 115-V ac power which is commonly used for household appliances is an rms value. Thus the power supplied to our homes is provided by a 60-Hz voltage having a maximum value of $115\sqrt{2} \approx 163$ V. On the other hand, maximum values are more commonly used in electronics and communications.

Finally, let us consider the rms value of the current in (12.14), which is made up of sinusoids of *different* frequencies. In terms of rms currents, (12.15) becomes

$$P = R(I_{dc}^2 + I_{1\,rms}^2 + I_{2\,rms}^2 + \ldots + I_{N\,rms}^2) \qquad (12.19)$$

Since $P = RI_{rms}^2$, we see that the rms value of a sinusoidal current consisting of *different* frequencies is

$$I_{rms} = \sqrt{I_{dc}^2 + I_{1\,rms}^2 + I_{2\,rms}^2 + \ldots + I_{N\,rms}^2}$$

Similarly,

$$V_{rms} = \sqrt{V_{dc}^2 + V_{1\,rms}^2 + V_{2\,rms}^2 + \ldots + V_{N\,rms}^2} \qquad (12.20)$$

These results are particularly important in the study of *noise* in electrical networks, a subject of later courses.

EXERCISES

12.3.1 Find the rms value of a periodic current for which one period is defined by

(a) $i = I,$ $0 \leq t \leq t_1$
         $= 0,$ $t_1 < t < t_2 \; (T = t_2).$
(b) $i = It,$ $0 < t \leq T = t_1.$
(c) $i = I_m \sin \omega t,$ $0 \leq t \leq \pi/\omega$
         $= 0,$ $\pi/\omega \leq t \leq 2\pi/\omega \; (T = 2\pi/\omega).$

Ans. (a) $\sqrt{t_1/t_2}\,I$, (b) $It_1/\sqrt{3}$, (c) $I_m/2$

12.3.2 Find the rms values of (a) $i = 10 \sin \omega t + 20 \cos (\omega t + 30°)$, (b) $i = 10 \sin \omega t + 20 \cos (2\omega t + 10°)$, and (c) $i = 15 \,(1 + \cos 377t)$.

Ans. (a) 12.25 A, (b) 15.81 A, (c) 18.37 A

12.3.3 Find I_{rms}.   *Ans.* 0.5 A

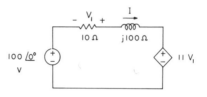

EXERCISE 12.3.3

12.4 POWER FACTOR

The average power delivered to a load in the ac steady state, repeating (12.17), is

$$P = V_{\mathrm{rms}} I_{\mathrm{rms}} \cos \theta$$

The power is thus equal to the product of the rms voltage, the rms current, and the cosine of the angle between the voltage and current phasors. In practice, the rms current and voltage are easily measured and their product, $V_{\mathrm{rms}} I_{\mathrm{rms}}$, is called the *apparent power*. The apparent power is usually referred to in terms of its units, volt-amperes (VA) or kilovoltamperes (kVA), in order to avoid confusing it with the unit of average power, the watt. It is clear that the average power can never be greater than the apparent power.

The ratio of the average power to the apparent power is defined as the *power factor*. Thus if we denote the power factor by *pf*, then in the sinusoidal case

$$pf = \frac{P}{V_{\mathrm{rms}} I_{\mathrm{rms}}} = \cos \theta \qquad\qquad (12.21)$$

which, of course, is dimensionless. The angle θ, in this case, is often referred to as the *pf angle*. We also recognize it as the angle of the impedance **Z** of the load.

In the case of purely resistive loads, the voltage and current are in phase. Therefore, $\theta = 0$, $pf = 1$, and the average and apparent powers are equal. A unity power factor ($pf = 1$) can also exist for loads which contain inductors and capacitors if the reactances of these elements are such that they cancel one another. Adjusting the reactance of loads so as to approximate this condition is very important in electrical power systems, as we shall see shortly.

In a purely reactive load, $\theta = \pm 90°$, $pf = 0$, and the average power is zero. In this case, the equivalent load is an inductance ($\theta = 90°$) or a capacitance ($\theta = -90°$), and the current and voltage differ in phase by $90°$.

A load for which $-90° < \theta < 0$ is equivalent to an *RC* combination, whereas one having $0 < \theta \leq 90°$ is an equivalent *RL* combination. Since $\cos \theta = \cos (-\theta)$, it is evident that the *pf* for an *RC* load having $\theta = -\theta_1$, where $0 < \theta_1 < 90°$, is equal to that of an *RL* load with $\theta = \theta_1$. To avoid this difficulty in identifying such loads, the *pf* is characterized as *leading* or *lagging* by the *phase of the current with respect to that of the voltage*. Therefore an *RC* load has a leading *pf* and an *RL* load has a lagging *pf*. For example, the series connection of a $100\,\Omega$ resistor and a 0.1-H inductor at 60 Hz has $\mathbf{Z} = 100 + j37.7 = 106.9\underline{/20.66°}\,\Omega$ and has a *pf* of $\cos 20.66° = 0.936$ lagging.

In practice, the power factor of a load is very important. In industrial applications, for instance, loads may require thousands of watts to operate, and the power factor greatly affects the electric bill. Suppose, for example, a mill consumes 100 kW from a 220-V rms line. At a *pf* of 0.85 lagging, we see that the rms current into the mill is

$$I_{rms} = \frac{P}{V_{rms}\,pf} = \frac{10^5}{(220)(0.85)} = 534.8 \text{ A}$$

which means that the apparent power supplied is

$$V_{rms}I_{rms} = (220)(534.8) \text{ VA} = 117.66 \text{ kVA}$$

Now suppose that the *pf* by some means is increased to 0.95 lagging. Then

$$I_{rms} = \frac{10^5}{(220)(0.95)} = 478.5 \text{ A}$$

and the apparent power is reduced to

$$V_{rms}I_{rms} = 105.3 \text{ kVA}$$

Comparing the latter case with the former, we see that I_{rms} was reduced by 56.3 A (10.5%). Therefore the generating station must generate a larger current in the case of the lower *pf*. Since the transmission lines supplying the power have resistance, the generator must produce a larger average power to supply the 100 kW to the load.

If the resistance is 0.1 Ω, for instance, then the power generated by the source must be

$$P_g = 10^5 + 0.1I^2_{rms}$$

Therefore we find

$$P_g = 128.6 \text{ kW}, \quad pf = 0.85$$
$$= 122.9 \text{ kW}, \quad pf = 0.95$$

which requires that the power station produce 5.7 kW (4.64%) more power to supply the lower *pf* load. It is for this reason that power companies encourage a *pf* exceeding say 0.9 and impose a penalty on large industrial users who do not comply.

Let us now consider a method of correcting the power factor of a load having an impedance

$$\mathbf{Z} = R + jX$$

We may alter the power factor by connecting an impedance $\mathbf{Z}_1$ in parallel with $\mathbf{Z}$, as shown in Fig. 12.6. For this connection, it is clear that the load voltage does not change. Since $\mathbf{Z}$ is fixed, $\mathbf{I}$ does not change, and the power delivered to the load is not affected. The current $\mathbf{I}_1$ supplied by the generator, however, does change.

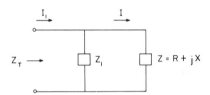

FIGURE 12.6 *Circuit for correcting the power factor*

Let us denote the impedance of the parallel combination by

$$\mathbf{Z}_T = \frac{\mathbf{ZZ}_1}{\mathbf{Z} + \mathbf{Z}_1}$$

In general, we select $\mathbf{Z}_1$ so that (1) $\mathbf{Z}_1$ absorbs zero average power, and (2) $\mathbf{Z}_T$ satisfies the desired power factor $pf = PF$. The first condition requires that $\mathbf{Z}_1$ be purely reactive. That is,

$$\mathbf{Z}_1 = jX_1$$

The second condition requires that

$$\cos\left[\tan^{-1}\left(\frac{\text{Im } \mathbf{Z}_T}{\text{Re } \mathbf{Z}_T}\right)\right] = PF$$

Substituting Z_T in terms of R, X, and X_1 into this equation, we find that (see Prob. 12.14)

$$X_1 = \frac{R^2 + X^2}{k_1 R \tan(\cos^{-1} PF) - X} \qquad (12.22)$$

where $k_1 = 1$ if the power factor is lagging upon correction and $k_1 = -1$ if the power factor is leading upon correction. In each case, $\tan(\cos^{-1} PF)$ is taken as positive.

As an example of the application of (12.22), let us change the power factor for the circuit of Fig. 12.4 to 0.95 lagging. We have already found that

$$\mathbf{Z} = 100 + j100 = 141.4\underline{/45°}$$

Therefore, before a parallel reactance is added across $\mathbf{Z}$, the power factor is

$$pf = \cos\theta = \cos 45° = 0.707 \text{ lagging}$$

Since we desire a power factor of 0.95 lagging, $k_1 = 1$ in (12.22), and

$$X_1 = \frac{100^2 + 100^2}{100 \tan(\cos^{-1} 0.95) - 100} = -297.92 \ \Omega$$

Since $X_1 < 0$, the reactance is a capacitance $C = -1/\omega X_1 = 33.6 \ \mu\text{F}$. The load impedance now becomes

$$\mathbf{Z}_T = \frac{(100 + j100)(-j297.92)}{100 + j100 - j297.92} = 190.0\underline{/18.2°}$$

Therefore the power to the corrected load is

$$P = \frac{100^2}{2(190.0)} \cos(18.2°) = 25 \text{ W}$$

which is the same as that delivered to $\mathbf{Z}$ in Fig. 12.4. The current, however, is

$$I_{rms} = \frac{100}{190\sqrt{2}} = 0.372 \text{ A}$$

as compared to that of Fig. 12.4, given by

$$I_{rms} = \frac{I_m}{\sqrt{2}} = 0.5 \text{ A}$$

We see, therefore, that the current has been reduced by 0.128 A or 25.6%.

EXERCISES

12.4.1 Find the apparent power for (a) a load that requires 30 A rms from a 230-V rms line and (b) a load consisting of a 100-Ω resistor in parallel with a 25-μF capacitor connected to a 110-V rms 60-Hz source.

Ans. (a) 6.9 kVA, (b) 166.3 VA

12.4.2 Find the power factor for (a) a load consisting of a series connection of a 100-Ω resistor and a 20-μF capacitor operating at 60 Hz, (b) one that is capacitive and requires 50 A rms and 5 kW at 110 V rms, and (c) one consisting of a 5-kW load at a power factor of 0.85 leading, connected in parallel with a 10-kW load at a power factor of 0.9 lagging.

Ans. (a) 0.602 leading, (b) 0.909 leading, (c) 0.993 lagging

12.4.3 A load consists of a 200-Ω resistor in series with a 0.1-H inductor. Find the parallel capacitance necessary to adjust to (a) a *pf* of 0.95 lagging and (b) a *pf* of 0.95 leading if $\omega = 1000$ rad/s. *Ans.* (a) 0.685 μF, (b) 3.31 μF

12.5 COMPLEX POWER

We shall now introduce a *complex power* in the ac steady state which is very useful for determining and correcting power factors associated with interconnected loads. Let us begin by defining rms phasors for general sinusoidal voltages and currents. The phasor representations for (12.7) and (12.8) are

$$\mathbf{V} = V_m e^{j\phi}$$
$$\mathbf{I} = I_m e^{j(\phi - \theta)}$$

The rms phasors for these quantities are defined as

$$\mathbf{V}_{\text{rms}} = \frac{\mathbf{V}}{\sqrt{2}} = V_{\text{rms}} e^{j\phi}$$

$$\mathbf{I}_{\text{rms}} = \frac{\mathbf{I}}{\sqrt{2}} = I_{\text{rms}} e^{j(\phi - \theta)}$$

$$(12.23)$$

Let us now consider the average power given in (12.17). Using Euler's formula, we may write

$$P = V_{\text{rms}} I_{\text{rms}} \cos \theta = \text{Re}(V_{\text{rms}} I_{\text{rms}} e^{j\theta})$$

Next, inspecting (12.23), we see that

$$\mathbf{V}_{\text{rms}} \mathbf{I}_{\text{rms}}^* = V_{\text{rms}} I_{\text{rms}} e^{j\theta}$$

where I_{rms}^* is the complex conjugate of I_{rms}. Thus

$$P = \mathrm{Re}\,(V_{rms}I_{rms}^*) \tag{12.24}$$

and the product $V_{rms}I_{rms}^*$ is a complex power whose real part is the average power. Denoting this complex power by S, we have

$$S = V_{rms}I_{rms}^* = P + jQ \tag{12.25}$$

where Q is the *reactive power*. Dimensionally, P and Q have the same units; however, the unit of Q is defined as the *var* (voltampere reactive) to distinguish it from the watt. The magnitude of the complex power is

$$|S| = |V_{rms}I_{rms}^*| = |V_{rms}||I_{rms}^*| = V_{rms}I_{rms}$$

which, of course, is equal to the apparent power.

From (12.25), we see that

$$Q = \mathrm{Im}\,S = V_{rms}I_{rms}\sin\theta \tag{12.26}$$

For an impedance Z, we know that $\sin\theta = (\mathrm{Im}\,Z)/|Z|$, so that

$$Q = V_{rms}I_{rms}\frac{\mathrm{Im}\,Z}{|Z|}$$

Therefore, since $V_{rms}/|Z| = I_{rms}$, we see that

$$Q = I_{rms}^2\,\mathrm{Im}\,Z$$

$$= V_{rms}^2\frac{\mathrm{Im}\,Z}{|Z|^2} \tag{12.27}$$

A phasor diagram of V_{rms} and I_{rms} is shown in Fig. 12.7. We see that the phasor current can be resolved into the two components $I_{rms}\cos\theta$ and $I_{rms}\sin\theta$. The component $I_{rms}\cos\theta$ is in phase with V_{rms}, and it produces the real power P. In contrast, $I_{rms}\sin\theta$ is 90° out of phase with V_{rms}, and it causes the reactive power Q. Since $I_{rms}\sin\theta$ is 90° out of phase with V_{rms}, it is called the *quadrature component* of I_{rms}. As a consequence, the reactive power is sometimes referred to as the *quadrature power*.

It is often convenient to view the complex power in terms of a diagram such as that of Fig. 12.8. It is evident that for an inductive load (lagging *pf*), $0 < \theta \leq 90°$, Q is positive, and S lies in the first quadrant. For a capacitive load (leading *pf*), $-90° \leq \theta < 0$, Q is negative, and S lies in the fourth quadrant. A load having a unity power factor requires that $Q = 0$ since $\theta = 0$. In general, we see that

$$\theta = \tan^{-1}\left(\frac{Q}{P}\right) \tag{12.28}$$

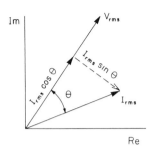

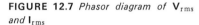

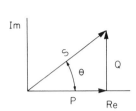

FIGURE 12.7 *Phasor diagram of* $\mathbf{V}_{rms}$
and $\mathbf{I}_{rms}$

FIGURE 12.8 *Diagram for the*
complex power

Let us now consider the complex power associated with a load consisting of two impedances $\mathbf{Z}_1$ and $\mathbf{Z}_2$, as shown in Fig. 12.9. The complex power delivered to the combined impedances is

$$\mathbf{S} = \mathbf{V}_{rms}\mathbf{I}_{rms}^* = \mathbf{V}_{rms}(\mathbf{I}_{1\,rms} + \mathbf{I}_{2\,rms})^*$$
$$= \mathbf{V}_{rms}\mathbf{I}_{1\,rms}^* + \mathbf{V}_{rms}\mathbf{I}_{2\,rms}^*$$

Therefore the complex power delivered by the source to the interconnected loads is the sum of that delivered to each individual load, and the complex power is thus conserved. This statement is true no matter how many individual loads there are or how they are interconnected, because it depends only on Kirchhoff's laws and the definition of complex power. This principle is known as *conservation of complex power.*

Conservation of complex power may be used in a straightforward manner to correct the power factor. As an example, let us consider the circuit of Fig. 12.6 once again. The complex power to the uncorrected load $\mathbf{Z}$ is

$$\mathbf{S} = P + jQ$$

Connecting a pure reactance $\mathbf{Z}_1$ in parallel with $\mathbf{Z}$ results in a complex power to $\mathbf{Z}_1$ of

$$\mathbf{S}_1 = jQ_1$$

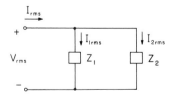

FIGURE 12.9 *Load consisting of* $\mathbf{Z}_1$ *and*
$\mathbf{Z}_2$ *in parallel*

Therefore, from conservation of complex power, for the composite load we have

$$S_T = S + S_1$$
$$= P + j(Q + Q_1)$$

It is evident that the addition of Z_1 does not affect the average power P delivered to the load. It does, however, affect the net reactive power. We can therefore select Q_1 to obtain our desired power factor, which, of course, is usually raised. This causes a reduction in the current required to produce P, as discussed previously.

Let us again consider the circuit of Fig. 12.4 and change the power factor to $PF = 0.95$ lagging. The complex power of the uncorrected load is

$$S = V_{rms}I_{rms}^* = P + jQ = 25 + j25$$

since

$$V_{rms} = 70.7 \text{ V}$$

$$I_{rms} = \frac{V_{rms}}{Z} = 0.3535(1 - j1) \text{ A}$$

From (12.28), we see that $Q_T = Q + Q_1$ must satisfy

$$\theta = \tan^{-1}\left(\frac{Q_T}{P}\right)$$

Therefore

$$\cos \theta = PF = \cos\left(\tan^{-1}\frac{Q_T}{P}\right)$$

and

$$Q_T = P \tan(\cos^{-1} PF)$$
$$= 25 \tan 18.2° = 8.22 \text{ vars}$$

The required Q_1 is

$$Q_1 = Q_T - Q = 8.22 - 25 = -16.78 \text{ vars}$$

Since $Q_1 = V_{rms}^2(\text{Im } Z_1)/|Z_1|^2 = V_{rms}^2/X_1$, then

$$X_1 = \frac{(70.7)^2}{-16.78} = -297.9 \text{ Ω}$$

which represents a capacitance $C = -1/\omega X_1 = 33.6 \ \mu\text{F}$. This is identical to our previous result in Sec. 12.4.

As a final example, let us find the power factor of two loads connected in parallel, as shown in Fig. 12.9. Suppose Z_1 represents a 10-kW load with a power factor $pf_1 = 0.9$ lagging and Z_2 a 5-kW load with $pf_2 = 0.95$ leading. For Z_1 we have

$$S_1 = P_1 + jQ_1$$

where $P_1 = 10^4$ W, $\theta_1 = \cos^{-1} pf_1 = 25.84°$, and, from (12.28),

$$Q_1 = P_1 \tan \theta_1 = 4843 \text{ vars}$$

Similarly, for $\mathbf{Z}_2$ we have

$$\mathbf{S}_2 = P_2 + jQ_2$$

where $P_2 = 5 \times 10^3$ W, $\theta_2 = -\cos^{-1} pf_2 = -18.2°$, and

$$Q_2 = 5 \times 10^3 \tan \theta_2 = -1643 \text{ vars}$$

The total complex power is

$$\mathbf{S}_T = P_1 + P_2 + j(Q_1 + Q_2)$$
$$= 1.5 \times 10^4 + j3200$$

Therefore, for the combined loads,

$$\theta = \tan^{-1} \left(\frac{3200}{1.5 \times 10^4} \right) = 12.04°$$

and

$$pf = 0.978 \text{ lagging}$$

EXERCISES

12.5.1 Find the complex power delivered to a load which has a 0.91 lagging power factor and (a) absorbs 1 kW, (b) 1 kvar, and (c) 1 kVA.

Ans. (a) $1099\underline{/24.5°}$, (b) $2412\underline{/24.5°}$, (c) $1000\underline{/24.5°}$

12.5.2 Find the impedance of the loads in Ex. 12.5.1 if $V_{\text{rms}} = 115$ V.

Ans. (a) $12.03\underline{/24.5°}$, (b) $5.48\underline{/24.5°}$, (c) $13.23\underline{/24.5°}$

12.5.3 Repeat Ex. 12.4.2 (c) using the concept of complex power.

12.6 POWER MEASUREMENT

A *wattmeter* is a device which measures the average power that is delivered to a load. It contains a rotating high-resistance *voltage*, or *potential, coil*, connected in parallel with the load and a fixed low-resistance *current coil* which is connected in series with the load. The device has four terminals, a pair to accommodate each coil. A typical connection is shown in Fig. 12.10. We see that the current coil responds to the load

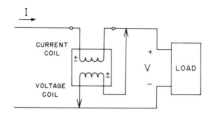

FIGURE 12.10 *Typical connection of a wattmeter*

current, whereas the voltage coil responds to the load voltage. For frequencies above a few hertz, the meter movement responds to the average power. Ideally, the voltage across the current coil and the current in the voltage coil are both zero, so that the presence of the meter does not influence the power it is measuring.

One terminal on each of the coils is marked $\pm$ so that if the current enters the $\pm$ terminal of the current coil and the $\pm$ terminal of the voltage coil is positive with respect to its other terminal, then the meter gives a positive, or upscale, reading. In Fig. 12.10 this corresponds to the load absorbing power. If the terminal connections of either the current coil or voltage coil (but not both) are reversed, a negative, or downscale, reading is indicated. Most meters cannot read downscale—the pointer simply rests on the downscale stop. Thus such a reading requires reversing the connections of one of the coils, usually the voltage coil. Reversing the connections of both coils does not affect the reading.

The wattmeter of Fig. 12.10, represented by the rectangle and the two coils, is connected to read

$$P = |\mathbf{V}| \cdot |\mathbf{I}| \cos \theta$$

where $\mathbf{V}$ and $\mathbf{I}$ are as indicated and θ is the angle between $\mathbf{V}$ and $\mathbf{I}$ (or, equivalently, the angle of the load impedance). This, of course, is the power delivered to the load.

Other types of meters are available for measuring the apparent power and the reactive power. An *apparent power* or VA meter simply measures the product of the rms current and rms voltage. The *varmeter*, on the other hand, measures the reactive power.

EXERCISE

12.6.1 Determine the power reading of each wattmeter after assigning the terminal markings required for a positive reading.   *Ans.* 87.5 W, 9.375 W, 0 W

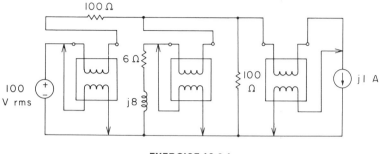

EXERCISE 12.6.1

PROBLEMS

12.1 One cycle of a periodic current is given by

$$i = 0.1(1 - e^{-t}), \qquad 0 \le t < 1$$

If the current flows in a 10-Ω resistor, find the average power.

12.2 Determine the average power for $p(t)$ of Fig. 12.1 (b).

12.3 Find the average power absorbed by the 10-Ω resistor.

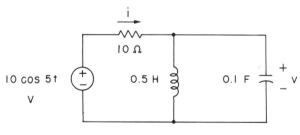

PROBLEM 12.3

12.4 Repeat Prob. 12.3 with the $\frac{1}{10}$-F capacitor replaced by a current-controlled voltage source of value $v = 10i$ V.

12.5 For a Thevenin equivalent circuit consisting of a voltage source $\mathbf{V}_g$ and an impedance $\mathbf{Z}_g = R_g + jX_g$, (a) show that the circuit delivers maximum average power to a load $\mathbf{Z}_L = R_L + jX_L$ when $R_L = R_g$ and $X_L = -X_g$, and (b) show that the maximum average power is delivered to a load R_L when

$R_L = |\mathbf{Z}_g|$. This is the *maximum power transfer theorem* for ac circuits. (In both cases, $\mathbf{V}_g$ and $\mathbf{Z}_g$ are fixed, and the load is variable.)

12.6   Find the average power delivered to R.

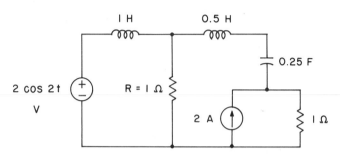

PROBLEM 12.6

12.7   Repeat Prob. 12.6 with the 2-A source replaced by a source having $i = 2\cos t$ A in the same direction.

12.8   We have defined the average power for an instantaneous power $p(t)$, which is not necessarily periodic, by

$$P = \lim_{\tau \to \infty} \frac{1}{\tau} \int_0^\tau p(t)\, dt$$

Show for v and i of (12.7)–(12.8) that this definition yields the same result as (12.6).

12.9   Given: $i = i_1 + i_2$, where

$$i_1 = I_{1m} \cos \omega_1 t$$
$$i_2 = I_{2m} \cos \omega_2 t$$

is the current flowing in a resistor R. Using the definition of Prob. 12.8, show for $\omega_1 \neq \omega_2$ that

$$P = P_1 + P_2$$

where P_1 and P_2 are the average powers associated with i_1 and i_2, respectively, acting alone. Note that this includes the case of $p(t)$ being nonperiodic.

12.10   Find the rms value of the voltage (a) $v = 100 \cos (100t + 35°)$ V and (b) $v = 8 \cos 2t + 4 \cos 3t + 6\sqrt{2} \cos (3t - 45°)$.

12.11   Determine the rms value of a periodic current for which one cycle is given by

(a) $i = \dfrac{I_m}{\pi} t$ A, $0 \leq t \leq \pi$

         $= 0,$ $\pi < t \leq 2\pi.$

(b) $i = I_m e^{-t}$ A, $0 \leq t < 1.$

12.12   Find the power factor of the parallel connection of a 1-kΩ resistor, a 2-H inductor, and a 10-μF capacitor at 60 Hz. At what frequency does a unity power factor occur?

12.13 Repeat Prob. 12.12 for the elements connected in series.

12.14 Derive (12.22).

12.15 Determine the power factor for the circuit. What parallel reactance will change the power factor to 0.95 lagging?

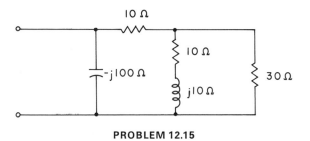

PROBLEM 12.15

12.16 Determine the power factor as seen from the terminals of the independent voltage source.

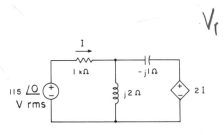

PROBLEM 12.16

12.17 Determine the reactive element needed to be added in parallel to the source to raise the power factor of Prob. 12.13 to 0.95 lagging. Assume a frequency of 100 Hz.

12.18 A load Z_L requires 10 kW at a power factor of 0.7 lagging for a load voltage of 230 V rms at 60 Hz. The transmission line connecting the generator to the load has a series impedance of $0.1 + j0.4 \, \Omega$. Determine the generator voltage and power factor.

12.19 A 230-V rms 60-Hz load draws 500 kW and 400 kvars. Compute the parallel capacitance necessary to give a power factor of (a) 1.0, (b) 0.9 lagging, and (c) 0.9 leading.

12.20 Two loads in parallel draw a total of 3 kW at a 0.9 lagging *pf* from a 115-V rms 60-Hz line. One load is known to absorb 1000 W at a 0.8 lagging power factor. Find (a) the power factor of the second load and (b) the parallel reactive element value necessary to correct the power factor to 0.95 lagging for the combined load.

12.21 A load Z_L requiring 5 kW at a 0.8 lagging power factor is supplied from two 60-Hz generators, as shown. Generator V_1 has a terminal voltage of 230 V rms and delivers 2.5 kW at a 0.8 lagging power factor through a transmission line having an impedance $Z_1 = 1.4 + j1.6\ \Omega$. The transmission line connecting V_2 has an impedance $Z_2 = 0.8 + j1\ \Omega$. Find (a) the load voltage V_L, (b) the generator voltage V_2, and (c) the complex power supplied by V_2.

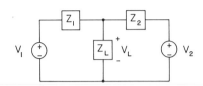

PROBLEM 12.21

13

THREE-PHASE CIRCUITS

As we have already noted, one very important use of ac steady-state analysis is its application to power systems, most of which are alternating current systems. One principal reason for this is that it is economically feasible to transmit power over long distances only if the voltages involved are very high, and it is easier to raise and lower voltages in ac systems than in dc systems. Alternating voltage can be stepped up for transmission and stepped down for distribution with transformers, as we shall see in Chapter 16. Transformers have no moving parts and are relatively simple to construct, whereas with the present technology, rotating machines are generally needed to raise and lower dc voltages.

Also, for reasons of economics and performance, almost all electric power is produced by *polyphase* sources (those generating voltages with more than one phase). In a single-phase circuit, the instantaneous power delivered to a load is pulsating, even if the current and voltage are in phase. A polyphase system, on the other hand, is somewhat like a multicylinder automobile engine in that the power delivered is steadier. Consequently, there is less vibration in the rotating machinery, which, in turn, performs more efficiently. An economic advantage is that the weight of the conductors and associated components required in a polyphase system is appreciably less than that required in a single-phase system that delivers the same power. Virtually all the power produced in the world is polyphase power at 50 or 60 Hz. In the United States 60 Hz is the standard frequency.

In this chapter we shall begin with single-phase, three-wire systems, but we shall concentrate on three-phase circuits, which are by far the most common of the polyphase systems. In the latter case the sources are three-phase generators that produce a *balanced* set of voltages, by which we mean three sinusoidal voltages having the same amplitude and frequency but displaced in phase by 120°. Thus the three-phase source is equivalent to three interconnected single-phase sources, each generating a voltage with a different phase. If the three currents drawn from the sources also constitute a balanced set, then the system is said to be a *balanced* three-phase system. This is the case to which we shall restrict ourselves, for the most part.

13.1 SINGLE-PHASE, THREE-WIRE SYSTEMS

Before proceeding to the three-phase case, let us digress in this section to establish our notation and consider an example of a single-phase system that is in common household use. This example will also serve to give us some practice with a single-phase system, with which we are already familiar and which will serve as an introduction to the three-phase systems to be considered next.

In this chapter we shall find extremely useful the double-subscript notation introduced in Chapter 2 for voltages. In the case of phasors, the notation is $\mathbf{V}_{ab}$ for the voltage of point a with respect to point b. We shall also use a double-subscript notation for current, taking, for example, $\mathbf{I}_{ab}$ as the current flowing in the *direct* path from point a to point b. These quantities are illustrated in Fig. 13.1, where the direct path from a to b is distinguished from the alternative path from a to b through c.

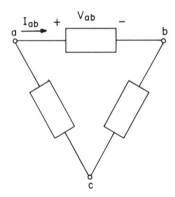

FIGURE 13.1 *Illustration of double-subscript notation*

Because of the simpler expressions for average power that result, we shall use rms values of voltage and current throughout this chapter. (These are also the values read by most meters.) That is, if

$$\mathbf{V} = |\mathbf{V}|\underline{/0°} \text{ V rms}$$
$$\mathbf{I} = |\mathbf{I}|\underline{/-\theta} \text{ A rms}$$

(13.1)

are the phasors associated with an element having impedance,

$$\mathbf{Z} = |\mathbf{Z}|\underline{/\theta} \ \Omega$$

(13.2)

the average power delivered to the element is

$$P = |\mathbf{V}| \cdot |\mathbf{I}| \cos \theta$$
$$= |\mathbf{I}|^2 \, \mathrm{Re} \, \mathbf{Z} \; \mathrm{W} \qquad (13.3)$$

In the time domain the voltage and current are

$$v = \sqrt{2} \, |\mathbf{V}| \cos \omega t \; \mathrm{V}$$
$$i = \sqrt{2} \, |\mathbf{I}| \cos (\omega t - \theta) \; \mathrm{A}$$

The use of double subscripts makes it easier to handle phasors both analytically and geometrically. For example, in Fig. 13.2(a), the voltage $\mathbf{V}_{ab}$ is

$$\mathbf{V}_{ab} = \mathbf{V}_{an} + \mathbf{V}_{nb}$$

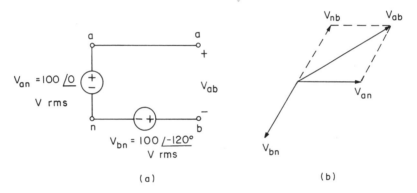

FIGURE 13.2 (a) Phasor circuit; (b) corresponding phasor diagram

This is evident without referring to a circuit since by KVL the voltage between two points a and b is the same regardless of the path, which in this case is the path a,n,b. Also, since $\mathbf{V}_{nb} = -\mathbf{V}_{bn}$, we have

$$\mathbf{V}_{ab} = \mathbf{V}_{an} - \mathbf{V}_{bn}$$
$$= 100 - 100\underline{/-120^\circ}$$

which, after simplification, is

$$\mathbf{V}_{ab} = 100\sqrt{3} \, \underline{/30^\circ} \; \mathrm{V \; rms}$$

These steps are shown graphically in Fig. 13.2(b).

A single-phase, three-wire source, as shown in Fig. 13.3, is one having three output terminals a, b, and a *neutral* terminal n, for which the terminal voltages are equal. That is,

$$\mathbf{V}_{an} = \mathbf{V}_{nb} = \mathbf{V}_1 \qquad (13.4)$$

This is a common arrangement in a normal household supplied with both 115 V and 230 V rms, since if $|\mathbf{V}_{an}| = |\mathbf{V}_1| = 115$ V, then $|\mathbf{V}_{ab}| = |2\mathbf{V}_1| = 230$ V.

Let us now consider the source of Fig. 13.3 loaded with two identical loads, both having an impedance $\mathbf{Z}_1$, as shown in Fig. 13.4. The currents in the lines aA and bB are

$$\mathbf{I}_{aA} = \frac{\mathbf{V}_{an}}{\mathbf{Z}_1} = \frac{\mathbf{V}_1}{\mathbf{Z}_1}$$

and

$$\mathbf{I}_{bB} = \frac{\mathbf{V}_{bn}}{\mathbf{Z}_1} = -\frac{\mathbf{V}_1}{\mathbf{Z}_1} = -\mathbf{I}_{aA}$$

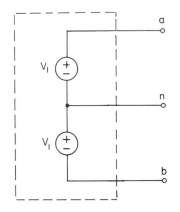

FIGURE 13.3 *Single-phase, three-wire source*

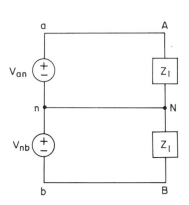

FIGURE 13.4 *Single-phase, three-wire system with two identical loads*

Therefore the current in the neutral wire, nN, by KCL is

$$\mathbf{I}_{nN} = -(\mathbf{I}_{aA} + \mathbf{I}_{bB}) = 0$$

Thus the neutral could be removed without changing any current or voltage in the system.

If the lines aA and bB are not perfect conductors but have equal impedances $\mathbf{Z}_2$, then $\mathbf{I}_{nN}$ is still zero because we may simply add the series impedances $\mathbf{Z}_1$ and $\mathbf{Z}_2$ and have essentially the same situation as in Fig. 13.4. Indeed, in the more general case shown in Fig. 13.5, the neutral current $\mathbf{I}_{nN}$ is still zero. This may be seen by writing the two mesh equations

$$(\mathbf{Z}_1 + \mathbf{Z}_2 + \mathbf{Z}_3)\mathbf{I}_{aA} + \mathbf{Z}_3\mathbf{I}_{bB} - \mathbf{Z}_1\mathbf{I}_3 = \mathbf{V}_1$$
$$\mathbf{Z}_3\mathbf{I}_{aA} + (\mathbf{Z}_1 + \mathbf{Z}_2 + \mathbf{Z}_3)\mathbf{I}_{bB} + \mathbf{Z}_1\mathbf{I}_3 = -\mathbf{V}_1$$

and adding the result, which yields

$$(\mathbf{Z}_1 + \mathbf{Z}_2 + \mathbf{Z}_3)(\mathbf{I}_{aA} + \mathbf{I}_{bB}) + \mathbf{Z}_3(\mathbf{I}_{aA} + \mathbf{I}_{bB}) = 0$$

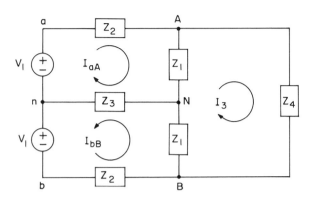

FIGURE 13.5 *Symmetrical single-phase, three-wire system*

or

$$\mathbf{I}_{aA} + \mathbf{I}_{bB} = 0 \qquad (13.5)$$

Since by KCL the left member of the last equation is $-\mathbf{I}_{nN}$, the neutral current is zero. This is, of course, a consequence of the symmetry of Fig. 13.5.

If the symmetry of Fig. 13.5 is destroyed by having unequal loads at terminals *A-N* and *N-B* or unequal line impedances in lines *aA* and *bB*, then there will be a neutral current. For example, let us consider the situation in Fig. 13.6, which has two loads operating at approximately 115 V and one at approximately 230 V. The mesh equations are

$$43\mathbf{I}_1 - 2\mathbf{I}_2 - 40\mathbf{I}_3 = 115$$
$$-2\mathbf{I}_1 + 63\mathbf{I}_2 - 60\mathbf{I}_3 = 115$$
$$-40\mathbf{I}_1 - 60\mathbf{I}_2 + (110 + j10)\mathbf{I}_3 = 0$$

Solving for the currents, we have

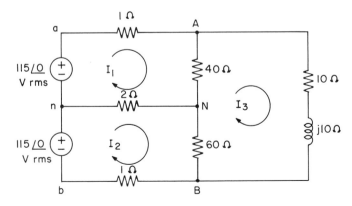

FIGURE 13.6 *Unsymmetrical single-phase, three-wire system*

$$\mathbf{I}_1 = 16.32\underline{/-33.7^\circ} \text{ A rms}$$

$$\mathbf{I}_2 = 15.73\underline{/-35.4^\circ} \text{ A rms}$$

$$\mathbf{I}_3 = 14.46\underline{/-39.9^\circ} \text{ A rms}$$

Therefore the neutral current is

$$\mathbf{I}_{nN} = \mathbf{I}_2 - \mathbf{I}_1 = 0.76\underline{/184.3^\circ} \text{ A rms}$$

and, of course, is not zero.

EXERCISES

13.1.1 Derive (13.5) by superposition applied to Fig. 13.5.

13.1.2 Find the power P_{40}, P_{60}, and P_{10+j10} delivered to the loads, 40 Ω, 60 Ω, and 10 + j10 Ω, respectively, of Fig. 13.6.   *Ans.* 249, 181, 2091 W

13.1.3 Find the power P_{aA}, P_{bB}, and P_{nN} lost in the lines in Fig. 13.6.
Ans. 266.3, 247.4, 1.2 W

13.1.4 Find the power P_{an} and P_{nb} delivered by the two sources in Fig. 13.6. Check the results in Exs. 13.1.2 and 13.1.3 for conservation of power.
Ans. 1561.4, 1474.5 W

13.2 THREE-PHASE Y-Y SYSTEMS

Let us consider the three-phase source of Fig. 13.7(a), which has *line* terminals *a*, *b*, and *c* and a *neutral* terminal *n*. In this case, the source is said to be **Y**-*connected* (connected in a **Y**, as shown). An equivalent representation is that of Fig. 13.7(b), which is somewhat easier to draw.

The voltages $\mathbf{V}_{an}$, $\mathbf{V}_{bn}$, and $\mathbf{V}_{cn}$ between the line terminals and the neutral terminal are called *phase voltages* and in the cases we shall consider are given by

$$\mathbf{V}_{an} = V_p\underline{/0^\circ}$$

$$\mathbf{V}_{bn} = V_p\underline{/-120^\circ} \tag{13.6}$$

$$\mathbf{V}_{cn} = V_p\underline{/120^\circ}$$

or

$$\mathbf{V}_{an} = V_p\underline{/0^\circ}$$

$$\mathbf{V}_{bn} = V_p\underline{/120^\circ} \tag{13.7}$$

$$\mathbf{V}_{cn} = V_p\underline{/-120^\circ}$$

In both cases, each phase voltage has the same rms magnitude V_p, and the phases are

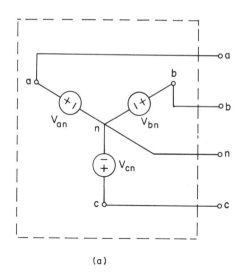 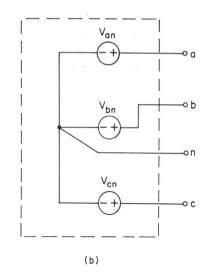

<table>
<tr><td>(a)</td><td>(b)</td></tr>
</table>

FIGURE 13.7 *Two representations of a Y-connected source*

displaced 120°, with $\mathbf{V}_{an}$ arbitrarily selected as the reference phasor. Such a set of voltages is called a *balanced* set and is characterized by

$$\mathbf{V}_{an} + \mathbf{V}_{bn} + \mathbf{V}_{cn} = 0 \tag{13.8}$$

as may be seen from (13.6) or (13.7).

The sequence of voltages in (13.6) is called the *positive sequence*, or *abc sequence*, while that of (13.7) is called the *negative*, or *acb*, sequence. Phasor diagrams of the two sequences are shown in Figs. 13.8(a) and (b), where we may see by inspection that

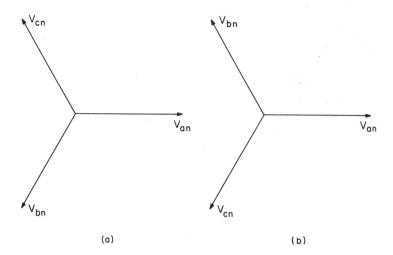

(a)   (b)

FIGURE 13.8 *(a) Positive and (b) negative phase sequence*

(13.8) holds. Evidently, the only difference between the positive and negative sequence is the arbitrary choice of the terminal labels, a, b, and c. Thus without loss in generality we shall consider only the positive sequence.

By (13.6), the voltages in the *abc* sequence may each be related to $\mathbf{V}_{an}$. The relationships, which will be useful later, are

$$\mathbf{V}_{bn} = \mathbf{V}_{an}/\underline{-120°},$$
$$\mathbf{V}_{cn} = \mathbf{V}_{an}/\underline{120°} \tag{13.9}$$

The *line-to-line* voltages, or simply *line* voltages, in Fig. 13.7 are $\mathbf{V}_{ab}$, $\mathbf{V}_{bc}$, and $\mathbf{V}_{ca}$, which may be found from the phase voltages. For example,

$$\mathbf{V}_{ab} = \mathbf{V}_{an} + \mathbf{V}_{nb}$$
$$= V_p/\underline{0°} - V_p/\underline{-120°}$$
$$= \sqrt{3}\,V_p/\underline{30°}$$

In like manner,

$$\mathbf{V}_{bc} = \sqrt{3}\,V_p/\underline{-90°}$$
$$\mathbf{V}_{ca} = \sqrt{3}\,V_p/\underline{-210°}$$

If we denote the magnitude of the line voltages by V_L, then we have

$$V_L = \sqrt{3}\,V_p \tag{13.10}$$

and thus

$$\mathbf{V}_{ab} = V_L/\underline{30°}, \qquad \mathbf{V}_{bc} = V_L/\underline{-90°}, \qquad \mathbf{V}_{ca} = V_L/\underline{-210°} \tag{13.11}$$

These results also may be obtained graphically from the phasor diagram shown in Fig. 13.9.

Let us now consider the system of Fig. 13.10, which is a *balanced* **Y-Y**, three-phase, four-wire system, if the source voltages are given by (13.6). The term **Y-Y** applies since both the source and the load are **Y**-connected. The system is said to be balanced since the source voltages constitute a balanced set and the load is balanced (each *phase impedance* is equal—in this case to $\mathbf{Z}_p$). The fourth wire is the neutral line *n-N*, which may be omitted to form a three-phase, three-wire system.

The line currents of Fig. 13.10 are evidently

$$\mathbf{I}_{aA} = \frac{\mathbf{V}_{an}}{\mathbf{Z}_p}$$

$$\mathbf{I}_{bB} = \frac{\mathbf{V}_{bn}}{\mathbf{Z}_p} = \frac{\mathbf{V}_{an}/\underline{-120°}}{\mathbf{Z}_p} = \mathbf{I}_{aA}/\underline{-120°} \tag{13.12}$$

$$\mathbf{I}_{cC} = \frac{\mathbf{V}_{cn}}{\mathbf{Z}_p} = \frac{\mathbf{V}_{an}/\underline{120°}}{\mathbf{Z}_p} = \mathbf{I}_{aA}/\underline{120°}$$

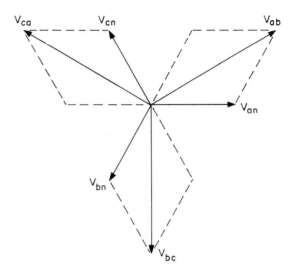

FIGURE 13.9 *Phasor diagram showing phase and line voltages*

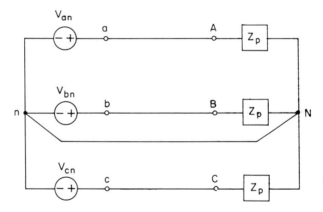

FIGURE 13.10 *Balanced Y-Y system*

The last two results are a consequence of (13.9) and show that the line currents also form a balanced set. Therefore their sum is

$$-\mathbf{I}_{nN} = \mathbf{I}_{aA} + \mathbf{I}_{bB} + \mathbf{I}_{cC} = 0$$

Thus the neutral carries no current in a balanced **Y-Y** four-wire system.

In the case of **Y**-connected loads, the currents in the lines aA, bB, and cC are also the *phase currents* (the currents carried by the phase impedances). If the magnitudes of the phase and line currents are I_p and I_L, respectively, then $I_L = I_p$, and (13.12) becomes

$$\mathbf{I}_{aA} = I_L\underline{/-\theta} = I_p\underline{/-\theta}$$
$$\mathbf{I}_{bB} = I_L\underline{/-\theta - 120°} = I_p\underline{/-\theta - 120°} \qquad (13.13)$$
$$\mathbf{I}_{cC} = I_L\underline{/-\theta + 120°} = I_p\underline{/-\theta + 120°}$$

where θ is the angle of $\mathbf{Z}_p$.

The average power P_p delivered to each phase of Fig. 13.10 is

$$P_p = V_p I_p \cos\theta$$
$$= I_p^2 \operatorname{Re} \mathbf{Z}_p \qquad (13.14)$$

and the total power delivered to the load is

$$P = 3P_p$$

The angle θ of the phase impedance is thus the power factor angle of the three-phase load as well as that of a single phase.

Suppose now that an impedance $\mathbf{Z}_L$ is inserted in each of the lines aA, bB, and cC and that an impedance $\mathbf{Z}_N$, not necessarily equal to $\mathbf{Z}_L$, is inserted in line nN. In other words, the lines are not to be perfect conductors but are to contain impedances. Evidently, the line impedances, except for $\mathbf{Z}_N$, are in series with the phase impedances, and the two sets of impedances may be combined to form perfect conducting lines aA, bB, and cC with a load impedance $\mathbf{Z}_p + \mathbf{Z}_L$ in each phase. Therefore, except for $\mathbf{Z}_N$ in the neutral, the equivalent system has perfect conducting lines and a balanced load. If the impedance $\mathbf{Z}_N$ were not present in the neutral, it would be a perfect conductor, and the system would be balanced as in Fig. 13.10. In this case, as we have seen, points n and N are at the same potential, and there is no neutral current. Thus it does not matter what is in the neutral line. It may be a short circuit or an open-circuit or contain an impedance such as $\mathbf{Z}_N$, and still no neutral current would flow and no voltage would appear across nN. Obviously, then, the presence of equal line imped-ances in aA, bB, and cC and an impedance in the neutral does not change the fact that the line currents form a balanced set.

As an example, let us find the line currents in Fig. 13.11. We may combine the 1-Ω line impedance and $(3 + j3)$-Ω phase impedance to obtain

$$\mathbf{Z}_p = 4 + j3 = 5\underline{/36.9°}\ \Omega$$

as the effective phase load. Since by the foregoing discussion there is no neutral current, we have

$$\mathbf{I}_{aA} = \frac{100\underline{/0°}}{5\underline{/36.9°}} = 20\underline{/-36.9°}\ \text{A}$$

The currents form a balanced, positive sequence set, so that we also have

$$\mathbf{I}_{bB} = 20\underline{/-156.9°}\ \text{A}, \qquad \mathbf{I}_{cC} = 20\underline{/-276.9°}\ \text{A}$$

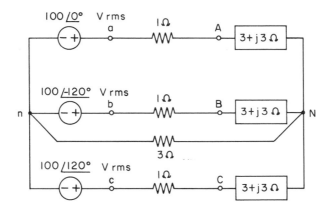

$100\,\underline{/0°}$ V rms

$100\,\underline{/-120°}$ V rms

$100\,\underline{/120°}$ V rms

FIGURE 13.11 *Balanced system with line impedances*

This example was solved on a "per-phase" basis. Since the impedance in the neutral is immaterial in a balanced **Y-Y** system, we may imagine the neutral line to be a short circuit. We may do this if it contains an impedance or even if the neutral wire is not present (a three-wire system). We may then look at only one phase, say "phase A," consisting of the source $\mathbf{V}_{an}$ in series with $\mathbf{Z}_L$ and $\mathbf{Z}_p$, as shown in Fig. 13.12. (The line nN is replaced by a short circuit.) The line current $\mathbf{I}_{aA}$, the phase voltage $\mathbf{I}_{aA}\mathbf{Z}_p$, and the voltage drop in the line $\mathbf{I}_{aA}\mathbf{Z}_L$ may all be found from this single-phase analysis. The other voltages and currents in the system may be found similarly, or from the previous results, since the system is balanced.

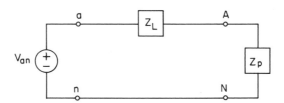

FIGURE 13.12 *Single phase for a per-phase analysis*

As another example, suppose we have a balanced **Y**-connected source, having line voltage $V_L = 200$ V rms, which is supplying a balanced **Y**-connected load with $P = 900$ W at a power factor of 0.9 lagging. Let us find the line current I_L and the phase impedance $\mathbf{Z}_p$. Since the power supplied to the load is 900 W, the power supplied to each phase is $P_p = \frac{900}{3} = 300$ W, and from

$$P_p = V_p I_p \cos \theta$$

we have

$$300 = \left(\frac{200}{\sqrt{3}}\right) I_p (0.9)$$

Therefore since for a Y-connected load the phase current is also the line current, we have

$$I_L = I_p = \frac{3\sqrt{3}}{2(0.9)} = 2.89 \text{ A}$$

The magnitude of $\mathbf{Z}_p$ is given by

$$|\mathbf{Z}_p| = \frac{V_p}{I_p} = \frac{200/\sqrt{3}}{3\sqrt{3}/(2)(0.9)} = 40 \ \Omega$$

and since $\theta = \cos^{-1}0.9 = 25.84°$ is the angle of $\mathbf{Z}_p$, we have

$$\mathbf{Z}_p = 40\underline{/25.84°} \ \Omega$$

If the load is unbalanced but there is a neutral wire which is a perfect conductor, then we may still use the per-phase method of solution for each phase. However, if this is not the case, this shortcut method does not apply. There is a very useful method employing so-called *symmetrical components* which is applicable to unbalanced systems and which the reader may encounter in a course on power systems. It is important to note, in any case, that a three-phase circuit is still a circuit and, balanced or unbalanced, may always be solved by general analysis procedures.

EXERCISES

13.2.1 Given: $\mathbf{V}_{ab} = 100\underline{/0°}$ V rms is a line voltage of a balanced Y-connected three-phase source. If the phase sequence is abc, find the phase voltages.

Ans. $57.7\underline{/-30°}$, $57.7\underline{/-150°}$, $57.7\underline{/-270°}$ V rms

13.2.2 In Fig. 13.10 the source voltages are determined by Ex. 13.2.1, and the load in each phase is a series combination of a 40-Ω resistor, a 100-μF capacitor, and a 0.1-H inductor. The frequency is $\omega = 500$ rad/s. Find the line currents and the power delivered to the load.

Ans. $1.15\underline{/-66.9°}$, $1.15\underline{/-186.9°}$, $1.15\underline{/-306.9°}$ A rms, 160 W

13.2.3 Show that if a balanced three-phase, three-wire system has two balanced three-phase loads connected in parallel, as shown, then the load is equivalent to that of Fig. 13.10 with

$$\mathbf{Z}_p = \frac{\mathbf{Z}_1\mathbf{Z}_2}{\mathbf{Z}_1 + \mathbf{Z}_2}$$

13.2.4 If in Ex. 13.2.3 $\mathbf{Z}_1$ draws 10 kW at a 0.8 power factor lagging, $\mathbf{Z}_2$ draws 15 kW at a 0.9 power factor leading, and the line voltage is $V_L = 300$ V rms, find the current I_L in each line.

Ans. 144.3 A rms

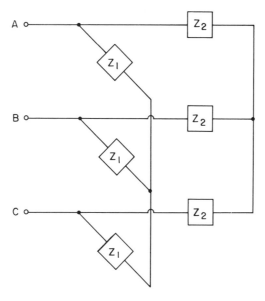

EXERCISE 13.2.3

13.3 THE DELTA CONNECTION

Another method of connecting a three-phase load to a line is the *delta* connection, or Δ connection. A balanced Δ-connected load (each of its phase impedances equal) is shown in Fig. 13.13(a), in a way that resembles a Δ, and in an equivalent way in Fig. 13.13(b). If the source is **Y**- or Δ-connected, then the system is a **Y**-Δ or a Δ-Δ system.

An advantage of a Δ-connected load over a **Y**-connected load is that loads may be added or removed more readily on a single phase of a Δ, since the loads are connected directly across the lines. This may not be possible in the **Y** connection, since the neutral may not be accessible. Also, for a given power delivered to the load the phase currents in a Δ are smaller than those in a **Y**. On the other hand, the Δ phase voltages are higher than those of the **Y** connection. Sources are rarely Δ-connected, because if the voltages are not perfectly balanced, there will be a net voltage, and consequently a circulating current, around the delta. This, of course, causes undesirable heating effects in the generating machinery. Also, the phase voltages are lower in the **Y**-connected generator, and thus less insulation is required. Obviously, systems with Δ-connected loads are three-wire systems, since there is no neutral connection.

From Fig. 13.13 we see that in the case of a Δ-connected load the line voltages are the same as the phase voltages. Therefore if the line voltages are given by (13.11), as before, then the phase voltages are

$$\mathbf{V}_{AB} = V_L \underline{/30°}, \qquad \mathbf{V}_{BC} = V_L \underline{/-90°}, \qquad \mathbf{V}_{CA} = V_L \underline{/-210°} \qquad (13.15)$$

where

$$V_L = V_p \qquad (13.16)$$

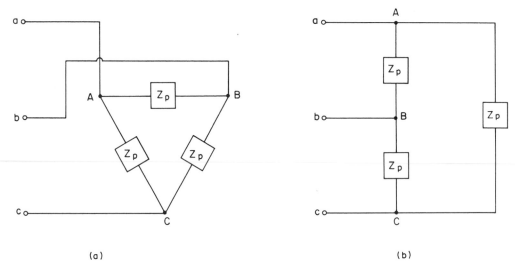

FIGURE 13.13 *Two versions of a Δ-connected load*

If $\mathbf{Z}_p = |\mathbf{Z}_p|\underline{/\theta}$, then the phase currents are

$$\mathbf{I}_{AB} = \frac{\mathbf{V}_{AB}}{\mathbf{Z}_p} = I_p\underline{/30° - \theta}$$

$$\mathbf{I}_{BC} = \frac{\mathbf{V}_{BC}}{\mathbf{Z}_p} = I_p\underline{/-90° - \theta} \qquad (13.17)$$

$$\mathbf{I}_{CA} = \frac{\mathbf{V}_{CA}}{\mathbf{Z}_p} = I_p\underline{/-210° - \theta}$$

where

$$I_p = \frac{V_L}{|\mathbf{Z}_p|} \qquad (13.18)$$

The current in line aA is

$$\mathbf{I}_{aA} = \mathbf{I}_{AB} - \mathbf{I}_{CA}$$

which after some simplification is

$$\mathbf{I}_{aA} = \sqrt{3}\, I_p\underline{/-\theta}$$

The other line currents, obtained similarly, are

$$\mathbf{I}_{bB} = \sqrt{3}\, I_p\underline{/-120° - \theta}$$
$$\mathbf{I}_{cC} = \sqrt{3}\, I_p\underline{/-240° - \theta}$$

Evidently the relation between the line and phase current magnitudes in the Δ case is

$$I_L = \sqrt{3}\, I_p \qquad (13.19)$$

and the line currents are thus

$$\mathbf{I}_{aA} = I_L\underline{/-\theta}, \qquad \mathbf{I}_{bB} = I_L\underline{/-120° - \theta}, \qquad \mathbf{I}_{cC} = I_L\underline{/-240° - \theta} \qquad (13.20)$$

Thus the currents and voltages are balanced sets, as expected. The relations between line and phase currents for the Δ-connected load are summed up in the phasor diagram of Fig. 13.14.

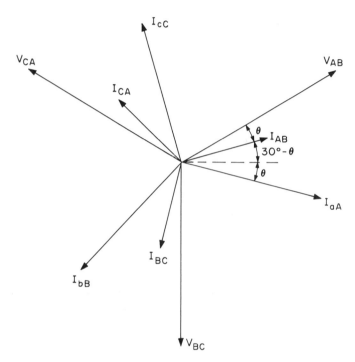

FIGURE 13.14 *Phasor diagram for a Δ-connected load*

As an example of a three-phase circuit with a Δ-connected load, let us find the line current I_L in Fig. 13.13 if the line voltage is 250 V rms and the load draws 1.5 kW at a lagging power factor of 0.8. For one phase, $P_p = \frac{1500}{3} = 500$ W, and thus

$$500 = 250I_p(0.8)$$

or

$$I_p = 2.5 \text{ A rms}$$

Therefore we have

$$I_L = \sqrt{3}\, I_p = 4.33 \text{ A rms}$$

Finally, in this section, let us derive a formula for the power delivered to a balanced three-phase load with a power factor angle θ. Whether the load is **Y**-connected or Δ-connected, we have

$$P = 3P_p = 3V_pI_p \cos\theta$$

In the Y-connected case, $V_p = V_L/\sqrt{3}$ and $I_p = I_L$, and in the Δ-connected case, $V_p = V_L$ and $I_p = I_L/\sqrt{3}$. In either case, then,

$$P = 3\frac{V_L I_L}{\sqrt{3}}\cos\theta$$

or

$$P = \sqrt{3}\, V_L I_L \cos\theta \qquad (13.21)$$

As a check on the previous example, (13.21) yields

$$1500 = \sqrt{3}\,(250)I_L(0.8)$$

or, as before,

$$I_L = 4.33 \text{ A rms}$$

EXERCISES

13.3.1 Solve Ex. 13.2.2 if the source and load are unchanged except that the load is Δ-connected. [*Suggestion*: Note that in (13.15), (13.17), and (13.20) 30° must be subtracted from every angle.]

Ans. $2\sqrt{3}\ \underline{/-66.9°}, 2\sqrt{3}\ \underline{/-186.9°}, 2\sqrt{3}\ \underline{/-306.9°}$, 480 W

13.3.2 A balanced Δ-connected load has $\mathbf{Z}_p = 4 + j3\ \Omega$, and the line voltage is $V_L = 200$ V rms at the load terminals. Find the total power delivered to the load.   *Ans.* 19.2 kW

13.3.3 If the lines in Ex. 13.3.2 each have a resistance of 0.1 Ω, find the power lost in the lines.   *Ans.* 1.44 kW

13.4 Y-Δ TRANSFORMATIONS

In many power systems applications it is important to be able to convert from a Y-connected load to an equivalent Δ-connected load and vice versa. For example, suppose we have a Y-connected load in parallel with a Δ-connected load, as shown in Fig. 13.15, and wish to replace the combination by an equivalent three-phase load. If both loads were Δ-connected, this would be relatively easy since corresponding phase impedances would be in parallel. Also, as we saw in Ex. 13.2.3, if both loads are Y-connected and balanced, the phase impedances may also be combined as parallel impedances.

To obtain Y-to-Δ or Δ-to-Y conversion formulas, let us consider the Y and Δ connections of Fig. 13.16. To effect a Y-Δ transformation we need expressions for

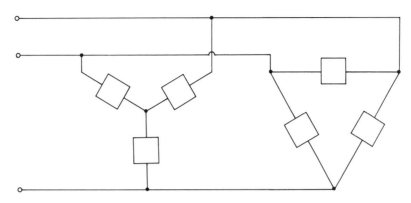

FIGURE 13.15 *Y-connected and Δ-connected load in parallel*

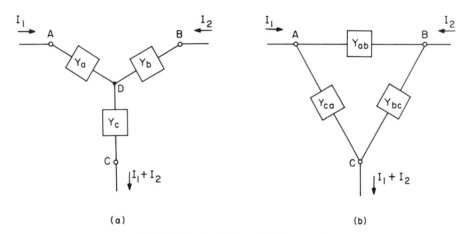

(a)   (b)

FIGURE 13.16 *(a)* **Y** *and (b)* **Δ** *connection*

$\mathbf{Y}_{ab}$, $\mathbf{Y}_{bc}$, and $\mathbf{Y}_{ca}$ of the Δ in terms of $\mathbf{Y}_a$, $\mathbf{Y}_b$, and $\mathbf{Y}_c$ of the **Y** so that the Δ connection is equivalent to the **Y** connection at the terminals *A*, *B*, and *C*. That is, if the **Y** is replaced by the Δ, the same node voltages $\mathbf{V}_A$, $\mathbf{V}_B$, and $\mathbf{V}_C$ will appear, and the same currents $\mathbf{I}_1$ and $\mathbf{I}_2$ will flow. Conversely, a Δ-**Y** transformation is an expression of the **Y** parameters in terms of the Δ parameters.

Let us begin by writing nodal equations for both circuits. If node *C* is taken as reference, in the case of the **Y** network we have

$$\mathbf{Y}_a\mathbf{V}_A - \mathbf{Y}_a\mathbf{V}_D = \mathbf{I}_1$$
$$\mathbf{Y}_b\mathbf{V}_B - \mathbf{Y}_b\mathbf{V}_D = \mathbf{I}_2$$
$$-\mathbf{Y}_a\mathbf{V}_A - \mathbf{Y}_b\mathbf{V}_B + (\mathbf{Y}_a + \mathbf{Y}_b + \mathbf{Y}_c)\mathbf{V}_D = 0$$

Solving for $\mathbf{V}_D$ in the third equation and substituting its value into the first two equations, we have, after simplification,

$$\left(\frac{Y_a Y_b + Y_a Y_c}{Y_a + Y_b + Y_c}\right)V_A - \left(\frac{Y_a Y_b}{Y_a + Y_b + Y_c}\right)V_B = I_1$$

$$-\left(\frac{Y_a Y_b}{Y_a + Y_b + Y_c}\right)V_A + \left(\frac{Y_a Y_b + Y_b Y_c}{Y_a + Y_b + Y_c}\right)V_B = I_2$$

(13.22)

The nodal equations for the Δ circuit are

$$(Y_{ab} + Y_{ca})V_A - Y_{ab}V_B = I_1$$

$$-Y_{ab}V_A + (Y_{ab} + Y_{bc})V_B = I_2$$

Equating coefficients of like terms in these equations and (13.22) and solving for the admittances of the Δ circuit, we have the Y-Δ transformation

$$Y_{ab} = \frac{Y_a Y_b}{Y_a + Y_b + Y_c}$$

$$Y_{bc} = \frac{Y_b Y_c}{Y_a + Y_b + Y_c}$$

(13.23)

$$Y_{ca} = \frac{Y_c Y_a}{Y_a + Y_b + Y_c}$$

If we imagine the Y and Δ circuits superimposed on a single diagram, then Y_a and Y_b are *adjacent* to Y_{ab}, Y_b and Y_c are *adjacent* to Y_{bc}, etc. Thus we may state (13.23) in words, as follows:

> The admittance of an arm of the Δ is equal to the product of the admittances of the adjacent arms of the Y divided by the sum of the Y admittances.

To obtain the Δ-Y transformation we may solve (13.23) for the Y admittances, a difficult task, or we may write two sets of loop equations for the Y and Δ circuits. In the latter case we shall have the dual of the procedure which led to (13.23). In either case, as the reader is asked to show in Ex. 13.4.4, the Δ-Y transformation is

$$Z_a = \frac{Z_{ab} Z_{ca}}{Z_{ab} + Z_{bc} + Z_{ca}}$$

$$Z_b = \frac{Z_{bc} Z_{ab}}{Z_{ab} + Z_{bc} + Z_{ca}}$$

(13.24)

$$Z_c = \frac{Z_{ca} Z_{bc}}{Z_{ab} + Z_{bc} + Z_{ca}}$$

where the Z's are the reciprocals of the Y's of Fig. 13.16. The rule is as follows:

> The impedance of an arm of the Y is equal to the product of the impedances of the adjacent arms of the Δ divided by the sum of the Δ impedances.

(By *adjacent* here we mean "on each side of and terminating on the same node as." For example, in the superimposed drawing of the **Y** and Δ, $\mathbf{Z}_a$ lies between $\mathbf{Z}_{ab}$ and $\mathbf{Z}_{ca}$ and all three have a common terminal A. Thus $\mathbf{Z}_{ab}$ and $\mathbf{Z}_{ca}$ are *adjacent* arms of $\mathbf{Z}_a$.)

As an example, let us find the input impedance **Z** of Fig. 13.17(a). This is a problem which would have required us to write loop or nodal equations in the past, because we cannot simplify the circuit by combining series and/or parallel impedances. Replacing the 1-, $\frac{1}{2}$-, and $\frac{1}{3}$-Ω resistors, which constitute a **Y**, by their equivalent Δ, as shown in Fig. 13.17(b), however, enables us to solve the problem readily.

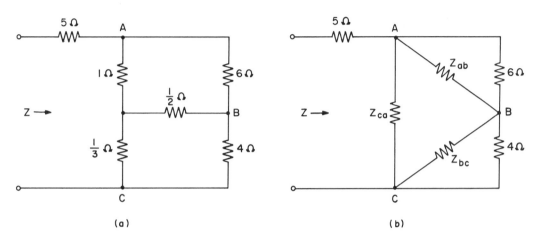

FIGURE 13.17 *Two equivalent circuits*

Comparing Figs. 13.17(a) and 13.16(a), we see that $\mathbf{Y}_a = 1$, $\mathbf{Y}_b = 1 \div \frac{1}{2} = 2$, and $\mathbf{Y}_c = 1 \div \frac{1}{3} = 3$ ℧. Therefore, from (13.23), we have

$$\mathbf{Y}_{ab} = \frac{1(2)}{1 + 2 + 3} = \frac{1}{3}$$

$$\mathbf{Y}_{bc} = \frac{2(3)}{1 + 2 + 3} = 1$$

$$\mathbf{Y}_{ca} = \frac{3(1)}{1 + 2 + 3} = \frac{1}{2}$$

Therefore in Fig. 13.17(b) we have

$$\mathbf{Z}_{ab} = 3\,\Omega, \qquad \mathbf{Z}_{bc} = 1\,\Omega, \qquad \mathbf{Z}_{ca} = 2\,\Omega$$

Thus Fig. 13.17(b) may be simplified by combining parallel and series resistors to obtain

$$\mathbf{Z} = \tfrac{37}{6}\,\Omega$$

As another example, suppose we have a balanced **Y**-connected load with phase impedance $\mathbf{Z}_y$ and wish to convert it to an equivalent Δ-connected load. By (13.23), since $\mathbf{Y}_a$, $\mathbf{Y}_b$, and $\mathbf{Y}_c$ are all equal to $\mathbf{Y}_y$, the reciprocal of $\mathbf{Z}_y$, then the equivalent Δ-connected load is also balanced because

$$\mathbf{Y}_{ab} = \mathbf{Y}_{bc} = \mathbf{Y}_{ca} = \frac{\mathbf{Y}_y^2}{3\mathbf{Y}_y} = \frac{\mathbf{Y}_y}{3}$$

Thus if $\mathbf{Z}_d$ is the phase impedance of the equivalent balanced Δ-connected load, then

$$\mathbf{Z}_d = 3\mathbf{Z}_y \tag{13.25}$$

which may be used to convert from Y to Δ and vice versa.

EXERCISES

13.4.1 Find the input impedance seen by the source in the figure for Prob. 11.8 using a Y-Δ or Δ-Y transformation to simplify the circuit. From this result find the average power delivered by the source.   *Ans.* $(1 + j2)/5\ \Omega$, 8 W

13.4.2 A balanced three-phase source with $V_L = 200$ V rms is delivering power to a balanced Y-connected load with phase impedance $\mathbf{Z}_1 = 6 + j8\ \Omega$ in parallel with a balanced Δ-connected load with phase impedance $\mathbf{Z}_2 = 9 + j12\ \Omega$. Find the power delivered by the source.   *Ans.* 7.2 kW

13.4.3 Show that the Y-Δ transformation of (13.23) is equivalent to

$$\mathbf{Z}_{ab} = \frac{\mathbf{Z}_a\mathbf{Z}_b + \mathbf{Z}_b\mathbf{Z}_c + \mathbf{Z}_c\mathbf{Z}_a}{\mathbf{Z}_c}$$

$$\mathbf{Z}_{bc} = \frac{\mathbf{Z}_a\mathbf{Z}_b + \mathbf{Z}_b\mathbf{Z}_c + \mathbf{Z}_c\mathbf{Z}_a}{\mathbf{Z}_a}$$

$$\mathbf{Z}_{ca} = \frac{\mathbf{Z}_a\mathbf{Z}_b + \mathbf{Z}_b\mathbf{Z}_c + \mathbf{Z}_c\mathbf{Z}_a}{\mathbf{Z}_b}$$

or in words:

The impedance of an arm of the Δ is equal to the sum of the products of the impedances of the Y, taken two at a time, divided by the impedance of the *opposite* arm of the Y.

13.4.4 Derive (13.24).

13.5 POWER MEASUREMENT

It appears to be a simple matter to measure the power delivered to a three-phase load by using one wattmeter for each of the three phases. This is illustrated for the three-wire **Y**-connected load in Fig. 13.18. Each wattmeter has its current coil in series with one phase of the load and its potential coil across one phase of the load. The connections are theoretically correct but may be useless in practice because the neutral point N may not be accessible (as, for example, in the case of a Δ-connected load). It would be better, in general, to be able to make the measurements using only the lines a, b, and c. In this section we shall show that this is, in fact, possible and that, moreover, only two wattmeters are required, instead of three. The method is general and is applicable to unbalanced as well as balanced systems.

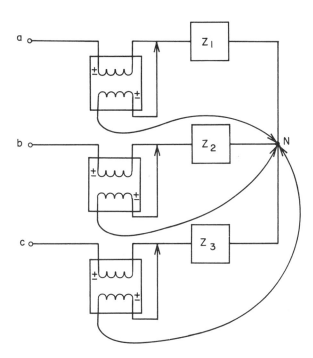

FIGURE 13.18 *Power measurement using three wattmeters*

Let us consider the three-wire **Y**-connected load of Fig. 13.19, which has three wattmeters connected so that each has its current coil in one line and its potential coil between that line and a common point x. If T is the period of the source voltages and i_a, i_b, and i_c are the time-domain line currents, directed into the loads, then the total power P_x indicated by the three meters is

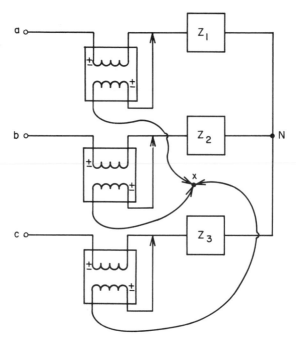

FIGURE 13.19 *Three wattmeters connected to a common point*

$$P_x = \frac{1}{T} \int_0^T (v_{ax}i_a + v_{bx}i_b + v_{cx}i_c)\,dt \qquad (13.26)$$

Regardless of the point x, which is completely arbitrary, we have

$$v_{ax} = v_{aN} + v_{Nx}$$
$$v_{bx} = v_{bN} + v_{Nx}$$
$$v_{cx} = v_{cN} + v_{Nx}$$

Substituting these results into (13.26) and rearranging, we obtain

$$P_x = \frac{1}{T} \int_0^T (v_{aN}i_a + v_{bN}i_b + v_{cN}i_c)\,dt + \frac{1}{T} \int_0^T v_{Nx}(i_a + i_b + i_c)\,dt$$

By KCL we have

$$i_a + i_b + i_c = 0$$

so that

$$P_x = \frac{1}{T} \int_0^T (v_{aN}i_a + v_{bN}i_b + v_{cN}i_c)\,dt \qquad (13.27)$$

Thus the sum of the three wattmeter readings is precisely the total average power delivered to the three-phase load, since the three terms in the integrand of (13.27) are the instantaneous phase powers.

Since the point x in Fig. 13.19 is arbitrary, we may place it on one of the lines. Then the meter whose current coil is in that line will read zero because the voltage across its potential coil is zero. Therefore the total power delivered to the load is measured by the other two meters, and the meter reading zero is unnecessary. For example, the point x is placed on line b in Fig. 13.20, and the total power delivered to the load is

$$P = P_A + P_C$$

where P_A and P_C are the readings of meters A and C. It is important to note that one or the other of the two wattmeters may indicate a negative reading, and thus the sum of the two readings is the *algebraic* sum.

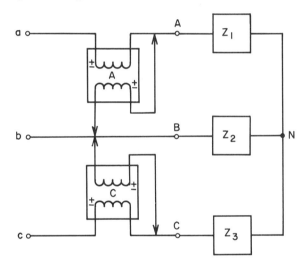

FIGURE 13.20 *Two wattmeters reading the total load power*

The proof of the two-wattmeter method of measuring the total three-phase power has been carried out for a **Y**-connected load. However, it holds also for a Δ-connected load, as the reader is asked to show in Prob. 13.19.

As an example, in Fig. 13.20 let the line voltages be a balanced *abc* sequence with

$$\mathbf{V}_{ab} = 100\sqrt{3}\;\underline{/0^\circ}\text{ V rms}$$

and phase impedances given by

$$\mathbf{Z}_1 = \mathbf{Z}_2 = \mathbf{Z}_3 = 10 + j10\;\Omega$$

Then we have

$$\mathbf{V}_{cb} = -\mathbf{V}_{bc} = -100\sqrt{3}\;\underline{/-120^\circ} = 100\sqrt{3}\;\underline{/60^\circ}\text{ V rms}$$

$$\mathbf{I}_{aA} = \frac{\mathbf{V}_{AN}}{\mathbf{Z}_1} = \frac{(100\sqrt{3}/\sqrt{3})\underline{/-30^\circ}}{10\sqrt{2}\;\underline{/45^\circ}} = 5\sqrt{2}\;\underline{/-75^\circ}\text{ A rms}$$

$$\mathbf{I}_{cC} = 5\sqrt{2}\;\underline{/-315^\circ}\text{ A rms}$$

The meter readings are thus

$$P_A = |\mathbf{V}_{ab}||\mathbf{I}_{aA}| \cos (\text{ang } \mathbf{V}_{ab} - \text{ang } \mathbf{I}_{aA})$$
$$= (100\sqrt{3})(5\sqrt{2}) \cos (0° + 75°)$$
$$= 317 \text{ W}$$

and

$$P_C = |\mathbf{V}_{cb}||\mathbf{I}_{cC}| \cos (\text{ang } \mathbf{V}_{cb} - \text{ang } \mathbf{I}_{cC})$$
$$= (100\sqrt{3})(5\sqrt{2}) \cos (60° + 315°)$$
$$= 1183 \text{ W}$$

with a total of

$$P = 1500 \text{ W}$$

As a check, the power delivered to phase A is

$$P_p = |\mathbf{V}_{AN}||\mathbf{I}_{aA}| \cos (\text{ang } \mathbf{V}_{AN} - \text{ang } \mathbf{I}_{aA})$$
$$= \left(\frac{100\sqrt{3}}{\sqrt{3}}\right)(5\sqrt{2}) \cos (-30° + 75°)$$
$$= 500 \text{ W}$$

Since the system is balanced, the total power is

$$P = 3P_p = 1500 \text{ W}$$

which agrees with the previous result.

EXERCISES

13.5.1 In Fig. 13.18, let $\mathbf{Z}_1 = \mathbf{Z}_2 = \mathbf{Z}_3 = 10\underline{/30°}$ Ω, and let the line voltages be a balanced *abc* sequence set, with $\mathbf{V}_{ab} = 200\underline{/0°}$ V rms. Find the reading of each meter.   *Ans.* $2/\sqrt{3}$ kW

13.5.2 If the power delivered to the load of Ex. 13.5.1 is measured by the two wattmeters A and C connected as shown in Fig. 13.20, find the wattmeter readings. Check for consistency with the answer of Ex. 13.5.1.
                                        Ans. $2/\sqrt{3}$, $4/\sqrt{3}$ kW

13.5.3 Find the wattmeter readings P_A and P_C and the total power P if the line voltages are as given in Ex. 13.5.1 and $\mathbf{Z}_1 = \mathbf{Z}_2 = \mathbf{Z}_3 = 10\underline{/75°}$ Ω. Check the answer by using $P = 3P_p$.   *Ans.* -598, 1633, 1035 W

PROBLEMS

13.1 If $\mathbf{V}_{an} = \mathbf{V}_{nb} = 100\underline{/0^\circ}$ V rms, $\mathbf{Z}_1 = 10\underline{/30^\circ}$ Ω, in A-N, and $\mathbf{Z}_2 = 20\underline{/60^\circ}$ Ω, in N-B, find the neutral current $\mathbf{I}_{nN}$ in Fig. 13.4.

13.2 In Fig. 13.5, let $\mathbf{V}_1 = 100\underline{/0^\circ}$ V rms, $\mathbf{Z}_1 = 5 + j5$ Ω, $\mathbf{Z}_2 = 0.5$ Ω, $\mathbf{Z}_3 = 1$ Ω, and $\mathbf{Z}_4 = 10 - j5$ Ω. Find the average power absorbed by the loads, lost in the lines, and delivered by the sources.

13.3 In Fig. 13.10, the line currents form a balanced, positive sequence set with $\mathbf{I}_{aA} = 10\underline{/0^\circ}$ A rms. If $\mathbf{Z}_p = \sqrt{3}\,\underline{/30^\circ}$ Ω, find the power delivered to the three-phase load and the line voltages.

13.4 A balanced Y-connected load is present on a 240-V rms (line-to-line) three-phase system. If the phase impedance is $4\underline{/60^\circ}$ Ω, find the total average power delivered to the load.

13.5 Repeat Prob. 13.4 if the load is Δ-connected and the phase impedance is $12\underline{/60^\circ}$ Ω.

13.6 A balanced Y-Y system has a positive sequence source with $\mathbf{V}_{an} = 100\underline{/0^\circ}$ V rms and $\mathbf{Z}_p = 10\underline{/30^\circ}$ Ω. Find the line voltage, the line current, and the power delivered to the load.

13.7 A balanced Y-connected source, $\mathbf{V}_{an} = 240\underline{/0^\circ}$ V rms, positive sequence, is connected by four perfect conductors (having zero impedance) to an unbalanced load, $\mathbf{Z}_{AN} = 10 - j5$ Ω, $\mathbf{Z}_{BN} = 10$ Ω, and $\mathbf{Z}_{CN} = j20$ Ω. Find the four line currents.

13.8 A balanced Y-Y three-wire, positive sequence system has $\mathbf{V}_{an} = 200\underline{/0^\circ}$ V rms and $\mathbf{Z}_p = 3 - j4$ Ω, and the lines each have an impedance of 1 Ω. Find the line current and the power delivered to the load.

13.9 A balanced three-phase Y-connected load draws 6 kW at a power factor of 0.8 lagging. A balanced Y of capacitors is to be placed in parallel with the load so that the power factor of the combination is 0.9 lagging. If the frequency is 60 Hz and the line voltages are a balanced 200-V rms set, find the capacitances required.

13.10 In Fig. 13.10, the source is balanced, with positive phase sequence, and $\mathbf{V}_{an} = 100\underline{/0^\circ}$ V rms. If the source provides 3 kVA at $pf = 0.8$ lagging, find $\mathbf{Z}_p$.

13.11 Repeat Prob. 13.10 if the load is a balanced Δ.

13.12 Repeat Prob. 13.11 if each line contains an impedance of 1 Ω.

13.13 A balanced Y-Δ system with $\mathbf{V}_{an} = 120\underline{/0^\circ}$ V rms, positive phase sequence, has $\mathbf{Z}_p = 6 - j9$ Ω and an impedance of 1 Ω in the lines. Find the power delivered to the load.

13.14 A balanced three-phase, positive-sequence source with $\mathbf{V}_{ab} = 200\underline{/0^\circ}$ V rms is supplying a Δ-connected load, $\mathbf{Z}_{AB} = 50$ Ω, $\mathbf{Z}_{BC} = 20 + j20$ Ω, and $\mathbf{Z}_{CA} = 30 - j40$ Ω. Find the phasor line currents. (Assume perfectly conducting lines.)

13.15 Repeat Prob. 13.14 if the impedance of each line is $1 \, \Omega$, $\mathbf{Z}_{AB} = 10 \, \Omega$, $\mathbf{Z}_{BC} = j10 \, \Omega$, and $\mathbf{Z}_{CA} = 10 - j10 \, \Omega$.

13.16 A balanced three-phase positive-sequence source with $\mathbf{V}_{ab} = 240 \underline{/0°}$ V rms is supplying a parallel combination of a Y-connected load and a Δ-connected load. If the Y and Δ loads are balanced with phase impedances of $8 + j8 \, \Omega$ and $12 - j24 \, \Omega$, respectively, find the line current I_L and the power supplied by the source, assuming perfectly conducting lines.

13.17 Find the power delivered by the source in Prob. 13.16 if the Δ load is replaced by an unbalanced Δ load of $\mathbf{Z}_{AB} = 6 - j24 \, \Omega$, $\mathbf{Z}_{BC} = \mathbf{Z}_{CA} = -j24 \, \Omega$.

13.18 For a balanced three-phase system, if the phase voltages are

$$v_a(t) = V_m \cos \omega t$$
$$v_b(t) = V_m \cos (\omega t - 120°)$$
$$v_c(t) = V_m \cos (\omega t - 240°)$$

then the phase currents are

$$i_a(t) = I_m \cos (\omega t - \theta)$$
$$i_b(t) = I_m \cos (\omega t - \theta - 120°)$$
$$i_c(t) = I_m \cos (\omega t - \theta - 240°)$$

Show that the total instantaneous power,

$$p(t) = v_a i_a + v_b i_b + v_c i_c$$

is a constant, given by

$$p(t) = \frac{3}{2} V_m I_m \cos \theta$$

which is also P, the total average power. [*Suggestion*: Recall that $\cos \alpha + \cos (\alpha - 120°) + \cos (\alpha - 240°) = 0$.]

13.19 Show that P_x given by (13.26) is equal to the total average power delivered to a Δ-connected load.

13.20 In the system of Fig. 13.20, the line voltages are a balanced, positive-sequence set, with $\mathbf{V}_{ab} = 100 \underline{/0°}$ V rms, and $\mathbf{Z}_1 = \mathbf{Z}_2 = \mathbf{Z}_3 = 10 \underline{/30°} \, \Omega$. Find the power delivered to the load (a) by finding the readings of the two wattmeters and (b) by $P = 3P_p$.

13.21 Repeat Prob. 13.20 if $\mathbf{Z}_1 = \mathbf{Z}_2 = \mathbf{Z}_3 = 10 \underline{/60°} \, \Omega$. Note that in this case wattmeter C reads the total power since wattmeter A reads zero.

13.22 Show that in Fig. 13.20 if $\mathbf{V}_{ab} = V_L \underline{/\alpha}$, $\mathbf{V}_{bc} = V_L \underline{/\alpha - 120°}$, $\mathbf{V}_{ca} = V_L \underline{/\alpha - 240°}$ V rms, and $\mathbf{Z}_1 = \mathbf{Z}_2 = \mathbf{Z}_3 = |\mathbf{Z}| \underline{/60°}$, then wattmeter A reads zero and wattmeter C reads the total average power delivered to the load,

$$P = \frac{V_L^2}{2|\mathbf{Z}|}$$

This is a generalization of the result of Prob. 13.21.

13.23 Show that if a balanced, positive-sequence source is connected by three perfectly conducting wires to a balanced load having phase impedance $\mathbf{Z}_p = |\mathbf{Z}_p|\underline{/\theta}$, then the two wattmeters, connected as in Fig. 13.20, read, respectively, P_A and P_C, where

$$\frac{P_A}{P_C} = \frac{\cos{(30° + \theta)}}{\cos{(30° - \theta)}}$$

From this result show that

$$\tan\theta = \sqrt{3}\,\frac{P_C - P_A}{P_C + P_A}$$

Thus we may find the power factor, $\cos\theta$, of the load from the two wattmeter readings.

13.24 In Prob. 13.23, find the *pf* of the load if (a) $P_A = P_C$, (b) $P_A = -P_C$, (c) $P_A = 0$, (d) $P_C = 0$, and (e) $P_A = 2P_C$.

14

COMPLEX FREQUENCY
AND NETWORK FUNCTIONS

In the previous chapters we have considered resistive circuit analysis, natural and forced responses of circuits containing storage elements, and, in particular, ac steady-state analysis. The excitations we have considered were, for the most part, constants, exponentials, and sinusoids. In this chapter we shall consider an excitation, the damped sinusoid, which includes all these excitations as special cases. From this function we shall develop generalized phasors and general network functions which include, as special cases, the phasors and impedances of Chapters 10–13.

The network functions will be expressed in terms of a complex frequency that includes the frequency $j\omega$ as a special case. The concepts of complex frequency and general network functions will enable us to combine all of our earlier results into one common procedure. Both the natural and forced responses of a circuit may be found from its excitation and its network function, as we shall see. In addition, the network function may be used to determine the frequency-domain properties of the circuit, which will be the subject of Chapter 15.

14.1 THE DAMPED SINUSOID

In this section we shall consider the *damped sinusoid*,

$$v = V_m e^{\sigma t} \cos(\omega t + \phi) \tag{14.1}$$

which is a sinusoid, like those of the previous chapters, multiplied by a damping factor $e^{\sigma t}$. The constant σ (the Greek letter sigma) is real and is usually negative or zero, which accounts for the term *damping factor*.

The damped sinusoid contains, as special cases, most of the functions we have considered thus far. For example, if $\sigma = 0$, we have the pure sinusoid

$$v = V_m \cos(\omega t + \phi) \tag{14.2}$$

of the previous chapters. If $\omega = 0$, we have the exponential function

$$v = V_0 e^{\sigma t} \tag{14.3}$$

and if $\sigma = \omega = 0$, we have the constant (dc) case

$$v = V_0 \tag{14.4}$$

where, in both (14.3) and (14.4), $V_0 = V_m \cos \phi$. These functions are sketched in Fig. 14.1 for the various cases of σ and ω.

As we see from Fig. 14.1, $\sigma > 0$ represents growing oscillations (b) or a growing exponential (e), and $\sigma < 0$ represents decaying oscillations (a) or a decaying exponential (d). Finally, $\sigma = 0$ represents ac steady state (c) or dc steady state (f).

The units of ω are radians per second and of ϕ are radians or degrees, as before. Since σt is dimensionless, σ is in units of 1 per sec (1/s). This unit was encountered earlier, in connection with natural frequencies, in Chapter 9, where the dimensionless unit, *neper* (Np), was used for σt. Thus σ is the *neper frequency* in nepers per second (Np/s). The neper was chosen in honor of the Scottish mathematician John Napier (1550–1617), who invented logarithms.

As an example of a circuit having a damped sinusoidal excitation, let us find the forced response i in Fig. 14.2. The loop equation is

$$2\frac{di}{dt} + 5i = 25e^{-t} \cos 2t \tag{14.5}$$

Trying as a forced response

$$i = e^{-t}(A \cos 2t + B \sin 2t)$$

which is the excitation and all its possible derivatives, we have

$$2e^{-t}(-2A \sin 2t - A \cos 2t + 2B \cos 2t - B \sin 2t)$$
$$+ 5e^{-t}(A \cos 2t + B \sin 2t) = 25e^{-t} \cos 2t$$

Since this must be an identity, we have

$$3A + 4B = 25$$
$$-4A + 3B = 0$$

or $A = 3$ and $B = 4$. Therefore the forced response is

$$i = e^{-t}(3 \cos 2t + 4 \sin 2t) \text{ A}$$

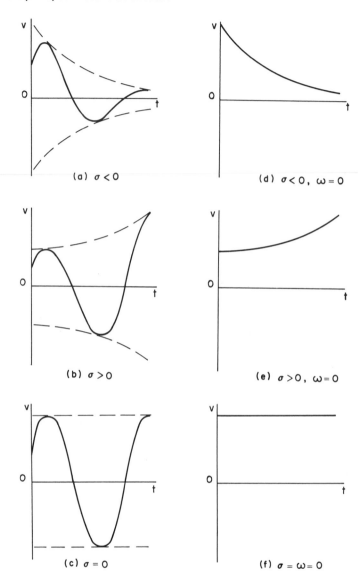

FIGURE 14.1 *Various cases of (14.1)*

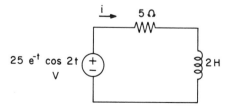

FIGURE 14.2 *Circuit excited by a damped sinusoid*

or, equivalently,

$$i = 5e^{-t} \cos(2t - 53.1°) \text{ A} \tag{14.6}$$

As expected, the forced response is, like the excitation, a damped sinusoid.

EXERCISES

14.1.1 Excite the circuit of Fig. 14.2 by the complex function

$$v_1 = 25e^{-t}e^{j2t}$$
$$= 25e^{(-1+j2)t}$$

and show that the forced response is

$$i_1 = (5\underline{/-53.1°})e^{(-1+j2)t}$$

14.1.2 Show that if i_1 is the response to v_1 in

$$2\frac{di_1}{dt} + 5i_1 = v_1$$

then Re i_1 is the response to Re v_1. Apply this result to the functions of Ex. 14.1.1 to obtain i given by (14.6).

14.1.3 Find the forced response v using the methods of Exs. 14.1.1 and 14.1.2 if

$$\frac{d^3v}{dt^3} + 6\frac{d^2v}{dt^2} + 11\frac{dv}{dt} + 6v = i$$

where

$$i = 4e^{-2t} \cos(t - 60°)$$

Ans. $2e^{-2t} \cos(t + 30°)$

14.2 COMPLEX FREQUENCY AND GENERALIZED PHASORS

In Chapter 10 we considered circuits with sinusoidal excitations such as

$$v_1 = V_m \cos(\omega t + \phi) \tag{14.7}$$

which we also wrote in the equivalent form

$$v_1 = \text{Re}(V_m e^{j\phi}e^{j\omega t})$$

Using the phasor representation of v_1,

$$\mathbf{V} = V_m e^{j\phi} = V_m \underline{/\phi} \tag{14.8}$$

we may also write

$$v_1 = \mathrm{Re}(\mathbf{V}e^{j\omega t}) \tag{14.9}$$

The phasor representation was extremely useful in solving ac steady-state circuits in which the voltages and currents were sinusoids of the form of (14.7).

In this chapter we wish to consider circuits in which the excitations and forced responses are damped sinusoids, such as

$$v = V_m e^{\sigma t} \cos(\omega t + \phi) \tag{14.10}$$

A good question at this point is, Can we define phasor representations of damped sinusoids that will work as well for us as the phasors, such as (14.8), of undamped sinusoids? It seems plausible that we can, because the properties of sinusoids that made the phasors possible are shared by damped sinusoids. That is, the sum or difference of two damped sinusoids is a damped sinusoid, and the derivative, indefinite integral, or constant multiple of a damped sinusoid is a damped sinusoid. In all these operations only V_m and ϕ may change. This is exactly the case with undamped sinusoids.

Let us see what follows if we write (14.10) in the form analogous to (14.9). That is,

$$\begin{aligned}
v &= V_m e^{\sigma t} \cos(\omega t + \phi) \\
&= \mathrm{Re}[V_m e^{\sigma t} e^{j(\omega t + \phi)}] \\
&= \mathrm{Re}[V_m e^{j\phi} e^{(\sigma + j\omega)t}]
\end{aligned}$$

If we define the quantity

$$s = \sigma + j\omega \tag{14.11}$$

then we have

$$v = \mathrm{Re}(\mathbf{V}e^{st}) \tag{14.12}$$

where $\mathbf{V}$ is the phasor of (14.8).

Evidently, since $\mathbf{V}$ is the same for both, the undamped sinusoid given by (14.9) is identical in form to the damped sinusoid of (14.12). The only difference is that the number $j\omega$ is used in one case and the number s is used in the other. Obviously, then, we may do anything with the damped sinusoids that we did with undamped sinusoids as long as we use s instead of $j\omega$. We may define the phasor $\mathbf{V}$ of (14.8) to be the phasor representation of v in (14.10) and use it for damped sinusoidal circuit problems in exactly the same way that we used it for the sinusoidal problems. In the damped sinusoidal case we may wish to write the phasor as $\mathbf{V}(s)$ to distinguish it from the undamped sinusoidal case, $\mathbf{V}(j\omega)$.

As an example, the damped sinusoid

$$v = 25e^{-t} \cos 2t \ V \tag{14.13}$$

of Fig. 14.2 has the phasor representation

$$\mathbf{V} = \mathbf{V}(s) = 25\underline{/0^\circ} \qquad\qquad (14.14)$$

where $s = -1 + j2$. Conversely, if $\mathbf{V}$ is given by (14.14) and s is as specified, then v is given by (14.13).

Some authors prefer to call $\mathbf{V}(s)$, corresponding to $v(t)$ in (14.10), a *generalized phasor*, even though it is identical to the phasors of sinusoidal functions. It is, however, a function of a *generalized frequency*, namely s, given by (14.11). Since s is a complex number, it is more often called a *complex frequency*. Its components are

$$\sigma = \text{Re } s \text{ Np/s}$$

$$\omega = \text{Im } s \text{ rad/s}$$

which have units of frequency, and, indeed, s is the coefficient of t in an exponential function, as was the case in Chapter 9 for the natural frequencies of a circuit. The units of s are 1/second and are sometimes called complex nepers per second or complex radians per second.

It might be worth noting at this point that a function which can be written in the form

$$f(t) = K_1 e^{s_1 t} + K_2 e^{s_2 t} + \cdots + K_n e^{s_n t}$$

where the K_i and s_i are independent of t, may be said to be characterized by the complex frequencies $s_1, s_2, \ldots, s_n$. For example, writing (14.10) in the form

$$v = V_m e^{\sigma t}\left(\frac{e^{j(\omega t + \phi)} + e^{-j(\omega t + \phi)}}{2}\right)$$

results in

$$v = K_1 e^{(\sigma + j\omega)t} + K_2 e^{(\sigma - j\omega)t}$$

where $K_1 = V_m e^{j\phi}/2$ and $K_2 = V_m e^{-j\phi}/2 = K_1^*$ (the complex conjugate). Thus v possesses not one but two complex frequencies, namely $s_1 = \sigma + j\omega$ and $s_2 = s_1^* = \sigma - j\omega$. This concept of complex frequency is consistent with that of the natural frequencies considered in Chapter 9.

EXERCISES

14.2.1 Find the complex frequencies associated with (a) $6 + 4e^{-2t}$, (b) $\cos \omega t$, (c) $\sin(\omega t + \theta)$, (d) $6e^{-3t}\sin(4t + 10^\circ)$, and (e) $e^{-t}(1 + \cos 2t)$.
       Ans. (a) $0, -2$; (b) $\pm j\omega$; (c) $\pm j\omega$; (d) $-3 \pm j4$; (e) $-1, -1 \pm j2$

14.2.2 Show that if i is a damped sinusoid,

$$i = I_m e^{\sigma t} \cos(\omega t + \phi)$$

then v, defined by

$$v = L\frac{di}{dt} + Ri$$

is also a damped sinusoid of the same complex frequency.

Ans. $v = I_m \sqrt{(R + \sigma L)^2 + \omega^2 L^2}\, e^{\sigma t} \cos\left(\omega t + \phi + \tan^{-1}\dfrac{\omega L}{R + \sigma L}\right)$

14.2.3 Find s and $V(s)$ if $v(t)$ is given by (a) 6, (b) $6e^{-t}$, (c) $6e^{-t} \cos(2t + 10°)$, and (d) $6 \cos(2t + 10°)$.

Ans. (a) 0, $6\underline{/0°}$; (b) -1, $6\underline{/0°}$; (c) $-1 + j2$, $6\underline{/10°}$; (d) $j2$, $6\underline{/10°}$

14.2.4 Find $v(t)$ if (a) $\mathbf{V} = 4\underline{/0°}$, $s = -3$; (b) $\mathbf{V} = 5\underline{/15°}$, $s = j2$; and (c) $\mathbf{V} = 6\underline{/-30°}$, $s = -2 + j4$.

Ans. (a) $4e^{-3t}$, (b) $5 \cos(2t + 15°)$, (c) $6e^{-2t} \cos(4t - 30°)$

14.3 IMPEDANCE AND ADMITTANCE

Because of the identical form of the phasor representations for sinusoids and damped sinusoids, we may find forced responses to damped sinusoidal excitations using phasors precisely as we did in the previous chapters. All the concepts and rules, such as impedance, admittance, KCL, KVL, Thevenin's theorem, Norton's theorem, super-position, etc., carry over to the damped sinusoidal case exactly. We need only to use $s = \sigma + j\omega$ rather than $j\omega$.

It follows that, in the s domain, the phasor current $\mathbf{I}(s)$ and voltage $\mathbf{V}(s)$, associated with a two-terminal device, are related by

$$\mathbf{V}(s) = \mathbf{Z}(s)\mathbf{I}(s)$$

where $\mathbf{Z}(s)$ is the *generalized impedance*, or simply impedance, of the device. We may obtain $\mathbf{Z}(s)$ from $\mathbf{Z}(j\omega)$, the impedance in the ac steady-state case, by simply replacing $j\omega$ by s.

For a resistance R, the impedance is, therefore,

$$\mathbf{Z}_R(s) = R$$

In the case of an inductance L, the impedance is

$$\mathbf{Z}_L(s) = sL$$

and for a capacitance C, it is

$$\mathbf{Z}_C(s) = \frac{1}{sC}$$

In a similar manner the admittances are, respectively,

$$Y_R(s) = \frac{1}{R} = G, \qquad Y_L(s) = \frac{1}{sL}, \qquad Y_C(s) = sC$$

Impedances in series or parallel are combined in exactly the same way as in the ac steady-state case, since we merely replace $j\omega$ by s. For example, let us reconsider the circuit of Fig. 14.2, whose phasor circuit is shown in Fig. 14.3, where $s = -1 + j2$. The impedance seen from the source terminals consists of the impedances $2s$ and 5 in series and thus is given by

$$Z(s) = 2s + 5\,\Omega$$

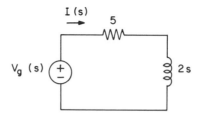

FIGURE 14.3 *Phasor circuit of Fig. 14.2*

Therefore since the input voltage phasor is

$$V_g(s) = 25\underline{/0°}\ V$$

we have

$$I(s) = \frac{V_g(s)}{2s + 5} = \frac{25\underline{/0°}}{2(-1 + j2) + 5} = 5\underline{/-53.1°}\ A$$

Thus in the time domain the forced response, as before, is

$$i(t) = 5e^{-t}\cos(2t - 53.1°)\ A$$

As another example, let us consider the time-domain circuit of Fig. 14.4(a), where it is required to find the forced response $v_0(t)$ for a given damped sinusoidal input $v_g(t)$. The phasor circuit is shown in Fig. 14.4(b), from which we have the nodal equations,

$$\left(\frac{1}{2} + 1 + \frac{1}{4}s\right)V_1 - \frac{1}{2}V_g - V_2 - \frac{1}{4}sV_o = 0$$

$$\left(1 + \frac{1}{4}s\right)V_2 - V_1 = 0$$

We note that $V_2 = V_o/2$ since the op amp and its two connected resistors constitute a VCVS with a gain of 2. Therefore eliminating V_1 and V_2, we have

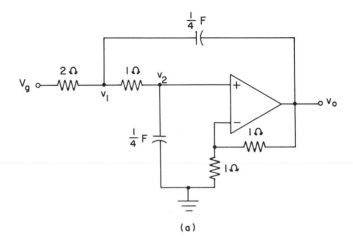

(a)

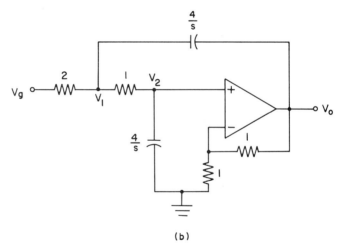

(b)

FIGURE 14.4 *(a) Time-domain circuit; (b) its phasor circuit*

$$\mathbf{V}_o(s) = \frac{16}{s^2 + 2s + 8} \mathbf{V}_g(s) \qquad (14.15)$$

If we have

$$v_g(t) = e^{-2t} \cos 4t \text{ V}$$

then $s = -2 + j4$ and $\mathbf{V}_g(s) = 1 \underline{/0°}$. Thus from (14.15) we have

$$\mathbf{V}_o(s) = \sqrt{2} \underline{/135°} \text{ V}$$

and therefore

$$v_o(t) = \sqrt{2} e^{-2t} \cos (4t + 135°) \text{ V}$$

As a final example in this section let us find the forced response i in Fig. 14.5 if

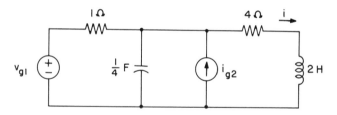

FIGURE 14.5 *Circuit with two sources*

$$v_{g1} = 8e^{-t} \cos t \text{ V}$$

and

$$i_{g2} = 2e^{-5t} \text{ A}$$

Since the complex frequencies of v_{g1} and i_{g2} are different, we must use superposition. That is,

$$i = i_1 + i_2$$

where i_1 is due to v_{g1} acting alone and i_2 is due to i_{g2} acting alone.

The phasor circuits for $i_{g2} = 0$ and $v_{g1} = 0$ are shown in Figs. 14.6(a) and (b),

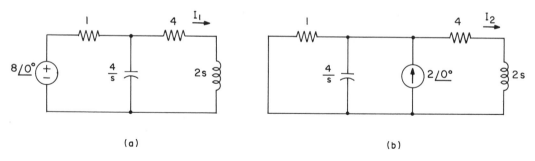

(a)   (b)

FIGURE 14.6 *Phasor circuits associated with Fig. 14.5*

respectively. The phasor currents are $\mathbf{I}_1$ and $\mathbf{I}_2$, as shown. In Fig. 14.6(a), using current division, we have

$$\mathbf{I}_1 = \frac{8\underline{/0°}}{1 + \{[(2s + 4)(4/s)]/(2s + 4 + 4/s)\}} \times \frac{4/s}{2s + 4 + 4/s}$$

which, since $s = -1 + j1$ for this case, becomes

$$\mathbf{I}_1 = 2\sqrt{2}\,\underline{/-45°}\text{ A}$$

Therefore the forced component due to v_{g1} alone is

$$i_1 = 2\sqrt{2}\,e^{-t} \cos(t - 45°)\text{ A}$$

In Fig. 14.6(b), again by current division, we obtain

$$I_2 = \left(\frac{[1(4/s)]/(1 + 4/s)}{\{[1(4/s)]/(1 + 4/s)\} + 2s + 4}\right)(2\underline{/0°})$$

which, since $s = -5$, is

$$I_2 = 0.8\underline{/0°}\ A$$

Therefore the forced component due to i_{g2} alone is

$$i_2 = 0.8e^{-5t}\ A$$

The complete forced response of Fig. 14.5 is, therefore,

$$i = 2\sqrt{2}\,e^{-t}\cos(t - 45°) + 0.8e^{-2t}\ A$$

EXERCISES

14.3.1 Change the voltage source in Fig. 14.5 to

$$v_{g1} = 6e^{-3t}\cos 2t\ V$$

and find the forced response i_1 due to v_{g1} alone.   *Ans.* $-4e^{-3t}\cos 2t$ A

14.3.2 Replace all the circuit of Fig. 14.6 (a), except the inductor, by its Thevenin equivalent, and find I_1.

$$Ans.\ V_{oc} = \frac{32}{s + 4},\ Z_{th} = \frac{4(s + 5)}{s + 4},\ \text{where } s = -1 + j1$$

14.3.3 Assume $I_1 = 1$ A in Fig. 14.6 (a), and use the method of proportionality to find the true I_1.

$$Ans.\ I_1 = \frac{2V_g}{s^2 + 6s + 10} = 2\sqrt{2}\,\underline{/-45°}\ A$$

14.4 NETWORK FUNCTIONS

A generalization of impedance and admittance is the so-called *network function*, which, in the case of a single excitation and response, is defined as the ratio of the response phasor to the excitation phasor. For example, if $V(s)$ and $I(s)$ are the voltage and current phasors associated with a two-terminal network, then the input impedance

$$Z(s) = \frac{V(s)}{I(s)}$$

is the network function if $I(s)$ is the excitation and $V(s)$ is the response. On the other hand, if $\mathbf{V}(s)$ is the input and $\mathbf{I}(s)$ is the output, then the input admittance

$$\mathbf{Y}(s) = \frac{\mathbf{I}(s)}{\mathbf{V}(s)}$$

is the network function.

The foregoing examples are special cases in that both the input and output are measured at the same pair of terminals. In general, for a given input current or voltage at a specified pair of terminals, the output may be a current or voltage at any place in the circuit. Thus the network function, which we designate in general as $\mathbf{H}(s)$, may be a ratio of a voltage to a current (in which case its units are ohms), a current to a voltage (with units of mhos), a voltage to a voltage, or a current to a current. In the last two cases, $\mathbf{H}(s)$ is a dimensionless quantity.

As an example, consider the circuit of Fig. 14.4(b), which was analyzed in the previous section. If $\mathbf{V}_o(s)$ is the output phasor and $\mathbf{V}_g(s)$ is the input phasor, then, by (14.15), the network function is

$$\mathbf{H}(s) = \frac{\mathbf{V}_o(s)}{\mathbf{V}_g(s)} = \frac{16}{s^2 + 2s + 8}$$

As another example, if the source $\mathbf{V}_g = 8\underline{/0°}$ V is the input and $\mathbf{I}_1$ is the output of Fig. 14.6(a), then from Ex. 14.3.3 we see that

$$\mathbf{H}(s) = \frac{\mathbf{I}_1(s)}{\mathbf{V}_g(s)} = \frac{2}{s^2 + 6s + 10} \tag{14.16}$$

In the former example $\mathbf{H}(s)$ is dimensionless and in the latter case its units are mhos.

The network function $\mathbf{H}(s)$ is independent of the input, being a function only of the network elements and their interconnections. Of course, when the input is specified this determines the value of s to be used in a given application. From a knowledge of the network function and the input function, we may then find the output phasor and subsequently the time-domain output. To be specific, suppose $\mathbf{V}_i(s)$ is the input and $\mathbf{V}_o(s)$ is the output. Then

$$\mathbf{H}(s) = \frac{\mathbf{V}_o(s)}{\mathbf{V}_i(s)} \tag{14.17}$$

from which

$$\mathbf{V}_o(s) = \mathbf{H}(s)\mathbf{V}_i(s) \tag{14.18}$$

In general, since s is complex, $\mathbf{H}(s)$ is complex. Therefore we may write, in polar form,

$$\mathbf{H}(s) = |\mathbf{H}(s)|\underline{/\theta} \tag{14.19}$$

where $|\mathbf{H}(s)|$ is the amplitude and θ the phase of $\mathbf{H}(s)$. Therefore if

$$\mathbf{V}_i(s) = V_m\underline{/\phi} \tag{14.20}$$

then

$$\mathbf{V}_o(s) = V_m|\mathbf{H}(s)|\underline{/\phi + \theta} \tag{14.21}$$

Thus the amplitude of $\mathbf{V}_o$ is $V_m|\mathbf{H}(s)|$, and its phase is $\phi + \theta$. That is, the amplitude of the output is that of the input times that of the network function, and the phase of the output is that of the input plus that of the network function.

As an example, $\mathbf{I}_1(s)$ in (14.16) is given by

$$\mathbf{I}_1(s) = \mathbf{H}(s)\mathbf{V}_g(s)$$

where $\mathbf{V}_g(s) = 8\underline{/0°}$ V and $s = -1 + j1$, as given in the previous section. Therefore

$$\mathbf{H}(-1 + j1) = \frac{2}{(-1 + j1)^2 + 6(-1 + j1) + 10}$$

$$= \frac{\sqrt{2}}{4}\underline{/-45°}$$

and

$$\mathbf{I}_1 = \left(\frac{\sqrt{2}}{4}\right)(8)\underline{/0 - 45°}$$

$$= 2\sqrt{2}\,\underline{/-45°}\ \text{A}$$

as was obtained earlier.

As a final note in this section, if there are two or more inputs, we may use superposition to define a network function relating the output to each individual input, with the other inputs made zero. For example, in the circuit of Fig. 14.5, we may find, from Fig. 14.6(a) (with $\mathbf{I}_{g2} = 0$), the network function

$$\mathbf{H}_1(s) = \frac{\mathbf{I}_1}{\mathbf{V}_{g1}}$$

where $s = -1 + j1$ and $\mathbf{V}_{g1} = 8\underline{/0°}$ V. Then from Fig. 14.6(b) (with $\mathbf{V}_{g1} = 0$), we have the network function

$$\mathbf{H}_2(s) = \frac{\mathbf{I}_2}{\mathbf{I}_{g2}}$$

where $s = -5$ and $\mathbf{I}_{g2} = 2\underline{/0°}$ A. We may then find $\mathbf{I}_1$ and $\mathbf{I}_2$ and subsequently the corresponding time-domain functions, i_1 and i_2. These are, by superposition, the components of i, as before.

EXERCISES

14.4.1   Given the network function

$$\mathbf{H}(s) = \frac{2(s + 4)}{s^2 + 2s + 2}$$

and the input

$$\mathbf{V}_i(s) = 4\underline{/0°}$$

find the forced response $v_o(t)$ if (a) $s = -1 + j3$, (b) $s = -3 + j1$, and (c) $s = -1$.   *Ans.* (a) $3\sqrt{2}\ e^{-t} \cos(3t - 135°)$, (b) $-2e^{-3t} \sin t$, (c) $24e^{-t}$

14.4.2   Find $\mathbf{H}(s)$ if the response is (a) $\mathbf{I}_1(s)$, (b) $\mathbf{I}_o(s)$, and (c) $\mathbf{V}_o(s)$.

$$\textit{Ans.}\ \text{(a)}\ \frac{s^2 + 3s + 1}{(s + 1)(s + 2)(s + 3)},\ \text{(b)}\ \frac{1}{(s + 1)(s + 2)(s + 3)},\ \text{(c)}\ \frac{1}{(s + 1)(s + 2)}$$

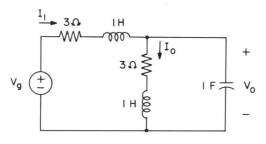

EXERCISE 14.4.2

14.5 POLES AND ZEROS

In general, the network function is a ratio of polynomials in s with real coefficients that are independent of the excitation. To illustrate this, let us consider the example of Fig. 14.2, described by

$$2\frac{di}{dt} + 5i = v$$

where i is the output and v is the input. Using the same technique with complex forcing functions that we used in Chapter 10, we note that if $v = \mathbf{V}e^{st}$, then the output must have the same form, namely $i = \mathbf{I}e^{st}$, where, of course, $\mathbf{V} = \mathbf{V}(s)$ and $\mathbf{I} = \mathbf{I}(s)$ are the phasor representations of v and i. Substituting these values into the differential equation, we have

$$(2s + 5)\mathbf{I}e^{st} = \mathbf{V}e^{st}$$

from which the network function is

$$\frac{\mathbf{I}}{\mathbf{V}} = \frac{1}{2s + 5}$$

In the general case, if the input and output of the circuit are $v_i(t)$ and $v_o(t)$, respectively, then the describing equation is

$$a_n\frac{d^n v_o}{dt^n} + a_{n-1}\frac{d^{n-1}v_o}{dt^{n-1}} + \cdots + a_1\frac{dv_o}{dt} + a_0 v_o$$

$$= b_m\frac{d^m v_i}{dt^m} + b_{m-1}\frac{d^{m-1}v_i}{dt^{m-1}} + \cdots + b_1\frac{dv_i}{dt} + b_0 v_i \qquad (14.22)$$

The a's and b's, of course, are real constants and are independent of v_i.

As before, if $v_i = V_i e^{st}$, then the output must have the form $v_o = V_o e^{st}$, where $V_i(s)$ and $V_o(s)$ are the phasor representations of v_i and v_o. Substituting these values into (14.22), we have

$$(a_n s^n + a_{n-1} s^{n-1} + \cdots + a_1 s + a_0) V_o e^{st}$$
$$= (b_m s^m + b_{m-1} s^{m-1} + \cdots + b_1 s + b_0) V_i e^{st} \qquad (14.23)$$

From this we obtain the network function

$$\frac{V_o(s)}{V_i(s)} = H(s) = \frac{b_m s^m + b_{m-1} s^{m-1} + \cdots + b_1 s + b_0}{a_n s^n + a_{n-1} s^{n-1} + \cdots + a_1 s + a_0} \qquad (14.24)$$

which is a ratio of polynomials in s.

We may also write the network function (14.24) in the factored form

$$H(s) = \frac{b_m (s - z_1)(s - z_2) \ldots (s - z_m)}{a_n (s - p_1)(s - p_2) \ldots (s - p_n)} \qquad (14.25)$$

In this case the numbers $z_1, z_2, \ldots, z_m$ are called *zeros* of the network function because they are values of s for which the function becomes zero. The numbers $p_1, p_2, \ldots, p_n$ are values of s for which the function becomes infinite and are called *poles* of the network function. The values of the poles and zeros, along with the values of the factors a_n and b_m, uniquely determine the network function.

As an example, the network function

$$H(s) = \frac{6(s + 1)(s^2 + 2s + 2)}{s(s + 2)(s^2 + 4s + 13)}$$
$$= \frac{6(s + 1)(s + 1 + j1)(s + 1 - j1)}{s(s + 2)(s + 2 + j3)(s + 2 - j3)} \qquad (14.26)$$

has zeros at -1, $-1 + j1$, and $-1 - j1$ and poles at 0, -2, $-2 + j3$, and $-2 - j3$. As is generally the case, because the a's and b's of (14.24) are real, complex poles or zeros always occur in conjugate pairs. Since the network function is a ratio of a third-degree to a fourth-degree polynomial, it approaches zero as s becomes infinite. Thus we also have a zero at $s = \infty$. If the numerator were higher in degree than the denominator, $s = \infty$ would be a pole.

The poles and zeros of a network function may be sketched on a *pole-zero plot*, which is simply the *s-plane*, consisting of σ- and $j\omega$-axes, with the poles and zeros located on it. Zeros are represented by a small circle and poles by a small cross. As an example, the pole-zero plot of (14.26) is shown in Fig. 14.7. The values of σ are on the horizontal (σ) axis and those of ω are on the vertical (so-called $j\omega$) axis.

As we shall see in Chapter 15, pole-zero plots are useful in considering frequency-domain properties of circuits.

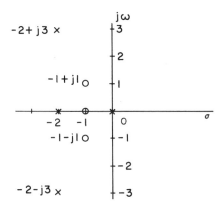

FIGURE 14.7 *Pole-zero plot*

EXERCISES

14.5.1 If the zeros of $H(s)$ are $s = -1$, $-1 \pm j1$, the poles are $s = -2$, $-1 \pm j2$, and $H(0) = 4$, find $H(s)$.

$$Ans. \quad \frac{20(s + 1)(s^2 + 2s + 2)}{(s + 2)(s^2 + 2s + 5)}$$

14.5.2 Draw the pole-zero plot of the network function of Ex. 14.5.1.

14.6 THE NATURAL RESPONSE
FROM THE NETWORK FUNCTION

As we know from our previous experience, an output of a circuit consists of the sum of a natural and a forced response. In the last few chapters we have been concerned entirely with finding the forced response, and we have seen that the phasor technique enables us to do this in a very easy, straightforward way in the cases of sinusoidal, damped sinusoidal, and exponential excitations. In power systems studies the forced response is, of course, an ac steady-state response and is always present. Therefore the forced response is usually of more interest than the natural response, which is transient and gone after a very short time.

   With damped sinusoidal excitations, on the other hand, both the forced and natural responses are transients. (In an actual circuit the natural response must be a transient, for otherwise we would have either a sustained or a growing response without an external excitation.) Therefore the natural response assumes more importance, relative to the forced response, than it does in ac steady-state cases. In finding natural responses in Chapters 8 and 9 we considered only first- and second-order circuits, for the simple reason that our methods were applied to the describing differential equations and became more difficult to use as the order of the circuit increased. However, as we shall see in this section, the natural response may be obtained relatively easily from the phasor representation.

The results of the previous section illustrate that we may obtain the network function quite readily from the describing equation. For example, if (14.22) is the describing equation, then the network function (14.24) is

$$H(s) = \frac{V_o(s)}{V_i(s)} = \frac{N(s)}{D(s)} \tag{14.27}$$

with the numerator

$$N(s) = b_m s^m + b_{m-1} s^{m-1} + \cdots + b_1 s + b_0 \tag{14.28}$$

and the denominator

$$D(s) = a_n s^n + a_{n-1} s^{n-1} + \cdots + a_1 s + a_0 \tag{14.29}$$

The last two expressions are merely the right member and left member, respectively, of (14.22) with the derivatives replaced by powers of s. For example, if the describing equation is

$$\frac{d^2 v_o}{dt^2} + 4\frac{dv_o(t)}{dt} + 3v_o = 2\frac{dv_i}{dt} + v_i \tag{14.30}$$

then the network function is

$$H(s) = \frac{V_o(s)}{V_i(s)} = \frac{2s + 1}{s^2 + 4s + 3} \tag{14.31}$$

The process is reversible if there has been no cancellation of common terms in the network function. For example, it is a simple matter to write down (14.30), given (14.31).

In general, from (14.28) and (14.29) we may reconstruct the describing equation (14.22). In the special case of (14.29) equated to zero

$$D(s) = 0 \tag{14.32}$$

replacing powers of s by corresponding derivatives of v_o results in the homogeneous equation of the system

$$a_n\frac{d^n v_o}{dt^n} + a_{n-1}\frac{d^{n-1} v_o}{dt^{n-1}} + \cdots + a_1\frac{dv_o}{dt} + a_0 v_o = 0$$

as discussed in Chapter 9. Therefore (14.32) is the characteristic equation, and its roots are the natural frequencies of the circuit. Since these roots, by (14.29), are also the poles of the network function, we see that the natural response of the circuit is

$$v_n = A_1 e^{p_1 t} + A_2 e^{p_2 t} + \cdots + A_n e^{p_n t} \tag{14.33}$$

where the natural frequencies $p_1, p_2, \ldots, p_n$ are the poles of the network function and $A_1, A_2, \ldots, A_n$ are arbitrary constants. Modifications, of course, must be made, as described in Chapter 9, if the natural frequencies are not distinct.

We now have a very simple method, based on phasors, for finding the complete response of a circuit. All we need to find is the network function from which by (14.27) we may obtain the output phasor. The forced response is found from the phasor response in the usual way, and the natural response is given by (14.33), where the natural frequencies are the poles of the network function.

If the input phasor is of the form

$$\mathbf{V}_i(s) = V_m \underline{/\phi}$$

where V_m and ϕ are constants, then, by (14.27),

$$\mathbf{V}_o(s) = (V_m \underline{/\phi}) \mathbf{H}(s)$$

Therefore since $\mathbf{V}_i(s)$ has no poles, the poles of $\mathbf{H}(s)$ are the poles of $\mathbf{V}_o(s)$. Thus the complete response $v_o(t)$ may be obtained from its phasor representation $\mathbf{V}_o(s)$. The forced response is obtained as before, and the natural frequencies are the poles of $\mathbf{V}_o(s)$, from which the natural response may be constructed.

As an example, let us find the complete response i of the circuit of Fig. 14.8. Using current division, the phasor $\mathbf{I}(s)$ of i is given by

$$\mathbf{I}(s) = \frac{\mathbf{V}_g(s)}{12 + \{[3s(2s + 6)]/(3s + 2s + 6)\}} \times \frac{3s}{3s + 2s + 6}$$

$$= \frac{s}{(s + 1)(s + 12)}$$

since $\mathbf{V}_g = 2\underline{/0^\circ}$. Since the poles of $\mathbf{I}$ are $s = -1, -12$, the natural response is

$$i_n(t) = A_1 e^{-t} + A_2 e^{-12t} \text{ A}$$

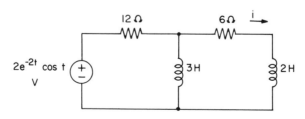

FIGURE 14.8 *Second-order circuit*

For $s = -2 + j1$, the phasor representation is

$$\mathbf{I} = 0.16\underline{/12.7°}$$

so that the forced response is

$$i_f(t) = 0.16e^{-2t} \cos(t + 12.7°) \text{ A}$$

The complete response is therefore

$$i(t) = A_1 e^{-t} + A_2 e^{-12t} + 0.16e^{-2t} \cos(t + 12.7°) \text{ A}$$

The arbitrary constants may be evaluated, as in Chapter 9, if we know the initial energy conditions.

EXERCISES

14.6.1 Find the natural response in Ex. 14.4.1, assuming that there is no cancellation in the network function.   *Ans.* $e^{-t}(A_1 \cos t + A_2 \sin t)$ V

14.6.2 Find the complete response in Ex. 14.4.1 (a) if the natural response is as given in Ex. 14.6.1 and

$$v_o(0^+) = \frac{dv_o(0^+)}{dt} = 0$$

Ans. $3e^{-t}[\cos t - 3 \sin t + \sqrt{2} \cos(3t - 135°)]$ V

14.6.3 Find the complete response $v_o(t)$ in Ex. 14.4.2 if $v_g(t) = e^{-t} \cos t$ V and

$$v_o(0^+) = \frac{dv_o(0^+)}{dt} = \frac{d^2 v_o(0^+)}{dt^2} = 0$$

(Note that the natural frequency $s = -3$ has been cancelled in the network function. Thus the corresponding term must be added to the natural response.)
Ans. $-\frac{1}{2}e^{-t} + \frac{3}{2}e^{-2t} - \frac{1}{2}e^{-3t} + (1/\sqrt{2})e^{-t} \cos(t - 135°)$ V

14.7 NATURAL FREQUENCIES*

As we have seen, the natural frequencies are the poles of the network function if there is no cancellation of a common pole and zero. Also, as discussed in Chapter 9, the natural frequencies are the same for any response in a given circuit unless one portion of the circuit is physically separated from the rest. Thus if we are looking only for the natural frequencies, we may consider any response, and it is obviously better to choose one which is easier to find.

For example, let us consider the circuit of Fig. 14.8 with the source killed (since the natural response corresponds to a zero source). The natural frequencies are the poles of any network function. Therefore we may excite the circuit in some proper manner and, for some chosen output, determine the network function. Figure 14.9 illustrates the two proper means of applying an excitation to the dead circuit. We may insert a voltage source in series with an element as in x-x' of Fig. 14.9(a), or we may place a current source across an element, as across y-y' of Fig. 14.9(b). The first entry into the circuit is sometimes called the pliers entry since we cut a wire and insert a voltage source. The second entry is a soldering entry since we solder the source across two nodes.

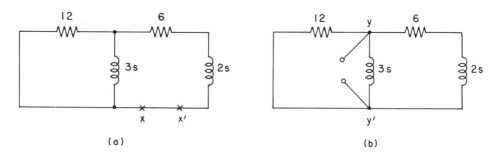

Figure 14.9 *Circuit with two possible entries*

In the pliers entry, if $V_x(s)$ is the voltage source inserted and $I_x(s)$, the current into the circuit at the source, is the response, then the network function is

$$Y_x(s) = \frac{I_x(s)}{V_x(s)}$$

Thus the natural frequencies are the poles of $Y_x(s)$ or the zeros of its reciprocal, $Z_x(s)$, the input impedance at x-x'. In this case we have

$$Z_x(s) = 2s + 6 + \frac{12(3s)}{12 + 3s}$$

$$= \frac{2(s + 1)(s + 12)}{s + 4}$$

Therefore the natural frequencies are $-1, -12$, as before.

In the case of Fig. 14.9(b), if a current source $I_y(s)$ is placed across y-y' and the voltage $V_y(s)$ across the source is the output, then the network function is

$$Z_y(s) = \frac{V_y(s)}{I_y(s)}$$

Therefore the natural frequencies are the poles of $Z_y(s)$ or the zeros of its reciprocal, $Y_y(s)$, given by

$$Y_y(s) = \frac{1}{12} + \frac{1}{3s} + \frac{1}{2s + 6}$$

$$= \frac{(s + 1)(s + 12)}{12s(s + 3)}$$

Again, the natural frequencies are -1, -12.

EXERCISES

14.7.1 Find the natural frequencies of the circuit of Fig. 14.5 by killing the sources and using (a) a pliers entry in series with the capacitor and (b) a soldering entry across the capacitor. *Ans.* $-3 \pm j1$

14.7.2 Find the natural frequencies of the circuit of Ex. 14.4.2 by killing the source and using (a) a pliers entry in series with the capacitor and (b) a soldering entry across the capacitor. (Note: Do not cancel the common pole and zero, $s = -3$.) *Ans.* $-1, -2, -3$

14.8 TWO-PORT NETWORKS

One of the most important applications of the network function concept is to networks for which the input and output signals are measured at different pairs of terminals. The simplest and most-often-encountered circuit for which this is possible is the *two-port* network, a port being defined as a pair of terminals at which a signal may enter or leave. A general two-port network is symbolized by Fig. 14.10(b), as contrasted with the *one-port* network of Fig. 14.10(a).

In general, as seen in Fig. 14.10(b), a two-port network has four terminals. It is possible, of course, for two of the terminals to be the same, in which case we have a three-terminal network. A general example of this case is shown in Fig. 14.11.

We may associate two pairs of currents and voltages with a general two-port network, as shown in the frequency-domain case of Fig. 14.12, with variables $V_1(s)$, $I_1(s)$, $V_2(s)$, and $I_2(s)$, as indicated. In case the network is linear, these variables may be related in a number of ways. For example, if I_1 and I_2 are inputs and V_1 and V_2 are outputs, we know by superposition that V_1 and V_2 each have components pro-

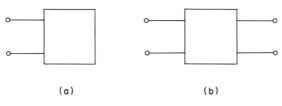

(a)   (b)

FIGURE 14.10 *(a) One-port and (b) two-port network*

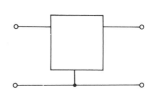

FIGURE 14.11 *Three-terminal two-port network*

portional to $\mathbf{I}_1$ and $\mathbf{I}_2$. That is,

$$\mathbf{V}_1 = \mathbf{z}_{11}\mathbf{I}_1 + \mathbf{z}_{12}\mathbf{I}_2$$
$$\mathbf{V}_2 = \mathbf{z}_{21}\mathbf{I}_1 + \mathbf{z}_{22}\mathbf{I}_2 \tag{14.34}$$

where the $\mathbf{z}$'s are proportionality factors, which in general are functions of s.

Since the $\mathbf{z}$'s multiply currents to yield voltages, they must be measured in ohms. Therefore they are impedance functions. We may find the $\mathbf{z}$'s from the network by open-circuiting either port 1 (with variables $\mathbf{V}_1$ and $\mathbf{I}_1$) or port 2 (with $\mathbf{V}_2$ and $\mathbf{I}_2$). For example, if port 2 is open-circuited ($\mathbf{I}_2 = 0$), then from (14.34) we have

$$\mathbf{z}_{11} = \left.\frac{\mathbf{V}_1}{\mathbf{I}_1}\right|_{\mathbf{I}_2=0}$$
$$\mathbf{z}_{21} = \left.\frac{\mathbf{V}_2}{\mathbf{I}_1}\right|_{\mathbf{I}_2=0} \tag{14.35}$$

Similarly, if port 1 is open-circuited ($\mathbf{I}_1 = 0$), we have

$$\mathbf{z}_{12} = \left.\frac{\mathbf{V}_1}{\mathbf{I}_2}\right|_{\mathbf{I}_1=0}$$
$$\mathbf{z}_{22} = \left.\frac{\mathbf{V}_2}{\mathbf{I}_2}\right|_{\mathbf{I}_2=0} \tag{14.36}$$

Accordingly, the $\mathbf{z}$'s are called *open-circuit impedances*, or *open-circuit parameters*, or simply *z-parameters*. In any case, they are examples of network functions.

As an example, let us consider the three-terminal network of Fig. 14.13. Because

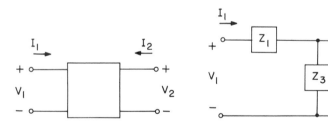

FIGURE 14.12 *General two-port network*

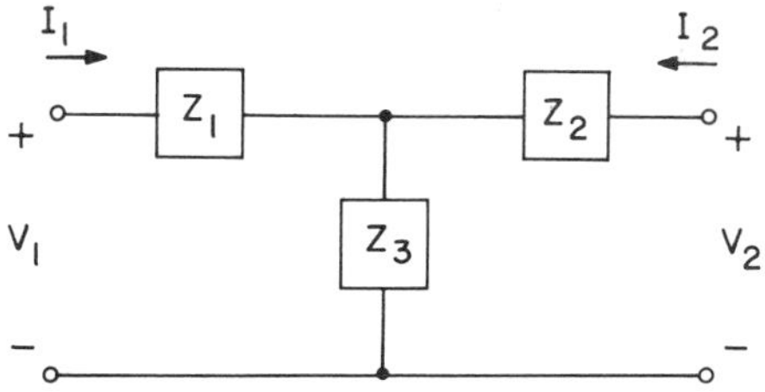

FIGURE 14.13 *T network*

of its shape, it is sometimes called a T network. Evidently, it is simply a $\mathbf{Y}$ network, as discussed in Chapter 13. The reader may easily show, using (14.35) and (14.36), that

$$\mathbf{z}_{11} = \mathbf{Z}_1 + \mathbf{Z}_3$$
$$\mathbf{z}_{12} = \mathbf{z}_{21} = \mathbf{Z}_3 \tag{14.37}$$
$$\mathbf{z}_{22} = \mathbf{Z}_2 + \mathbf{Z}_3$$

If in Fig. 14.12 $\mathbf{V}_1$ and $\mathbf{V}_2$ are inputs and $\mathbf{I}_1$ and $\mathbf{I}_2$ are outputs, we may obtain, in an analogous manner,

$$I_1 = y_{11}V_1 + y_{12}V_2$$
$$I_2 = y_{21}V_1 + y_{22}V_2$$
(14.38)

where, evidently,

$$y_{11} = \frac{I_1}{V_1}\Big|_{V_2=0}$$

$$y_{12} = \frac{I_1}{V_2}\Big|_{V_1=0}$$

$$y_{21} = \frac{I_2}{V_1}\Big|_{V_2=0}$$

$$y_{22} = \frac{I_2}{V_2}\Big|_{V_1=0}$$

Accordingly, the **y**'s are *short-circuit admittances* ($\mathbf{V}_1 = 0$ or $\mathbf{V}_2 = 0$), or *short-circuit parameters*, or simply *y-parameters*.

As an example, let us find the *y*-parameters of the three-terminal circuit of Fig. 14.14. Because of its shape, it is sometimes called a π network. Evidently, it is simply a Δ network, as considered in Chapter 13. Using the expressions for the *y*-parameters, we may write, by inspection,

$$y_{11} = \mathbf{Y}_a + \mathbf{Y}_b$$
$$y_{12} = y_{21} = -\mathbf{Y}_b$$
$$y_{22} = \mathbf{Y}_b + \mathbf{Y}_c$$
(14.39)

The *z*- and *y*-parameters are but two sets of parameters associated with a two-port network. Others include the *hybrid parameters* and the *transmission parameters*, which will be defined and considered in the exercises and problems.

The two-port parameters may be used to obtain various network functions. For example, if port 2 is open ($\mathbf{I}_2 = 0$), then from (14.34) we may find the voltage ratio function

$$\frac{\mathbf{V}_2}{\mathbf{V}_1} = \frac{\mathbf{z}_{21}}{\mathbf{z}_{11}}$$
(14.40)

As another example, let us find the voltage ratio function for the two-port loaded with 1 Ω, as shown in Fig. 14.15. Evidently, we have $\mathbf{V}_2 = -\mathbf{I}_2$, which substituted into

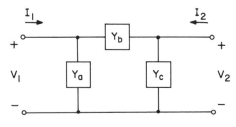

FIGURE 14.14 π *network*

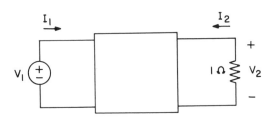

FIGURE 14.15 *Two-port loaded with a 1-Ω resistor*

the second equation of (14.38) yields, after some rearrangement,

$$\frac{\mathbf{V}_2}{\mathbf{V}_1} = \frac{-\mathbf{y}_{21}}{1 + \mathbf{y}_{22}} \tag{14.41}$$

Another use for the two-port parameters is in the construction of equivalent circuits. For example, in the first equation of (14.34), $\mathbf{V}_1$ is a sum of two terms, $\mathbf{z}_{11}\mathbf{I}_1$ and $\mathbf{z}_{12}\mathbf{I}_2$. The first may be obtained by an impedance $\mathbf{z}_{11}$ carrying a current $\mathbf{I}_1$, and the second may be obtained by a dependent voltage controlled by $\mathbf{I}_2$. In a similar way we may interpret the second equation of (14.34) and draw the equivalent circuit. The result, as may be verified by inspection, is shown in Fig. 14.16. We say that this circuit is equivalent (that is, at the terminals) to that of Fig. 14.12, because they both have the same two-port parameters. Thus if the port currents are the same, then the port voltages are the same, and vice versa, for both networks.

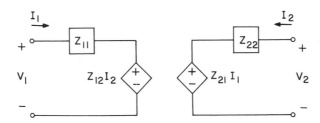

FIGURE 14.16 *Equivalent circuit of Fig. 14.12*

If $\mathbf{z}_{12} = \mathbf{z}_{21}$ (or, equivalently, $\mathbf{y}_{12} = \mathbf{y}_{21}$), the network is said to be a *reciprocal* network. If the elements comprising the network are resistors, inductors, capacitors, or independent sources, then the network is always reciprocal. An equivalent circuit for a general reciprocal three-terminal network is shown in Fig. 14.17. There are, of course, many other equivalent circuits that may be drawn using the various network parameters.

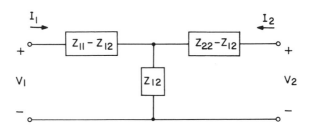

FIGURE 14.17 *Equivalent circuit of a general reciprocal three-terminal network*

EXERCISES

14.8.1   Draw an equivalent circuit for a two-port network with parameters $z_{11} = 6\ \Omega$, $z_{12} = z_{21} = 4\ \Omega$, and $z_{22} = 10\ \Omega$.

14.8.2   Show that the y-parameters may be obtained from the z-parameters by

$$y_{11} = \frac{z_{22}}{\Delta}, \qquad y_{12} = -\frac{z_{12}}{\Delta}$$

$$y_{21} = -\frac{z_{21}}{\Delta}, \qquad y_{22} = \frac{z_{11}}{\Delta}$$

where $\Delta = z_{11}z_{22} - z_{12}z_{21}$. Use this result to find the y-parameters of the circuit of Ex. 14.8.1. [*Suggestion*: Solve (14.34) for the voltages, and compare the result with (14.38).]   *Ans.* $y_{11} = \frac{5}{22}$, $y_{12} = y_{21} = -\frac{1}{11}$, $y_{22} = \frac{3}{22}\ \mho$

14.8.3   Two sets of hybrid parameters, $h_{11}, h_{12}, h_{21}, h_{22}$ and $g_{11}, g_{12}, g_{21}, g_{22}$, are defined by

$$V_1 = h_{11}I_1 + h_{12}V_2$$
$$I_2 = h_{21}I_1 + h_{22}V_2$$

and

$$I_1 = g_{11}V_1 + g_{12}I_2$$
$$V_2 = g_{21}V_1 + g_{22}I_2$$

Find the h-parameters and the g-parameters of the network of Ex. 14.8.1. [*Suggestion*: Solve (14.34) for V_1 and I_2 in terms of I_1 and V_2, etc., and compare results.]   *Ans.* h's: $\frac{44}{10}, \frac{4}{10}, -\frac{4}{10}, \frac{1}{10}$
g's: $\frac{1}{6}, -\frac{4}{6}, \frac{4}{6}, \frac{44}{6}$

PROBLEMS

14.1   Find the phasor representation $V(s)$ of (a) $v(t) = 6e^{-t}\cos(4t + 30°)$, (b) $v(t) = 8e^{-9t}$, and (c) $v(t) = e^{-2t}(4\cos 3t + 3\sin 3t)$.

14.2   Find $v(t)$ if (a) $V(s) = 6\underline{/10°}$ with $s = -3 + j8$, (b) $V(s) = 5\underline{/0°}$ with $s = -10$, and (c) $V(s) = 4 + j3$ with $s = -1 + j2$.

14.3   Find the complex frequencies which characterize (a) $v = e^{-t}(\cos 2t + \sin 2t)$ and (b) $v = e^{-2t} + e^{-4t} + 2\cos t$.

14.4   Find the impedance $Z(s)$ and locate its poles and zeros. If an excitation $v_g = 6e^{-t}\cos 2t$ V is applied at the terminals, find the forced component of the current it delivers.

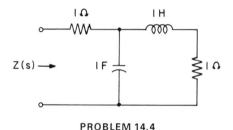

PROBLEM 14.4

14.5 Repeat Prob. 14.4 for the circuit shown.

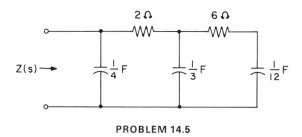

PROBLEM 14.5

14.6 Find in Prob. 9.6 the network functions

$$\mathbf{H}_1 = \frac{\mathbf{I}_1(s)}{\mathbf{V}_g(s)}, \qquad \mathbf{H}_2(s) = \frac{\mathbf{I}_2(s)}{\mathbf{V}_g(s)}$$

Use these results to find the forced responses i_1 and i_2 if

$$v_g(t) = 10e^{-t} \cos 5t \text{ V}$$

14.7 If the current source in Prob. 9.13 is $i_g(t) = 9e^{-t} \cos 3t$ A, find for $t > 0$ the network function $\mathbf{I}(s)/\mathbf{I}_g(s)$ and the forced response $i(t)$. (*Suggestion*: Use the proportionality principle.)

14.8 If the voltage source in Prob. 9.18 is $v_g(t) = 6e^{-t} \cos 2t$ V, find the network function $\mathbf{I}(s)/\mathbf{V}_g(s)$ and the forced response i in each case.

14.9 In Prob. 14.7, replace everything in the circuit except the 5-H inductor by its Thevenin equivalent circuit. Use the result to find the network function and the forced response.

14.10 In Prob. 9.19, let the current source be $i_g(t) = 4e^{-2t} \cos 4t$ A and $C = \frac{1}{6}$ F. Replace the corresponding phasor circuit, except for the series combination of 1 H and 1 Ω, by its equivalent Thevenin circuit. Use the result to obtain the network function $\mathbf{I}(s)/\mathbf{I}_g(s)$ and the forced response $i(t)$.

14.11 Find the network function $\mathbf{V}(s)/\mathbf{V}_g(s)$ in Prob. 9.21. Use the result to find the forced response $v(t)$.

14.12 Find the network function $\mathbf{V}(s)/\mathbf{V}_g(s)$ in Prob. 9.22. Use this result to find the forced response $v(t)$ if $v_g(t) = 6e^{-2t} \cos 4t$ V.

14.13 Find the network function and the forced response $v(t)$ if $v_g(t) = 6e^{-2t}$ V.

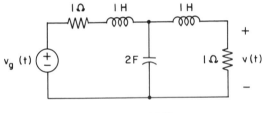

PROBLEM 14.13

14.14   Given the network function

$$H(s) = \frac{V_o(s)}{V_i(s)} = \frac{4s(s+2)}{(s+1)(s+3)}$$

If no cancellation has occurred, find the complete response $v_o(t)$, for $t > 0$, if $v_i(t) = 6e^{-t} \cos 2t$ V and $v_o(0^+) = dv_o(0^+)/dt = 0$.

14.15   Repeat Prob. 14.14 if

$$H(s) = \frac{4s(s+2)}{s^2 + 2s + 2}$$

14.16   Repeat Prob. 14.14 if $v_i(t) = 6e^{-t}$ V. (*Suggestion*: Observe the describing differential equation.)

14.17   Find the complete response $v(t)$, for $t > 0$, if there is no initial stored energy.

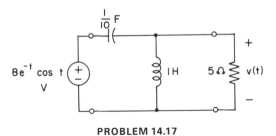

PROBLEM 14.17

14.18   Find the network function $V_o(s)/V_i(s)$ and the forced response if (a) $v_i(t) = 6e^{-2t} \cos t$ V and (b) $v_i(t) = 6 \cos t$ V.

14.19   Find the complete response $v(t)$, for $t > 0$, if $v(0) = 0$.

14.20   Find the network function, the natural frequencies, and the forced response v if $i_g = 5e^{-2t} \cos 2t$ A.

14.21   Find the network function $V(s)/V_g(s)$ in Prob. 9.23 and the forced response $v(t)$ if $v_g(t) = 6e^{-2t} \cos t$ V.

14.22   Find the complete response $v(t)$, for $t > 0$, in Prob. 14.21 if there is no initial stored energy.

14.23   Considering the figure for Prob. 14.17 as a two-port network with terminals as shown, find the z-parameters as functions of s.

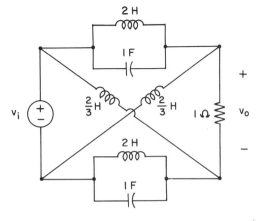

PROBLEM 14.18

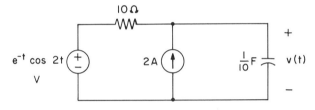

PROBLEM 14.19

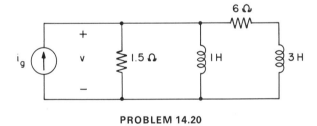

PROBLEM 14.20

14.24 Show that the hybrid parameters defined in Ex. 14.8.3 may be obtained from the z-parameters by

$$\mathbf{h}_{11} = \frac{\Delta}{\mathbf{z}_{22}}, \qquad \mathbf{h}_{12} = \frac{\mathbf{z}_{12}}{\mathbf{z}_{22}}$$

$$\mathbf{h}_{21} = -\frac{\mathbf{z}_{21}}{\mathbf{z}_{22}}, \qquad \mathbf{h}_{22} = \frac{1}{\mathbf{z}_{22}}$$

$$\mathbf{g}_{11} = \frac{1}{\mathbf{z}_{11}}, \qquad \mathbf{g}_{12} = -\frac{\mathbf{z}_{12}}{\mathbf{z}_{11}}$$

$$\mathbf{g}_{21} = \frac{\mathbf{z}_{21}}{\mathbf{z}_{11}}, \qquad \mathbf{g}_{22} = \frac{\Delta}{\mathbf{z}_{11}}$$

where $\Delta = \mathbf{z}_{11}\mathbf{z}_{22} - \mathbf{z}_{12}\mathbf{z}_{21}$.

14.25 Find the y-, h-, and g-parameters of the two-port network of Prob. 14.23.

14.26 A set of *transmission* parameters, A, B, C, and D, is defined for the two-port network of Fig. 14.12 by

$$V_1 = AV_2 - BI_2$$
$$I_1 = CV_2 - DI_2$$

Show that

$$A = \frac{z_{11}}{z_{21}}, \qquad B = \frac{\Delta}{z_{21}}$$

$$C = \frac{1}{z_{21}}, \qquad D = \frac{z_{22}}{z_{21}}$$

where Δ is given in Prob. 14.24. Note that the transmission parameters relate the output variables V_2, I_2 to the input variables V_1, I_1.

14.27 Find the transmission parameters of the two-port network of Prob. 14.23.

14.28 Derive the **Y**-Δ transformation of Chapter 13 by making Figs. 14.13 and 14.14 equivalent at the terminals (that is, equal two-port parameters).

14.29 Find the *y*-parameters of the network shown. Terminate the output port with a 1-Ω resistor, and find the resulting network function V_2/V_1.

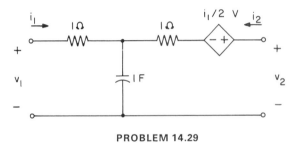

PROBLEM 14.29

14.30 Show that for Fig. 14.15, in terms of the transmission parameters we have

$$\frac{V_2}{V_1} = \frac{1}{A + B}$$

14.31 Check the result for the terminated network of Prob. 14.29 by using the result of Prob. 14.30.

15

FREQUENCY RESPONSE

Frequency-domain functions are very useful, as we have seen, for finding corresponding time-domain functions. The frequency response of a circuit, however, is extremely useful in its own right, as we shall see in this chapter. For example, if we are interested in which frequencies are dominant in an output signal, say $\mathbf{V}(j\omega)$, then we need only consider the amplitude $|\mathbf{V}(j\omega)|$. The dominant frequencies correspond to relatively large amplitudes, and frequencies that are virtually suppressed correspond to relatively small amplitudes.

There are many applications in which frequency responses are important. One very common application is in the design of electric filters, which are networks that pass signals of certain frequencies and block signals of other frequencies. That is, if the output signal of the filter has amplitude $|\mathbf{V}(j\omega)|$, then ω_1 passes if $|\mathbf{V}(j\omega_1)|$ is relatively large and is blocked if $|\mathbf{V}(j\omega_1)|$ is relatively small. There are many examples of electric filters in our modern society, some of the more common being those in our television sets, which allow us to tune in a certain channel by passing its band of frequencies while filtering out those of the other channels.

In this chapter we shall consider frequency responses, both amplitude and phase. We shall also define resonance and quality factor and show how they are related to the frequency responses. Finally, we shall consider methods of scaling networks to yield a given frequency response with practical circuit element values.

15.1 AMPLITUDE AND PHASE RESPONSES

A network function $\mathbf{H}(j\omega)$, as well as any phasor quantity, is in general a complex function, having a real and an imaginary part. That is,

$$H(j\omega) = \text{Re } H(j\omega) + j \text{ Im } H(j\omega) \tag{15.1}$$

As we know, we may also write the network function in the polar form

$$H(j\omega) = |H(j\omega)| e^{j\phi(\omega)} \tag{15.2}$$

where $|H(j\omega)|$ is the *amplitude*, or *magnitude, response* and $\phi(\omega)$ is the *phase response*, given, respectively, by

$$|H(j\omega)| = \sqrt{\text{Re}^2\, H(j\omega) + \text{Im}^2\, H(j\omega)} \tag{15.3}$$

and

$$\phi(\omega) = \tan^{-1} \frac{\text{Im } H(j\omega)}{\text{Re } H(j\omega)} \tag{15.4}$$

The amplitude and phase responses are, of course, special cases of *frequency responses*.
    As an example, suppose the network function of the *RLC* parallel circuit of Fig. 15.1 is the input impedance

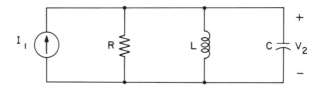

FIGURE 15.1 *RLC parallel circuit*

$$H(s) = \frac{V_2(s)}{I_1(s)} = Z(s) = \frac{1}{(1/R) + sC + (1/sL)} \tag{15.5}$$

or

$$H(s) = \frac{(1/C)s}{s^2 + (1/RC)s + (1/LC)} \tag{15.6}$$

For $s = j\omega$ we have, from (15.5),

$$H(j\omega) = \frac{1}{(1/R) + j[\omega C - (1/\omega L)]} \tag{15.7}$$

so that the amplitude and phase responses are

$$|H(j\omega)| = \frac{1}{\sqrt{(1/R^2) + [\omega C - (1/\omega L)]^2}} \tag{15.8}$$

and

$$\phi(\omega) = -\tan^{-1} R\left(\omega C - \frac{1}{\omega L}\right) \tag{15.9}$$

    Since R, L, and C are constants, the maximum amplitude occurs at the frequency $\omega = \omega_0$ for which the denominator in (15.8) is a minimum. Evidently this occurs when

$$\omega C - \frac{1}{\omega L} = 0$$

or

$$\omega_0 = \frac{1}{\sqrt{LC}} \tag{15.10}$$

Thus

$$|\mathbf{H}(j\omega)|_{\max} = |\mathbf{H}(j\omega_0)| = R$$

Also, it is clear that $|\mathbf{H}(j\omega)| \longrightarrow 0$ as $\omega \longrightarrow 0$ and $\omega \longrightarrow \infty$. Therefore the amplitude response has the form shown in Fig. 15.2(a). In a like manner we may sketch the phase response, shown in Fig. 15.2(b), since $\phi(\omega_0) = 0$, $\phi(\omega) \longrightarrow \pi/2$ as $\omega \longrightarrow 0$, and $\phi(\omega) \longrightarrow -\pi/2$ as $\omega \longrightarrow \infty$.

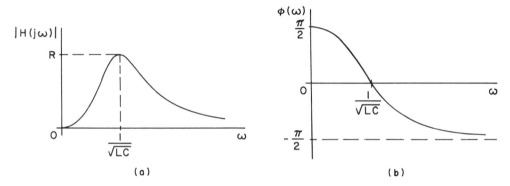

(a)   (b)

FIGURE 15.2 *(a) Amplitude and (b) phase response of (15.7)*

If the input of Fig. 15.1 is the time-domain function

$$i(t) = I_m \cos \omega t$$

then the input phasor is $\mathbf{I} = I_m\underline{/0^\circ}$, and the output phasor is $\mathbf{V}_2 = I_m\mathbf{Z} = I_m\mathbf{H}$. Thus the amplitude of the output is simply that of the network function multiplied by a constant. Therefore we may obtain as much information from the network function response as from the output response. For this reason and the reason that the network function depends only on the network and not on how it is excited, we shall usually consider the frequency response of the network function.

EXERCISES

15.1.1   Let $R = 2\,\Omega$, $L = \frac{1}{9}\,$H, and $C = \frac{1}{4}\,$F in Fig. 15.1, and find the maximum amplitude and the point where it occurs. Also, sketch the amplitude and phase responses.   *Ans.* $|\mathbf{H}|_{\max} = 2$, $\omega_0 = 6$

15.1.2   For the *RLC* series circuit with a voltage source v_g, let the network function be $\mathbf{H} = \mathbf{I}/\mathbf{V}_g$, where $\mathbf{I}$ is the phasor current. Show that the amplitude and phase responses are similar to those of Fig. 15.2 with $|\mathbf{H}|_{\max} = 1/R$ and $\omega_0 = 1/\sqrt{LC}$.

15.1.3   Let the network function of Fig. 15.1 be $\mathbf{H} = \mathbf{I}_L/\mathbf{I}_1$, where $\mathbf{I}_L$ is the inductor phasor current directed downward. Show that

$$\mathbf{H}(s) = \frac{1/LC}{s^2 + (1/RC)s + (1/LC)}$$

and that $|\mathbf{H}|_{\max} = 1$, occurring at $\omega_0 = 0$, provided

$$2R^2C \leq L$$

15.2   FILTERS

With reference to Fig. 15.2(a) we see that frequencies clustered around $\omega_0 = 1/\sqrt{LC}$ rad/s, or $f_0 = 1/(2\pi\sqrt{LC})$ Hz, correspond to relatively large amplitudes, while those near zero and larger than ω_0 correspond to relatively small amplitudes. Thus Fig. 15.1 is an example of a *bandpass filter*, which passes the *band* of frequencies centered around ω_0. In the general amplitude case, shown in Fig. 15.3, we say that ω_0, the frequency at

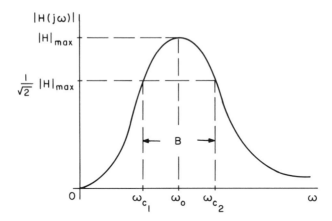

FIGURE 15.3 *General bandpass amplitude response*

which the maximum amplitude occurs, is the *center frequency*. The band of frequencies that passes, or the *passband*, is *defined* to be

$$\omega_{c_1} \leq \omega \leq \omega_{c_2}$$

where ω_{c_1} and ω_{c_2} are called the *cutoff* points and are defined as the frequencies at which the amplitude is $1/\sqrt{2} = 0.707$ times the maximum amplitude. The width of the passband, given by

$$B = \omega_{c_2} - \omega_{c_1} \tag{15.11}$$

is called the *bandwidth*.

As we shall see in Sec. 15.4, the bandwidth in the case of Fig. 15.1 is $B = 1/RC$. As another example, let us consider the circuit of Fig. 15.4. Analyzing the circuit,

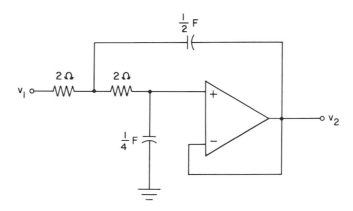

FIGURE 15.4 *Low-pass filter*

we may readily obtain the voltage-ratio function,

$$\mathbf{H}(s) = \frac{\mathbf{V}_2(s)}{\mathbf{V}_1(s)} = \frac{2}{s^2 + 2s + 2}$$

Letting $s = j\omega$ and calculating the amplitude response, we have

$$|\mathbf{H}(j\omega)| = \frac{2}{\sqrt{(2 - \omega^2)^2 + 4\omega^2}}$$

or, after simplification,

$$|\mathbf{H}(j\omega)| = \frac{1}{\sqrt{1 + (\omega^4/4)}} \tag{15.12}$$

The amplitude function continuously decreases as ω increases, because its numerator is constant and its denominator continuously increases with frequency. Therefore the amplitude response attains its maximum of $|\mathbf{H}|_{\max} = 1$ at $\omega_0 = 0$ and thus has the shape of Fig. 15.5(a). From this we see that the circuit of Fig. 15.4 is a *low-pass* filter. That is, it passes low frequencies (relatively large amplitudes) and rejects high frequencies (relatively small amplitudes).

There is only one cutoff point, as indicated in the figure. This follows from the definition

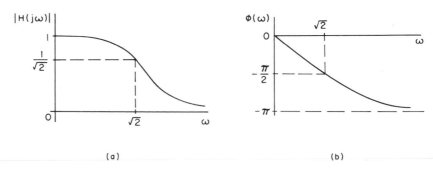

(a)   (b)

FIGURE 15.5 *Low-pass frequency responses*

$$|\mathbf{H}(j\omega_c)| = \frac{1}{\sqrt{2}}|\mathbf{H}(j\omega)|_{max}$$

$$= \frac{1}{\sqrt{2}} \quad (1)$$

$$= \frac{1}{\sqrt{1 + (\omega_c^4/4)}}$$

whose only real answer is $\omega_c = \sqrt{2}$. Thus the band of frequencies which passes is the low-frequency band

$$0 \le \omega \le \sqrt{2}$$

The phase response for Fig. 15.4 may be easily shown from $\mathbf{H}(j\omega)$ to be

$$\phi(\omega) = -\tan^{-1}\frac{2\omega}{2 - \omega^2}$$

which is sketched in Fig. 15.5(b).

An example of a passive, low-pass filter is that of Fig. 15.1, where the network function is as defined in Ex. 15.1.3. In fact, if $R = 1\,\Omega$, $L = 1$ H, and $C = \frac{1}{2}$ F, the network function is the same as that of Fig. 15.4.

There are many types of filters other than low-pass and bandpass. Two of the more common are *high-pass* filters, which pass high frequencies and reject low frequencies, and *band-reject* filters, which pass all frequencies except a single band. Typical

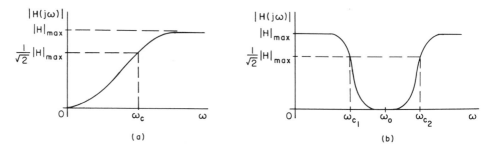

(a)   (b)

FIGURE 15.6 *(a) High-pass and (b) band-reject amplitude response*

amplitude responses are shown in Fig. 15.6, and examples are considered in Exs. 15.2.2 and 15.2.3. In Fig. 15.6(a), ω_c is the cutoff point, and the passband is $\omega > \omega_c$. In Fig. 15.6(b), ω_0 is the center frequency of the rejected band of bandwidth $B = \omega_{c_2} - \omega_{c_1}$.

EXERCISES

15.2.1 Show that

$$H(s) = \frac{2s}{s^2 + 0.2s + 1}$$

is the network function of a bandpass filter and find ω_0, ω_{c_1}, ω_{c_2}, and B.

Ans. 1, $(\mp 0.2 + \sqrt{4.04})/2 = 0.905, 1.105; 0.2$

15.2.2 Show that

$$H(s) = \frac{2s^2}{s^2 + s + 0.5}$$

is the network function of a high-pass filter, and find $|H(j\omega)|_{\max}$ and ω_c.

Ans. $|H|_{\max} = 2$, $\omega_c = 1/\sqrt{2}$

15.2.3 Show that

$$H(s) = \frac{3(s^2 + 25)}{s^2 + s + 25}$$

is the network function of a band-reject filter, and find $|H(j\omega)|_{\max}$, ω_0, ω_{c_1}, and ω_{c_2}.

Ans. $|H|_{\max} = 3$, $\omega_0 = 5$, $\omega_{c_1, c_2} = (\mp 1 + \sqrt{101})/2 = 4.525, 5.525$

15.3 RESONANCE

A physical system which has a sinusoidal type of natural response reacts vigorously, and sometimes violently, when it is excited at, or near, one of its natural frequencies. This effect may have been noticed by the reader in Sec. 9.6, particularly in the case of Ex. 9.6.3. The system in this respect is somewhat like all of us. When urged to do what it naturally wants to do, it responds with enthusiasm.

This phenomenon is known as *resonance*, and its side effects may be good or they may be bad. As an example, a singer may break a crystal goblet with his voice alone by properly producing a note at precisely the right frequency. Also, a bridge may be destroyed if it is subjected to a periodic force with the same frequency as one of its natural frequencies. This is why no thoughtful troop commander will march his men in step across a bridge. On the other hand, without resonance there could be no electric filters.

We shall define a sinusoidally excited network to be in resonance when the amplitude of the network function attains a pronounced maximum or minimum value. The frequency at which this occurs is called the *resonant frequency*. As an example, the *RLC* parallel circuit of Fig. 15.1 is in resonance when the frequency of the driving function is $\omega_0 = 1/\sqrt{LC}$. This was shown in Sec. 15.1 where it was demonstrated that the maximum network function amplitude occurred at ω_0. The amplitude response of Fig. 15.2(a) is typical, with its relatively high peak at the resonant frequency. The parallel *RLC* circuit is so important that the term *parallel resonance* is reserved for its resonant condition.

The reader may recall encountering the term *resonant frequency* earlier in Sec. 9.8 in connection with the underdamped case of the parallel *RLC* circuit. The two frequencies, there and here, are exactly the same.

The natural frequencies of the parallel *RLC* circuit are the poles of the network function, given from (15.6) by

$$s_{1,2} = -\alpha \pm j\omega_d \tag{15.13}$$

where

$$\alpha = \frac{1}{2RC} \tag{15.14}$$

and

$$\omega_d = \sqrt{\omega_0^2 - \alpha^2} \tag{15.15}$$

From the pole-zero plot, shown in Fig. 15.7, we see that the resonant frequency ω_0, or $s = j\omega_0$, is very near the natural frequency $s = -\alpha + j\omega_d$, since, by (15.15), ω_0 is the radius of the dashed semicircle. If R is made larger, then α is made smaller and the

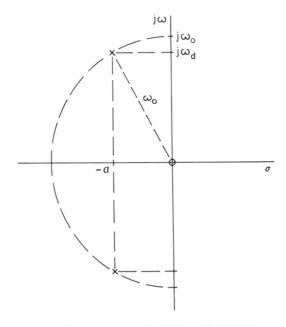

FIGURE 15.7 *Pole-zero plot of the parallel RLC circuit*

resonant frequency is even closer to the natural frequency. In this case, as is clear in Fig. 15.2(a), the peak is more pronounced. Of course, if R were infinite (open-circuited), then the resonant frequency would coincide with the natural frequency and the amplitude would be infinite.

Before leaving the parallel RLC circuit, let us note that the network function is actually the input impedance, as stated in (15.5). The resonant frequency $\omega = \omega_0 = 1/\sqrt{LC}$ is the frequency for which the input impedance is purely real, as seen in (15.7). Indeed, many authors define the resonant frequency precisely this way in the case of a two-terminal network. In the general case, maximum amplitude does not always occur exactly at the frequency of real impedance, but usually there is very little difference.

In the case of the RLC series circuit excited by a voltage source $\mathbf{V}_g$, if the phasor current $\mathbf{I}$ is the output, then

$$H(j\omega) = \frac{\mathbf{I}(j\omega)}{\mathbf{V}_g(j\omega)} = \mathbf{Y}(j\omega) = \frac{1}{R + j[\omega L - (1/\omega C)]}$$

where $\mathbf{Y}$ is the input admittance seen at the source terminals. Evidently series resonance occurs at $\omega_0 = 1/\sqrt{LC}$, yielding a maximum amplitude of $1/R$. As in the parallel resonance case, the resonant frequency is also the frequency of real input impedance or admittance. At resonance the effect of the storage elements exactly cancels, and the source sees only the resistance.

EXERCISES

15.3.1 Show that the resonant frequency in the case of the function of Ex. 15.2.1 coincides with the frequency for which the function is real.

15.3.2 Find the resonant frequency for the parallel RLC circuit described by (a) $R = 2\text{ k}\Omega$, $L = 4\text{ mH}$, and $C = 0.1\ \mu\text{F}$; (b) $\omega_d = 5\text{ rad/s}$ and $\alpha = 12\text{ Np/s}$; and (c) $\alpha = 1\text{ Np/s}$, $R = 4\ \Omega$, and $L = 2\text{ H}$.

Ans. (a) 50,000, (b) 13, (c) 2 rad/s

15.3.3 In Ex. 15.3.2 (a), find the amplitude of the voltage across the combination if the current source is $i_g = \cos \omega t$ mA and ω is (a) 10^4, (b) 5×10^4, and (c) 25×10^4 rad/s.

Ans. (a) 0.0417, (b) 2, (c) 0.0417 V

15.4 BANDPASS FUNCTIONS AND QUALITY FACTOR

The network function (15.6) is a special case of the general *second-order* bandpass function

$$H(s) = \frac{Ks}{s^2 + as + b} \tag{15.16}$$

where K, $a > 0$, and $b > 0$ are real constants. (The function is called second-order because of the quadratic denominator.) To see that the function is of the bandpass type, let us consider its amplitude

$$|\mathbf{H}(j\omega)| = \frac{|K\omega|}{\sqrt{(b - \omega^2)^2 + a^2\omega^2}}$$

$$= \frac{|K|}{\sqrt{a^2 + [(b - \omega^2)/\omega]^2}}$$

Evidently the maximum value is

$$|\mathbf{H}(j\omega)|_{max} = \frac{|K|}{a}$$

occurring at the center, or resonant, frequency ω_0, satisfying

$$b = \omega_0^2 \tag{15.17}$$

At a cutoff frequency ω_c, we must have

$$|\mathbf{H}(j\omega_c)| = \frac{1}{\sqrt{2}}|\mathbf{H}(j\omega)|_{max}$$

or

$$\frac{|K|}{\sqrt{a^2 + [(b - \omega_c^2)/\omega_c]^2}} = \frac{|K|}{\sqrt{2}\,a}$$

which evidently holds if

$$\frac{b - \omega_c^2}{\omega_c} = \pm a$$

Thus we have

$$\omega_c^2 \pm a\omega_c - b = 0 \tag{15.18}$$

which, because of the double-sign possibility, has four solutions. Using the positive sign, we have, by the quadratic formula,

$$\omega_{c_1} = \frac{-a + \sqrt{a^2 + 4b}}{2} \tag{15.19}$$

We have discarded the negative sign on the radical because this gives a negative cutoff frequency. Using the negative sign in (15.18) and again suppressing the negative root, we have the other cutoff frequency,

$$\omega_{c_2} = \frac{a + \sqrt{a^2 + 4b}}{2} \tag{15.20}$$

Evidently, from (15.19) and (15.20) we have the bandwidth,

$$B = \omega_{c_2} - \omega_{c_1} = a \tag{15.21}$$

Thus in view of (15.17) and (15.21) we may write the network function as

$$H(s) = \frac{Ks}{s^2 + Bs + \omega_0^2} \tag{15.22}$$

which is the general network function of a second-order bandpass filter having center frequency ω_0 and bandwidth B. The amplitude response is shown, of course, in Fig. 15.3.

Another result worth noting from (15.17), (15.19), and (15.20) is

$$\omega_0^2 = \omega_{c_1}\omega_{c_2} \tag{15.23}$$

which demonstrates that the center frequency ω_0 is the geometric mean $\sqrt{\omega_{c_1}\omega_{c_2}}$ of the cutoff points.

A good measure of *selectivity* or *sharpness of peak* in a resonant circuit is the so-called *quality factor Q*, which is defined as the ratio of the resonant frequency to the bandwidth. That is,

$$Q = \frac{\omega_0}{B} \tag{15.24}$$

(The letter Q is also our symbol for reactive power, as the reader will recall. However, the two quantities will never be used in the same context so there should be no confusion.) Evidently, since $B = \omega_0/Q$, a low Q corresponds to a relatively large bandwidth, and a high Q (sometimes arbitrarily taken as 5 or more) indicates a small bandwidth, or a more selective circuit.

With this definition of Q we may write (15.22) in the form

$$H(s) = \frac{Ks}{s^2 + (\omega_0/Q)s + \omega_0^2} \tag{15.25}$$

so that we see at a glance the center frequency, the quality factor, and the bandwidth. As an example, the filter described in Ex. 15.2.1 is a bandpass filter with $\omega_0 = 1$ rad/s, $B = 0.2$ rad/s, and $Q = 5$. Another example is the parallel RLC circuit with the network function given in (15.6). In that case, $\omega_0 = 1/\sqrt{LC}$, $B = 1/RC$, and $Q = \omega_0/B$, which has the equivalent values

$$Q = \omega_0 RC$$
$$= R\sqrt{\frac{C}{L}} \tag{15.26}$$
$$= \frac{R}{\omega_0 L}$$

We should mention that the quantity we have called Q is defined as *selectivity* by some authors, who reserve for Q the definition

$$Q = 2\pi \, \frac{\text{Total energy stored at resonance}}{\text{Energy dissipated per cycle at resonance}} \tag{15.27}$$

In the examples we have considered, the two definitions are the same, as the reader is asked to show in Exs. 15.4.1 and 15.4.2. However, in general, there is a slight difference.

Finally, let us consider the effect of Q on resonance. Incorporating (15.19) and (15.20) into one equation and replacing a and b by their values, we have

$$\omega_{c_1,c_2} = \mp \frac{\omega_0}{2Q} + \sqrt{\omega_0^2 + \left(\frac{\omega_0}{2Q}\right)^2}$$

or

$$\omega_{c_1,c_2} = \left(\mp \frac{1}{2Q} + \sqrt{\left(\frac{1}{2Q}\right)^2 + 1}\right)\omega_0 \tag{15.28}$$

Evidently if Q is high, then we may neglect $(1/2Q)^2$ in comparison with 1, and write

$$\omega_{c_1,c_2} \approx \mp \frac{\omega_0}{2Q} + \omega_0$$

or, approximately,

$$\omega_{c_1} = \omega_0 - \frac{B}{2}$$
$$\omega_{c_2} = \omega_0 + \frac{B}{2} \tag{15.29}$$

Thus as Q increases, the amplitude response approaches arithmetic symmetry. That is, the cutoff points are half the bandwidth above and below the center frequency.

In the example of Ex. 15.2.1, we have $\omega_0 = 1$ and $Q = 5$, which we consider as high. Thus by (15.29), the approximate cutoff frequencies are $\omega_{c_1} = 0.9$ and $\omega_{c_2} = 1.1$ rad/s. The exact values are

$$\omega_{c_1,c_2} = 0.905, \quad 1.105$$

EXERCISES

15.4.1   For the RLC parallel circuit in resonance, show that the total energy stored is

$$w(t) = \frac{1}{2} R^2 C I_m^2 \cos^2 \omega_0 t + \frac{R^2 I_m^2}{2\omega_0^2 L} \sin^2 \omega_0 t$$
$$= \tfrac{1}{2} R^2 C I_m^2$$

where the excitation is $i = I_m \cos \omega_0 t$.

15.4.2 Show that for the circuit of Ex. 15.4.1 the energy dissipated per cycle is

$$\int_0^{2\pi/\omega_0} RI_m^2 \cos^2 \omega_0 t \, dt = \frac{\pi RI_m^2}{\omega_0}$$

and thus by definition (15.27) and the result of Ex. 15.4.1 that

$$Q = \omega_0 RC$$

15.4.3 For the *RLC* series circuit with excitation $v_g = V_m \cos \omega t$ V, show that $\omega_0 = 1/\sqrt{LC}$ and $Q = \omega_0 L/R$.

15.5 USE OF POLE-ZERO PLOTS

The pole-zero plot of a network function may often be used to readily sketch the frequency responses. To see how this may be done, let us write the network function in the form

$$H(s) = \frac{K(s - z_1)(s - z_2)\ldots(s - z_m)}{(s - p_1)(s - p_2)\ldots(s - p_n)} \tag{15.30}$$

where, as before, the z's and p's are the zeros and poles. Each of the factors is a complex number of the form $s - s_1$ and may be represented in the s-plane by a vector, drawn from s_1 to s, as shown in Fig. 15.8. This is true since by vector addition the vector s is clearly the sum of the vectors s_1 and $s - s_1$.

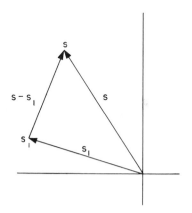

FIGURE 15.8 *Vector representation of* $s - s_1$

The typical vector $s - s_1$, drawn from s_1 to s, may be written in polar form where its magnitude is its length and its phase is the angle it makes with the positive real axis. If $s = j\omega$, the point s in Fig. 15.8 is on the $j\omega$-axis, and the factors of (15.30) may be written

$$jω - z_i = N_i e^{jα_i}, \qquad i = 1, 2, \ldots, m$$
$$jω - p_k = M_k e^{jβ_k}, \qquad k = 1, 2, \ldots, n$$

Therefore the network function is

$$\mathbf{H}(jω) = |\mathbf{H}(jω)| e^{jϕ(ω)}$$

where the amplitude is

$$|\mathbf{H}(jω)| = \frac{KN_1 N_2 \ldots N_m}{M_1 M_2 \ldots M_n} \tag{15.31}$$

and the phase is

$$ϕ(ω) = (α_1 + α_2 + \ldots + α_m) - (β_1 + β_2 + \ldots + β_n) \tag{15.32}$$

both of which may be measured directly from the pole-zero plot.

Thus for any point $jω$ we simply draw vectors from all the poles and zeros to $jω$, measure their lengths and angles, and calculate the amplitude and phase from (15.31) and (15.32). Quite often we need only a few points to sketch the responses, as we shall see.

As an example, suppose we have

$$\mathbf{H}(s) = \frac{4s}{s^2 + 2s + 401}$$

which has a zero at 0 and poles at $-1 \pm j20$. These are shown on the pole-zero plots of Fig. 15.9. By (15.31) and (15.32) we have

$$|\mathbf{H}(jω)| = \frac{4N}{M_1 M_2}$$

$$ϕ(ω) = α - (β_1 + β_2)$$

whose components are identified in Fig. 15.9(a). First we note that if $ω$ varies from 0^+ to $+∞$, then $α = 90°$. For $ω = 0$ we have, from Fig. 15.9(b),

$$|\mathbf{H}| = \frac{4(0)}{\sqrt{401}\sqrt{401}} = 0$$

$$ϕ = 90° - (-\tan^{-1} 20 + \tan^{-1} 20) = 90°$$

For $ω = 20$ [Fig. 15.9(c)] we have

$$|\mathbf{H}| = \frac{4(20)}{(1)\sqrt{1601}} ≈ 2$$

$$ϕ = 90° - (0 + \tan^{-1} 40) ≈ 0$$

This point is in the region where the amplitude changes the fastest, since $M_1 = 1$ is a

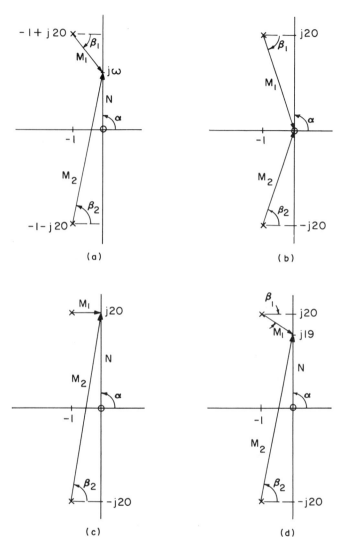

FIGURE 15.9 *Steps in the construction of the frequency responses (Figures not drawn to scale)*

minimum and is changing percentage-wise much faster than N or M_2. The amplitude therefore will reach its peak near this point. Actually we know from our previous work that $\omega_0 = \sqrt{401}$ yields the peak amplitude of 2.

If $\omega = \omega_h$, a very high value, say, such as 10^6, then all three vectors will be essentially vertical, so that

$$N \approx M_1 \approx M_2 \approx \omega_h$$

$$\alpha \approx \beta_1 \approx \beta_2 \approx 90°$$

Therefore we have $|\mathbf{H}| \approx 4/\omega_h$, a very small value, and $\phi \approx 90°$.

Sketching the functions, we have the forms of Fig. 15.2, as expected, with $\omega_0 \approx 20$ and $|\mathbf{H}|_{\max} \approx 2$.

We may get a rough idea of the cutoff points from Fig. 15.9(d). In this figure $M_1 = \sqrt{2}$, which is $\sqrt{2}$ times its value at the approximate peak, represented by Fig. 15.9(c). Since M_2 and N have changed by much smaller percentages, the amplitude is then approximately $1/\sqrt{2}$ times its peak, so that $\omega_{c_1} \approx 19$. By a similar argument at $\omega = 21$, we have $\omega_{c_2} \approx 21$. The exact values, by (15.19) and (15.20), are $\omega_{c_1, c_2} = 19.05,\ 21.05$. In this example, by comparing the network function with (15.25), we see that $Q = 10.01$. Thus we have a high Q so that the poles are very near the $j\omega$-axis. Thus ω_d is very near ω_0, and the approximations we have made are very near the exact values.

EXERCISES

15.5.1   Use the method of this section to sketch the amplitude and phase responses for

$$H(s) = \frac{16s}{s^2 + 4s + 2504}$$

Ans. $\omega_0 \approx 50$, $Q \approx 12.5$, $B = 4$, $\omega_{c_1, c_2} \approx 48,\ 52$

15.5.2   Find the exact values for the answers given in Ex. 15.5.1.

Ans. 50.04, 12.51, 4, 48.08, 52.08

15.6   SCALING THE NETWORK FUNCTION

The reader may have noticed that in many of our examples we have used network elements such as 1 Ω, 1 F, 2 H, etc., which are extremely nice numbers to have when we are analyzing a network. For example, in Fig. 15.4 we had elements of 2 Ω, $\frac{1}{4}$ F, and $\frac{1}{2}$ F, and the network was a low-pass filter with a cutoff frequency of $\sqrt{2}$ rad/s, or 0.23 Hz. However, such element values as these are not very practical, to say the least, and there is very little demand for filters that pass frequencies between 0 and 0.23 Hz. (As an illustration, a 1-F capacitor, constructed of two parallel plates 1 cm apart with air as the dielectric, would require a face area of 1.13×10^9 m².)

It would be an ideal situation if the circuits we analyze or design contain simple element values such as 1 F, etc., while the circuits we build have practical element values such as 0.047 μF, and useful characteristics such as cutoff points of 100 Hz. As we shall see in this section, *network scaling* allows us to have it both ways.

We shall consider two types of network scaling, namely *impedance scaling* and *frequency scaling*. To illustrate the former, let us suppose that the network function is an impedance given by

$$\mathbf{Z}'(s) = sL' + R' + \frac{1}{sC'}$$

The network has been impedance-scaled by an *impedance scale factor* k_i if the impedance $\mathbf{Z}(s)$ of the scaled network is $\mathbf{Z}(s) = k_i \mathbf{Z}'(s)$. In other words, we must have

$$\mathbf{Z}(s) = k_i \left(sL' + R' + \frac{1}{sC'} \right)$$

$$= s\,(k_i L') + k_i R' + \frac{1}{s(C'/k_i)} \tag{15.33}$$

If the impedance of the scaled network is

$$\mathbf{Z}(s) = sL + R + \frac{1}{sC}$$

then we see by comparison with (15.33) that

$$\begin{aligned} L &= k_i L' \\ R &= k_i R' \\ C &= \frac{C'}{k_i} \end{aligned} \tag{15.34}$$

In summary, to impedance-scale a network by the factor k_i, we multiply the L's and R's by k_i and divide the C's by k_i. We have illustrated this for a special case, but it may be shown to hold in general. Also, if there are dependent sources, the scaling is accomplished by multiplying gain constants having units of ohms by k_i, dividing those with units of mhos by k_i, and leaving unchanged those that are dimensionless. The scaling multiplies by k_i, divides by k_i, or leaves unchanged network functions that have units of ohms or mhos or are dimensionless.

To illustrate, let us impedance-scale the network of Fig. 15.4 by $k_i = 5$. The 2-Ω resistors become $2 \times 5 = 10\ \Omega$, the $\frac{1}{2}$-F capacitor becomes $\frac{1}{2} \div 5 = 0.1$ F, and the $\frac{1}{4}$-F capacitor becomes $\frac{1}{4} \div 5 = 0.05$ F. The op amp, an infinite gain device, remains an infinite gain device, and thus is unchanged. The network function $\mathbf{V}_2(s)/\mathbf{V}_1(s)$, being dimensionless, is also unchanged.

To frequency-scale a network function by a *frequency scale factor* k_f, we simply replace s by s/k_f. That is, if the unscaled network has the network function $\mathbf{H}'(S)$, then the scaled network function $\mathbf{H}(s)$ is obtained by letting $S = s/k_f$, resulting in

$$\mathbf{H}(s) = \mathbf{H}'\left(\frac{s}{k_f}\right) \tag{15.35}$$

Thus if the unscaled network had a property, such as center frequency, when $S = j1$, then the scaled network has this property at $s = jk_f$. This is clear from (15.35), which gives

$$H(jk_f) = H'(j1)$$

Another way to consider frequency scaling is to note that $s = k_f S$, so that if $s = j\omega$ corresponds to $S = j\Omega$, then $\omega = k_f \Omega$. Thus the values on the frequency axis have been multiplied by the scale factor k_f, without affecting values on the vertical axis of a frequency response.

The scaling of the network to effect the transformation of (15.35) is quite simple. If $Z'(S)$, given by

$$Z'(S) = SL' + R' + \frac{1}{SC'}$$

is any impedance in the unscaled network, then the corresponding impedance in the scaled network is

$$Z(s) = sL + R + \frac{1}{sC}$$

$$= Z'\left(\frac{s}{k_f}\right)$$

$$= s\left(\frac{L'}{k_f}\right) + R' + \frac{1}{s(C'/k_f)}$$

Comparing these results, we have

$$L = \frac{L'}{k_f}$$

$$R = R' \tag{15.36}$$

$$C = \frac{C'}{k_f}$$

Therefore, to frequency-scale a network by a factor k_f, we divide the L's and C's by k_f and leave the R's unchanged. If there are dependent sources with constant gains, these also are left unchanged.

To illustrate, the circuit of Fig. 15.4 is a low-pass filter with a cutoff frequency of $\omega_c = \sqrt{2}$ rad/s, as we have previously seen. Suppose we want to frequency-scale the network so that the cutoff frequency is 2 rad/s. Then the scale factor k_f is given by

$$\sqrt{2}\,k_f = 2$$

or $k_f = \sqrt{2}$. Dividing the C's in Fig. 15.4 by k_f, we have, in the scaled network, capacitances of $1/2\sqrt{2}$ F and $1/4\sqrt{2}$ F. The rest of the circuit remains unchanged, there being no inductors to scale.

Of course, we may perform both impedance and frequency scaling on a network. To obtain a network with a practical property such as center or cutoff frequency, we may first apply frequency scaling with the proper factor k_f. Then to make the resulting element values more practical we may impedance-scale by a factor k_i. For example, if the parallel *RLC* network of Fig. 15.1 contains a 1-Ω resistor, a 2-H inductor, and a $\frac{1}{2}$-F capacitor, then for the network function of (15.5) we have an amplitude response as in Fig. 15.2(a), with a resonant frequency of 1 rad/s and a peak amplitude of 1 Ω. Suppose we amplitude- and frequency-scale the network to obtain a resonant frequency of 10^6 rad/s using a capacitor of 1 nF. Then we have $k_f = 10^6$, and the new capacitance is

$$C = 10^{-9} = \frac{1/2}{k_i k_f} = \frac{1}{2k_i \times 10^6}$$

Therefore $k_i = 500$, and the new inductance and resistance are

$$L = \frac{2k_i}{k_f} = 10^{-3} \text{ H} = 1 \text{ mH}$$

$$R = 1k_i = 500 \ \Omega$$

The scaled network is shown in Fig. 15.10(a), and its amplitude response, a scaled version of Fig. 15.2(a), is shown in Fig. 15.10(b).

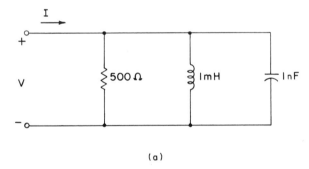

(a)

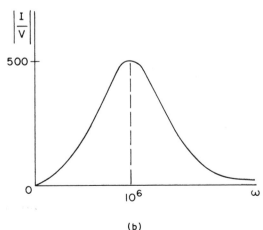

(b)

FIGURE 15.10 *(a) Network and (b) its amplitude response*

EXERCISE

15.6.1 Frequency- and impedance-scale the circuit of Fig. 15.4 to obtain $\omega_c = 2000\pi$
rad/s ($f_c = 1000$ Hz), using capacitors of 0.01 and 0.005 μF.
Ans. $k_f = (1000\sqrt{2})\pi$, $k_i = 10^5/2\sqrt{2}\pi$, R's $= 22.5$ kΩ each

15.7 THE DECIBEL

If the output of a network is used to provide some quantity that is to be sensed by a
human being, the function is sensed in a noncontinuous manner. For example, if the
network provides sound, as in the case of the telephone, the listener cannot detect
continuous changes in intensity. Moreover, if the sound intensity is 1 (on some
arbitrary scale) and it must be increased to 1.1 before the listener detects any change,
then the same listener can detect no change, if the original level is 2, until it is
increased to 2.2. In other words, the ear is not a linear device but more like a *logarithmic* device, since the differences,

$$\log 1.1 - \log 1 = \log 1.1$$

and

$$\log 2.2 - \log 2 = \log\left(\frac{2.2}{2}\right)$$

in the two cases are the same.

For this reason, among others, the amplitude response,

$$|\mathbf{H}(j\omega)| = \left|\frac{\mathbf{V}_2(j\omega)}{\mathbf{V}_1(j\omega)}\right|$$

is more commonly expressed in *decibels*, abbreviated dB and defined by

$$dB = 20 \log_{10}|\mathbf{H}(j\omega)| \tag{15.37}$$

Historically, the logarithmic unit, now known as the *bel*, was defined originally by
Alexander Graham Bell (1847–1922), who, of course, invented the telephone, as the
power unit,

$$\text{bels} = \log_{10}\frac{P_2}{P_1}$$

However, this proved to be a large unit, so that the unit decibel ($\frac{1}{10}$ bel) became
common. That is,

$$dB = 10 \log\frac{P_2}{P_1}$$

If the two average powers P_1 and P_2 are referred to equal impedances, the last expression may be given in terms of the corresponding voltages by

$$dB = 10 \log_{10} \left| \frac{V_2(j\omega)}{V_1(j\omega)} \right|^2$$

which, of course, is equivalent to (15.37). In any case, (15.37) is taken as the standard definition.

In practice, frequencies are not ideally blocked or *attenuated* in the filtering process, as may be seen in the low-pass response of Fig. 15.5(a). A zero amplitude would correspond to absolute, or infinite, attenuation, and any approach to such an ideal situation would be difficult to appreciate on a linear sketch. For example, if $|H(j\omega)| = 0.001$ in Fig. 15.5(a) ($\frac{1}{10}$ of 1 % of its peak value of 1), this would correspond to -60 dB (or 60 dB below its peak value of 0 dB). The latter figure means much more to a telephone engineer than the linear figure.

Even more often in practice it is of interest to consider the *attenuation* or *loss*, defined by

$$\alpha(\omega) = -20 \log_{10} |H(j\omega)|$$
$$= -20 \log_{10} \left| \frac{V_2(j\omega)}{V_1(j\omega)} \right|$$
$$= 20 \log_{10} \left| \frac{V_1(j\omega)}{V_2(j\omega)} \right| dB \tag{15.38}$$

In this case, a frequency ω_1 passes if $\alpha(\omega_1)$ is relatively small and is attenuated if $\alpha(\omega_1)$ is relatively large. The decibel units enable us to tell with some standard precision the degree to which the frequency is attenuated.

As an example, suppose

$$H(s) = \frac{1}{s^2 + \sqrt{2}s + 1}$$

As the reader may verify, the amplitude response is

$$|H(j\omega)| = \frac{1}{\sqrt{1 + \omega^4}}$$

which is that of a low-pass filter with $\omega_c = 1$ rad/s. The linear sketch looks somewhat like that of Fig. 15.5(a). The attenuation is given in decibels by

$$\alpha(\omega) = 20 \log_{10} \sqrt{1 + \omega^4}$$
$$= 10 \log_{10} (1 + \omega^4) \tag{15.39}$$

and is shown in Fig. 15.11 for $0 \le \omega \le 5$.

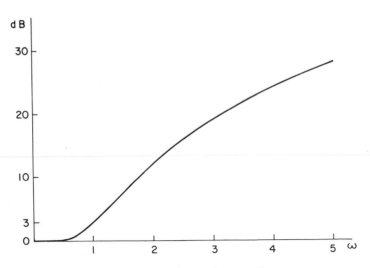

FIGURE 15.11 *Attenuation of a low-pass filter*

EXERCISES

15.7.1 Let the amplitude response $|H(j\omega)|$ be such that $|H(j\omega)|_{max} = |H(j0)| = K$, so that at cutoff $|H(j\omega_c)| = K/\sqrt{2}$. Find the loss, given by (15.38), at $\omega = 0$ and at $\omega = \omega_c$. Note that ω_c corresponds to the "3-dB point," meaning that at $\omega = \omega_c$ the loss is approximately 3 dB more than at the point of minimum loss, $\omega = 0$ in this case.

Ans. $\alpha(0) = -20 \log K$, $\alpha(\omega_c) = -20 \log K + 3$

15.7.2 Given the bandpass filter function

$$H(s) = \frac{0.2s}{s^2 + 0.2s + 1}$$

find the loss in decibels at $\omega = 0.0001, 0.5, 0.905, 1.0, 1.105, 10$, and 100 rad/s.

Ans. 94, 18, 3, 0, 3, 34, 54

PROBLEMS

15.1 For the circuit shown, $R = 2\,\Omega$, $L = 1$ H, and $C = 0.01$ F. If the input and output are V_1 and V_2, respectively, find the network function and sketch the amplitude and phase responses. Show that the peak amplitude and zero phase occur at $\omega = 10$ rad/s.

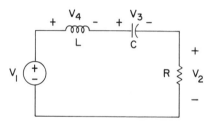

PROBLEM 15.1

15.2 For the circuit shown, $R = 1\,\Omega$, $L = 0.01\,\text{H}$, and $C = 1\,\text{F}$. If the input and output are $\mathbf{V}_1$ and $\mathbf{V}_2$, respectively, find the network function and sketch the amplitude and phase responses. Show that the peak amplitude and zero phase occur at $\omega = 10\,\text{rad/s}$.

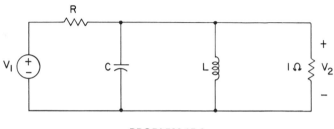

PROBLEM 15.2

15.3 For the circuit shown, $R_1 = R_2 = 0.5\,\Omega$ and $R_3 = 0.01\,\Omega$. If the input and output are $\mathbf{V}_1$ and $\mathbf{V}_2$, respectively, find the network function and sketch the

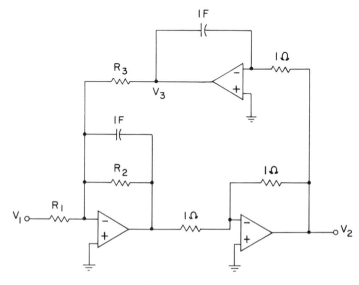

PROBLEM 15.3

amplitude and phase responses. Show that the peak amplitude and zero phase occur at $\omega = 10$ rad/s.

15.4 For the circuit shown, find the network function, $H(s) = V_2(s)/V_1(s)$, and sketch the amplitude and phase responses. Show that the peak amplitude and zero phase occur at $\omega = 0$.

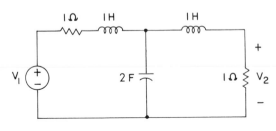

PROBLEM 15.4

15.5 Show that in the general case the network function in the figure for Prob. 15.1 is given by

$$H(s) = \frac{V_2(s)}{V_1(s)} = \frac{(R/L)s}{s^2 + (R/L)s + (1/LC)}$$

Thus by comparison with (15.22) and (15.25), show that the circuit is a bandpass filter with resonant, or center, frequency $\omega_0 = 1/\sqrt{LC}$, bandwidth $B = R/L$, and quality factor $Q = \omega_0/B = (1/R)\sqrt{L/C}$. Show also that the *gain* K of the filter, defined as its peak amplitude value, is given in this case by $K = 1$.

15.6 Show that in Prob. 15.5 if $L = Q$, $C = 1/Q$, and $R = 1$, then

$$H(s) = \frac{(1/Q)s}{s^2 + (1/Q)s + 1}$$

which is a bandpass network function with quality factor Q, center frequency $\omega_0 = 1$ rad/s, bandwidth $B = 1/Q$ rad/s, and gain $K = 1$.

15.7 Show that in the general case the network function in the figure for Prob. 15.2 is given by

$$H(s) = \frac{V_2(s)}{V_1(s)} = \frac{(1/RC)s}{s^2 + [(R+1)/RC]s + (1/LC)}$$

Thus show that the circuit is a bandpass filter with center frequency, bandwidth, quality factor, and gain given, respectively, by

$$\omega_0 = \frac{1}{\sqrt{LC}}$$

$$B = \frac{R+1}{RC}$$

$$Q = \frac{R}{R+1}\sqrt{\frac{C}{L}}$$

$$K = \frac{1}{R+1}$$

15.8 Show that in Prob. 15.7 if $L = (1 - K)/Q$, $C = Q/(1 - K)$, and $R = (1 - K)/K$, where $0 < K < 1$, then

$$H(s) = \frac{(K/Q)s}{s^2 + (1/Q)s + 1}$$

Thus the circuit has resonant frequency $\omega_0 = 1$ rad/s, quality factor Q, gain K, and bandwidth $B = 1/Q$ rad/s.

15.9 Show that in the general case of the figure for Prob. 15.3

$$H(s) = \frac{V_2(s)}{V_1(s)} = \frac{(1/R_1)s}{s^2 + (1/R_2)s + (1/R_3)}$$

Thus show that the circuit is a bandpass filter with center frequency $1/\sqrt{R_3}$ rad/s, bandwidth $1/R_2$ rad/s, gain R_2/R_1, and quality factor $R_2/\sqrt{R_3}$. Let $R_1 = Q/K$, $R_2 = Q$, and $R_3 = 1$ to obtain the network function of Prob. 15.8.

15.10 Using the results of Prob. 15.9, obtain a bandpass filter having the form of the figure for Prob. 15.3 with $\omega_0 = 1$ rad/s, $Q = 10$, and $K = 2$. Find the approximate cutoff points and the bandwidth.

15.11 Show that the network function V_3/V_1 in the figure for Prob. 15.1 is given by

$$\frac{V_3}{V_1} = \frac{1/LC}{s^2 + (R/L)s + (1/LC)}$$

Let $R = 1\ \Omega$, $L = 1/\sqrt{2}$ H, and $C = \sqrt{2}$ F, and show that the result is a low-pass filter with $\omega_c = 1$ rad/s by finding the amplitude response.

15.12 Show that the circuit of Prob. 15.4 is a low-pass filter with $\omega_c = 1$ rad/s. Compare its amplitude response with that of Prob. 15.11 and note the improvement around the cutoff point.

15.13 One type of low-pass filter is the nth-order *Butterworth* filter, whose amplitude response is

$$|H(j\omega)| = \frac{K}{\sqrt{1 + \omega^{2n}}}, \qquad n = 1, 2, 3, \ldots$$

where K is a constant. Ideally, a filter would pass all frequencies in its passband equally well ($|H| > 0$ would be constant) and perfectly block all other frequencies ($|H|$ would be zero). As the accompanying figure shows, for $n = 2, 3$, and 8, the Butterworth filter improves (approaches ideal) as the order n increases. Show that $\omega_c = 1$ rad/s for any n and that the filters of Probs. 15.11 and 15.12 are Butterworth filters of second and third orders, respectively. Finally, sketch a fourth-order Butterworth response and compare it with the second, third, and eighth orders.

15.14 Show that for the figure for Prob. 15.3

$$H(s) = \frac{V_3(s)}{V_1(s)} = -\frac{1/R_1}{s^2 + (1/R_2)s + (1/R_3)}$$

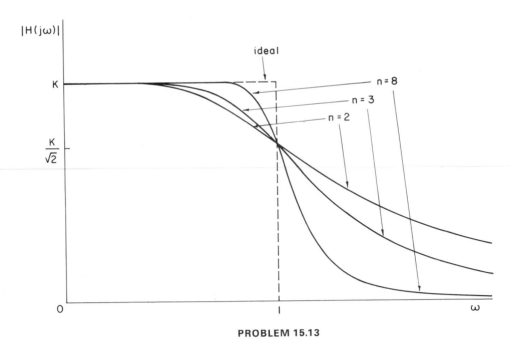

PROBLEM 15.13

Choose values of the resistances so that the circuit is a low-pass Butterworth filter with $\omega_c = 1$ rad/s and $H(0)$ -2. [*Suggestion*: By Prob. 15.11, the denominator of $H(s)$ is required to be $s^2 + \sqrt{2}s + 1$.]

15.15 If the *gain* of a low-pass filter with the network function $H(s)$ is defined to be $K = |H(0)|$, compare the gains in the general cases of Probs. 15.11 and 15.14, which are, respectively, passive and active circuits that perform low-pass filtering. Note that gains higher than 1 are possible with active elements present. This feature and the absence of inductors, which are undesirable at lower frequencies, are advantages of active filters over passive filters.

15.16 Show that in the figure for Prob. 15.1

$$\frac{V_4}{V_1} = \frac{s^2}{s^2 + (R/L)s + (1/LC)}$$

Let $R = 1\,\Omega$, $L = 1/\sqrt{2}$ H, and $C = \sqrt{2}$ F, and show that the result is a high-pass filter with cutoff $\omega_c = 1$ rad/s.

15.17 Apply impedance and frequency scaling to the circuit of Prob. 15.1 to obtain a bandpass filter with a center frequency $f_0 = 10^5$ Hz, with a quality factor $Q = 5$, and using a capacitor of 1 nF. (*Suggestion*: See Prob. 15.6.)

15.18 Obtain an active bandpass filter using the configuration of Prob. 15.3 with $f_0 = 1000$ Hz, $Q = 10$, and $K = 2$, using capacitors of 0.01 μF. (*Suggestion*: See Prob. 15.9.)

15.19 Show that the network function of the given circuit is

$$H(s) = \frac{V_2}{V_1} = \frac{K(s^2 + 1)}{s^2 + (1/Q)s + 1}$$

Thus the circuit is a band-reject filter with center frequency (rejected) $\omega_0 = 1$ rad/s. Also, as in the bandpass case, Q is the quality factor, and $B = \omega_0/Q = 1/Q$ is the bandwidth. Note that the *gain* is $H(0) = K$, where $0 < K < 1$.

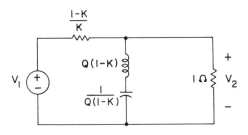

PROBLEM 15.19

15.20 Scale the network in the figure for Prob. 15.19 so that $\omega_0 = 10^6$ rad/s, $Q = 5$, $K = 0.5$, and the capacitance is 10 nF.

15.21 Show that the network of Prob. 11.18 is a band-reject filter with a gain of 1, $Q = 1$, and a center frequency $\omega_0 = 1$ rad/s. Scale the network to obtain a center frequency $f_0 = 60$ Hz, using capacitors of 1 and 2 nF.

15.22 The given circuit is a fourth-order bandpass filter (the denominator of the network function is fourth degree), with $\omega_0 = \sqrt{2}$ rad/s and $Q = 5$. Verify this by finding

$$H(s) = \frac{V_2}{V_1} = \frac{\frac{2}{25}s^2}{s^4 + \frac{2}{5}s^3 + \frac{102}{25}s + \frac{4}{5}s + 4}$$

and so forth. [*Suggestion*: Find the amplitude response and verify that $\omega_0 = \sqrt{2}$ yields the peak point and that $\omega_{c_1,c_2} \approx \omega_0 \mp (\omega_0/10)$ satisfy the appropriate equation. Since $|H| \geq 0$, its maximum occurs when the maximum of $|H|^2$ occurs. Also $|H|$ and $|H|^2$ are functions of ω^2; thus the maximum occurs when

$$\frac{d}{d\omega^2}|H(j\omega)|^2 = 0$$

Since $(d/d\omega)|H|^2 = (d/d\omega^2)|H|^2 (d\omega^2/d\omega) = 2\omega (d/d\omega^2)|H|^2 = 0$, we must check $\omega = 0$ separately.]

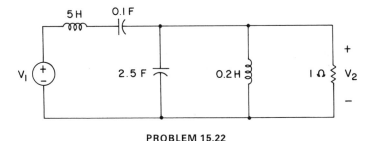

PROBLEM 15.22

16

TRANSFORMERS

In our study of inductance in Chapter 7, we found that a changing current produces a changing magnetic flux which induces a voltage in a coil. In this chapter, we shall consider the effect of a changing magnetic flux that is common to two or more distinct coils. Neighboring coils which share a common magnetic flux are said to be *mutually coupled*. In mutually coupled circuits, a changing current in one coil winding produces an induced voltage in the remaining mutually coupled windings. The induced voltage is characterized by a mutual inductance which exists between the neighboring coils.

A system of mutually coupled coils which are wound on a composite form, or *core*, is commonly called a *transformer*. Transformers are available in a wide variety of sizes and shapes that are designed for numerous applications. Devices as small as an aspirin tablet, for instance, are common in radios, television sets, and stereos for connecting various amplifier stages of the systems. On the other hand, transformers are designed for 60-Hz power applications which range in size from that of a ping-pong ball to those which are larger than an automobile.

We shall restrict ourselves to *linear* transformers (those whose constituent coils are linear) and begin by considering the properties of *self-* and *mutual* inductances. Energy-storage and impedance properties are then analyzed, and an important special case of the linear transformer, the ideal transformer, is introduced. We shall conclude the chapter with a discussion of equivalent circuits that are useful for representing linear transformers.

The circuit analyses presented in the chapter will include both time-domain and frequency-domain cases. Particular emphasis will be given to the frequency-domain case because the most important transformer applications occur in the ac steady state.

16.1 MUTUAL INDUCTANCE

In Sec. 7.4 we found that the inductance L of a linear inductor, as shown in Fig. 16.1, is related to the flux linkage λ by the expression

$$\lambda = N\phi = Li$$

From Faraday's law, the terminal voltage is given by

$$v = \frac{d\lambda}{dt} = L\frac{di}{dt}$$

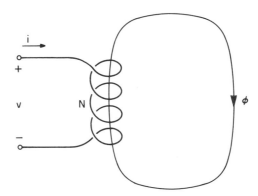

FIGURE 16.1 *Simple inductor*

It is evident from Faraday's law that a voltage is induced in a coil which contains a time-varying magnetic flux, regardless of the source of the flux. Let us therefore consider a second coil, having N_2 turns, positioned in the neighborhood of the first coil, having $N = N_1$ turns, as shown in Fig. 16.2(a). In this case, we have formed a simple transformer having two pairs of terminals in which coil 1 is referred to as the *primary winding* and coil 2 as the *secondary winding*.

To introduce several important inductive quantities, let us begin by examining the open-circuited secondary case of Fig. 16.2(a). The current i_1 produces a magnetic flux ϕ_{11} given by

$$\phi_{11} = \phi_{L1} + \phi_{21}$$

where ϕ_{L1} is the flux of i_1 that links coil 1 and not coil 2, called the *leakage flux*, and ϕ_{21} is the flux of i_1 that links coil 2 and coil 1, called the *mutual flux*.

We shall assume, as in the case of the linear inductor, that the flux in each coil links all turns of the coil. Since the secondary is an open circuit, no current flows in coil 2, and the flux linkage of this coil is

$$\lambda_2 = N_2\phi_{21}$$

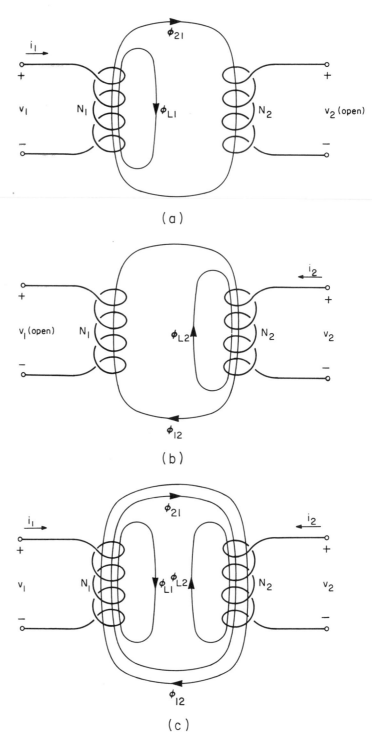

FIGURE 16.2 *Mutually coupled coils in which (a) $i_2 = 0$, (b) $i_1 = 0$, and (c) i_1 and i_2 are nonzero*

Therefore the voltage v_2 is given by

$$v_2 = \frac{d\lambda_2}{dt} = N_2 \frac{d\phi_{21}}{dt}$$

(We shall justify later that the polarity is as shown.)
In a linear transformer, the flux ϕ_{21} is linearly proportional to i_1. Thus we may write

$$N_2\phi_{21} = M_{21}i_1 \qquad (16.1)$$

where M_{21} is a *mutual inductance* in henrys (H). In terms of this mutual inductance, the open-circuit secondary voltage becomes

$$v_2 = M_{21} \frac{di_1}{dt}$$

Let us now find the primary voltage v_1. We know that

$$v_1 = \frac{d\lambda_1}{dt}$$

where λ_1, the flux linkage of coil 1, is given by

$$\lambda_1 = N_1\phi_{11}$$

As in the case of a single linear inductor, the inductance L_1 of the primary winding, sometimes called the *self-inductance* of the primary to distinguish it from the mutual inductance, is defined by

$$L_1 i_1 = N_1\phi_{11} \qquad (16.2)$$

Hence we see that

$$v_1 = L_1 \frac{di_1}{dt}$$

in the case under consideration.
Next let us consider Fig. 16.2(b) in which the primary is an open circuit and a current i_2 flows in the secondary. Proceeding as before, we have

$$\phi_{22} = \phi_{L2} + \phi_{12}$$

where ϕ_{L2} is the leakage flux of i_2 that links coil 2 but not coil 1, and ϕ_{12} is the mutual flux of i_2 that links coil 1 and coil 2. The flux linkage of coil 1 is

$$\lambda_1 = N_1\phi_{12}$$

Therefore the primary voltage is given by

$$v_1 = \frac{d\lambda_1}{dt} = N_1 \frac{d\phi_{12}}{dt}$$

If we define

$$N_1\phi_{12} = M_{12}i_2 \qquad (16.3)$$

where M_{12} is a mutual inductance, then the open-circuit primary voltage is

$$v_1 = M_{12}\frac{di_2}{dt}$$

In the next section we shall show that the mutual inductances M_{12} and M_{21} are equal; therefore we shall write

$$M = M_{12} = M_{21} \qquad (16.4)$$

and refer to M as the mutual inductance.

The secondary voltage is given by

$$v_2 = \frac{d\lambda_2}{dt}$$

where the flux linkage is

$$\lambda_2 = N_2\phi_{22}$$

We now define the relation

$$L_2 i_2 = N_2\phi_{22} \qquad (16.5)$$

where L_2 is the self-inductance of coil 2 in henrys (H). Therefore

$$v_2 = L_2\frac{di_2}{dt}$$

in the open-circuit primary case.

Let us now consider the general case of Fig. 16.2(c) in which both i_1 and i_2 are nonzero. The fluxes in coil 1 and coil 2, as shown in the figure, are

$$\phi_1 = \phi_{L1} + \phi_{21} + \phi_{12} = \phi_{11} + \phi_{12}$$
$$\phi_2 = \phi_{L2} + \phi_{12} + \phi_{21} = \phi_{21} + \phi_{22}$$

respectively. Therefore the flux linkages of the primary and secondary coils are

$$\lambda_1 = N_1\phi_{11} + N_1\phi_{12}$$
$$\lambda_2 = N_2\phi_{21} + N_2\phi_{22}$$

Substituting from (16.1)–(16.5), we find upon differentiation that the primary and secondary voltages are

$$v_1 = L_1\frac{di_1}{dt} + M\frac{di_2}{dt}$$

$$v_2 = M\frac{di_1}{dt} + L_2\frac{di_2}{dt} \qquad (16.6)$$

It is clear that the voltages consist of self-induced voltages due to the inductances L_1 and L_2 and mutual voltages due to the mutual inductance M.

In the preceding discussion, the coil windings in Fig. 16.2 are such that the algebraic sign of the mutual voltage terms $M\, di_1/dt$ and $M\, di_2/dt$ are positive for the terminal voltage and current assignments as shown. In practice, it is, of course, undesirable to show a detailed sketch of the windings. This is avoided by the use of a *dot convention* which designates the polarity of the mutual voltage. Equivalent circuit symbols for the transformer of Fig. 16.2 are shown in Fig. 16.3. The polarity markings are assigned so that a positively increasing current flowing into a dotted (undotted) terminal in one winding induces a positive voltage at the dotted (undotted) terminal of the other winding.

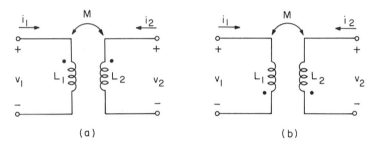

FIGURE 16.3 *Circuit symbols for the transformer of Fig. 16.2: (a) dots on upper terminals; (b) dots on lower terminals*

For the purpose of writing the describing equations, we may state the following rule:

> *A current* i *entering a dotted (undotted) terminal in one winding induces a voltage* M di/dt *with positive polarity at the dotted (undotted) terminal of the other winding.*

We note that with this rule it is unimportant whether the current i is increasing or not, since the sign of the induced voltage is accounted for by di/dt. That is, if i is increasing, the induced voltage $M\, di/dt$ is positive, and if i is decreasing, the induced voltage is negative. If i is a dc current, then, of course, the induced voltage is zero.

As an example, let us write the loop equations for Fig. 16.3(a). We see that i_2 enters a dotted terminal. Thus the mutual voltage $M\, di_2/dt$ has positive polarity at the dotted terminal of the primary. Similarly, since i_1 enters a dotted terminal, the mutual voltage $M\, di_1/dt$ has positive polarity at the dotted terminal of the secondary. Application of KVL around the primary and secondary circuits gives (16.6).

In a like manner, identical statements apply to the undotted terminals of Fig. 16.3(b), and KVL once again yields (16.6).

Let us now establish a method for placing polarity markings on a transformer. We begin by arbitrarily assigning a dot to a terminal, such as terminal a of Fig. 16.4(a).

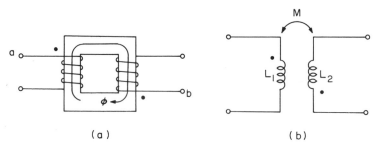

(a)   (b)

FIGURE 16.4 *(a) Model of a transformer to determine polarity mark-ings; (b) circuit symbol*

A current into this terminal produces a flux ϕ as shown. (The direction of ϕ is deter-mined by the right-hand rule, which states that if the fingers of the right hand encircle a coil in the direction of the current, the thumb indicates the direction of the flux.) The dotted terminal in the remaining coil is the one which a current enters to produce a flux in the same direction as ϕ. (This statement, which may be used to obtain the polarity of the induced voltage, is a consequence of a rule known as *Lenz's law*. The reader may study this law, named for the German scientist Heinrich F. E. Lenz, in a later course in electromagnetic field theory.) We note that a current flowing into b in Fig. 16.4(a) produces a flux in the same direction as ϕ. Thus terminal b receives the polarity dot. The circuit symbol for this transformer is shown in Fig. 16.4(b).

Evidently the polarity markings on a transformer are independent of the terminal voltage and current assignments. Consider, for instance, the assignment of Fig. 16.5(a). Applying KVL around the primary and secondary loops, we find

$$v_1 = L_1 \frac{di_1}{dt} - M \frac{di_2}{dt}$$

$$v_2 = -M \frac{di_2}{dt} + L_2 \frac{di_2}{dt}$$

In Fig. 16.5(b), the voltage and current assignments have been changed. Application of KVL for these assignments yields

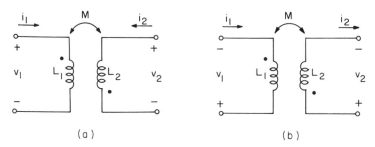

(a)   (b)

FIGURE 16.5 *Transformer with different terminal voltage and current assignments*

$$v_1 = -L_1\frac{di_1}{dt} - M\frac{di_2}{dt}$$

$$v_2 = M\frac{di_1}{dt} + L_2\frac{di_2}{dt}$$

The above results can also be obtained by an alternative method for selecting the sign of the mutual voltages, as follows:

> *If both currents enter (or leave) the dotted terminals of the coils, the mutual and self-inductance terms for each terminal pair have the same sign; otherwise, they have opposite signs.*

The student can easily verify the use of this method for the cases of Figs. 16.3 and 16.5.

As an example, let us find the open-circuit voltage v_2 in the circuit of Fig. 16.6, given that $i_1(0^-) = 0$. For $t > 0$, since no current flows in the secondary, applying KVL around the primary yields

$$\frac{di_1}{dt} + 10i_1 = 20$$

This is a first-order differential equation with a general solution of

$$i_1 = 2 + Ae^{-10t}$$

Since $i(0^+) = i(0^-) = 0$, $A = -2$ and

$$i_1 = 2(1 - e^{-10t})\text{ A}$$

Therefore, in the secondary,

$$v_2 = -0.25\frac{di_1}{dt}$$

$$= -5e^{-10t}\text{ V}$$

Let us now consider the transformer of Fig. 16.3(a) in the case of an excitation having a complex frequency s. In this case, the phasor equations corresponding to (16.6) are

$$\mathbf{V}_1(s) = sL_1\mathbf{I}_1(s) + sM\mathbf{I}_2(s)$$
$$\mathbf{V}_2(s) = sM\mathbf{I}_1(s) + sL_2\mathbf{I}_2(s)$$

The phasor circuit for this network is shown in Fig. 16.7. In the case of a purely sinusoidal excitation, we simply replace s by $j\omega$.

For a final example, consider finding the phasor voltage $\mathbf{V}_2(s)$ in the network of Fig. 16.8 due to the complex forcing function $\mathbf{V}_1(s)$. The loop equations in the primary

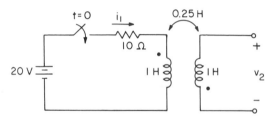
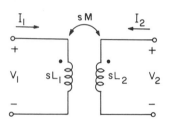

FIGURE 16.6 *Circuit with an open-circuit secondary*   **FIGURE 16.7** *Phasor circuit for a transformer*

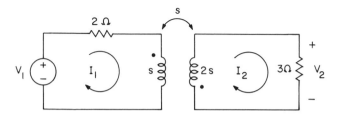

FIGURE 16.8 *Circuit with a complex excitation*

and secondary are

$$V_1 = (s + 2)I_1 + sI_2$$
$$0 = sI_1 + (2s + 3)I_2$$

Therefore

$$I_2 = \frac{\begin{vmatrix} s+2 & V_1 \\ s & 0 \end{vmatrix}}{\begin{vmatrix} s+2 & s \\ s & 2s+3 \end{vmatrix}}$$

$$= \frac{-sV_1}{s^2 + 7s + 6}$$

and if $H(s)$ is the voltage ratio function, then

$$H(s) = \frac{V_2}{V_1} = \frac{3I_2}{V_1} = \frac{-3s}{s^2 + 7s + 6}$$

$$= \frac{-3s}{(s + 1)(s + 6)}$$

A pole-zero plot and a sketch of $|H(j\omega)|$ are shown in Fig. 16.9. We may show that the maximum ac steady-state response occurs for $\omega = \sqrt{6}$ rad/s.
    Suppose, as an example, that $v_1 = 100 \cos 10t$ V. Then

$$V_1 = 100 \text{ V}, \qquad s = j10 \text{ rad/s}$$

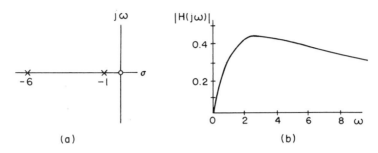

FIGURE 16.9 *(a) Pole-zero and (b)* $|H(j\omega)|$ *plot for the network of Fig. 16.8*

and

$$\mathbf{V}_2 = 25.6\underline{/126.7°}\ \text{V}$$

Thus the ac steady-state response is

$$v_2(t) = 25.6\cos(10t + 126.7°)\ \text{V}$$

EXERCISES

16.1.1   In the circuit of Fig. 16.3 (a), $L_1 = L_2 = 0.1$ H and $M = 10$ mH. Find v_1 and v_2 if (a) $i_1 = 10$ mA and $i_2 = 0$, (b) $i_1 = 0$ and $i_2 = 10\sin 100t$ mA, and (c) $i_1 = 0.1\cos t$ A and $i_2 = 0.3\sin(t + 30°)$ A.

Ans. (a) 0, 0; (b) $10\cos 100t$ mV, $0.1\cos 100t$ V; (c) $-10\sin t + 3\cos(t + 30°)$ mV, $-\sin t + 30\cos(t + 30°)$ mV

16.1.2   Verify that the coil windings of Fig. 16.2 are consistent with the polarity markings of Fig. 16.3.

16.1.3   Find I_2 in Fig. 16.8 for a sinusoidal voltage of $\mathbf{V}_1 = 10\underline{/0°}$ V having $\omega = 10$ rad/s if the 2-Ω resistor is replaced by the series combination of a 10-Ω resistor and a 1-F capacitor.   *Ans.* $0.417\underline{/163.5°}$ A

16.2 ENERGY STORAGE

We have previously shown that the energy stored in an inductor at time t is

$$w(t) = \tfrac{1}{2}Li^2(t)$$

Evidently, for a given inductance L, the energy is completely specified in terms of $i(t)$.

Let us now determine the energy stored in a pair of mutually coupled inductors, such as those of Figs. 16.2–16.4. In so doing, we shall show that $M_{12} = M_{21} = M$ and establish limits for the magnitude of M.

In Fig. 16.3, the stored energy is the sum of the energies supplied to the primary and secondary terminals. The instantaneous powers delivered to these terminals, from (16.6), are

$$p_1 = v_1 i_1 = \left(L_1 \frac{di_1}{dt} + M_{12} \frac{di_2}{dt}\right) i_1$$

$$p_2 = v_2 i_2 = \left(M_{21} \frac{di_1}{dt} + L_2 \frac{di_2}{dt}\right) i_2$$

(16.7)

where M has been replaced by M_{12} and M_{21} in the appropriate terms.

Let us now perform a simple experiment. Suppose we start at time t_0 with $i_1(t_0) = i_2(t_0) = 0$. Since the magnetic flux is zero, no energy is stored in the magnetic field; that is, $w(t_0) = 0$. Next, assume, beginning at time t_0, that we maintain $i_2 = 0$ and increase i_1 until, at time t_1, $i_1(t_1) = I_1$ and $i_2(t_1) = 0$. During this interval, $i_2 = 0$ and $di_2/dt = 0$. Thus the energy accumulated during this time is

$$w_1 = \int_{t_0}^{t_1} (p_1 + p_2)dt = \int_{t_0}^{t_1} L_1 i_1 \frac{di_1}{dt} dt$$

$$= \int_0^{I_1} L_1 i_1 \, di_1 = \tfrac{1}{2} L_1 I_1^2$$

As a final step, let us maintain $i_1 = I_1$ and increase i_2 until, at time t_2, $i(t_2) = I_2$. During this time interval, $di_1/dt = 0$, and the energy accumulated, using (16.7), is

$$w_2 = \int_{t_1}^{t_2} \left(M_{12} I_1 \frac{di_2}{dt} + L_2 i_2 \frac{di_2}{dt}\right) dt$$

$$= \int_0^{I_2} (M_{12} I_1 + L_2 i_2) \, di_2$$

$$= M_{12} I_1 I_2 + \tfrac{1}{2} L_2 I_2^2$$

Thus the energy stored in the transformer at time t_2 is

$$w(t_2) = w(t_0) + w_1 + w_2$$

$$= \tfrac{1}{2} L_1 I_1^2 + M_{12} I_1 I_2 + \tfrac{1}{2} L_2 I_2^2$$

Let us now repeat our experiment but reverse the order in which we increase i_1 and i_2. That is, in the interval from t_0 to t_1, we increase i_2 so that $i_2(t_1) = I_2$ while holding $i_1 = 0$. Finally, we maintain $i_2 = I_2$ while increasing i_1 so that $i_1(t_2) = I_1$. Using the same steps as before, we find in this case that

$$w(t_2) = \tfrac{1}{2} L_1 I_1^2 + M_{21} I_1 I_2 + \tfrac{1}{2} L_2 I_2^2$$

Since $i_1(t_2) = I_1$ and $i_2(t_2) = I_2$ in both experiments, then $w(t_2)$ should be the same in both cases. Comparing our results, we see that this requires

$$M_{12} = M_{21} = M$$

If we repeat our experiment with the polarity markings of Fig. 16.4, the sign of the mutual voltage is negative, which causes the mutual term in the energy MI_1I_2 to be negative. Thus a general expression for the energy at any time t is given by

$$w(t) = \tfrac{1}{2}L_1i_1^2 \pm Mi_1i_2 + \tfrac{1}{2}L_2i_2^2 \tag{16.8}$$

where the sign in the mutual term is positive if both currents enter dotted (or undotted) terminals; otherwise, it is negative.

The *coefficient of coupling* between the inductors indicates the amount of coupling and is defined by

$$k = \frac{M}{\sqrt{L_1L_2}} \tag{16.9}$$

Clearly, $k = 0$ if no coupling exists between the coils, since $M = 0$. We can establish the upper limit of k by substituting from (16.1)–(16.5), which yields

$$k = \frac{M}{\sqrt{L_1L_2}} = \frac{\sqrt{M_{12}M_{21}}}{\sqrt{L_1L_2}} = \sqrt{\frac{\phi_{12}}{\phi_{11}}}\sqrt{\frac{\phi_{21}}{\phi_{22}}} \le 1$$

since $\phi_{12}/\phi_{11} \le 1$ and $\phi_{21}/\phi_{22} \le 1$. Thus we must have

$$0 \le k \le 1$$

or, equivalently,

$$0 \le M \le \sqrt{L_1L_2}$$

If $k = 1$, all of the flux links all of the turns of both windings, which is a *unity-coupled* transformer.

The value of k (and hence M) depends on the physical dimensions and number of turns of each coil, their relative positions to one another, and the magnetic properties of the core on which they are wound. Coils are said to be *loosely coupled* if $k < 0.5$, whereas those in which $k > 0.5$ are *tightly coupled*. Most air-core transformers are loosely coupled, in contrast to iron-core devices for which k can approach 1.

Let us now examine the values of i_1 and i_2 in (16.8) for which $w(t)$ is zero. From the quadratic formula, we may write

$$i_1 = \mp\frac{Mi_2}{L_1} \pm \frac{i_2}{L_1}\sqrt{M^2 - L_1L_2}$$

For real values of i_1 and i_2, we see that for $w(t) = 0$ we must have

$$i_1 = i_2 = 0, \qquad M < \sqrt{L_1 L_2}$$

$$i_1 = \pm \frac{M}{L_1} i_2, \qquad M = \sqrt{L_1 L_2} \qquad\qquad (16.10)$$

where the sign of the mutual term is negative if both currents enter dotted (or undotted) terminals; otherwise, it is positive. The second equation of (16.10) shows that $w(t)$ can be zero in a unity-coupled ($k = 1$) transformer even though i_1 and i_2 are nonzero.

EXERCISE

16.2.1   Determine the energy stored in each case of Ex. 16.1.1 at $t = 0$.

Ans. (a) 5 μJ, (b) 0, (c) 1.775 mJ

16.3 CIRCUITS WITH LINEAR TRANSFORMERS

A two-winding transformer is in general a four-terminal device in which the reference potential in the primary can be different from that of the secondary without altering the values of v_1, v_2, i_1, or i_2. Consider, for example, the circuit of Fig. 16.10. KVL

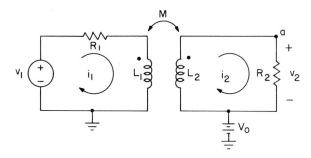

FIGURE 16.10 *Circuit showing different reference potentials in primary and secondary*

around the primary and secondary circuits gives

$$v_1(t) = R_1 i_1 + L_1 \frac{di_1}{dt} - M \frac{di_2}{dt}$$

$$0 = -M \frac{di_1}{dt} + R_2 i_2 + L_2 \frac{di_2}{dt}$$

Evidently, the voltages and currents are not affected by V_0. For this reason, the secondary of the transformer is said to have *dc isolation* from the primary. Point a, of course, is at an absolute potential of $V_0 + i_2R_2$ volts with respect to the ground reference. If we now let $V_0 = 0$, it is seen that the bottom terminals of the transformer are connected and that the primary and secondary circuits have a common reference point. The transformer, in this case, is a three-terminal device.

As a second example, let us find the complete response i_2 for $t > 0$ in Fig. 16.11,

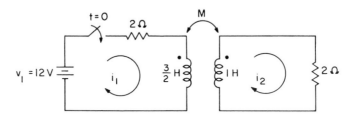

FIGURE 16.11 *Switching circuit containing a transformer*

given $M = 1/\sqrt{2}$ H and $i_1(0^-) = i_2(0^-) = 0$. Since the forced response for i_1 is a dc current (by inspection), no steady-state voltage is produced in the secondary; thus $i_{2f} = 0$. To find the natural response, let us first obtain the network function for $t > 0$. Application of KVL for a complex excitation $\mathbf{V}_1(s)$ yields

$$\mathbf{V}_1(s) = \left(\frac{3}{2}s + 2\right)\mathbf{I}_1 - \frac{1}{\sqrt{2}}s\mathbf{I}_2$$

$$0 = -\frac{1}{\sqrt{2}}s\mathbf{I}_1 + (s + 2)\mathbf{I}_2$$

from which we find

$$\mathbf{H}(s) = \frac{\mathbf{I}_2}{\mathbf{V}_1} = \frac{1}{\sqrt{2}}\frac{s}{(s+1)(s+4)}$$

The poles of $\mathbf{H}(s)$ are the natural frequencies, $-1, -4$, of the natural response. Thus, for $t > 0$,

$$i_2 = i_{2f} + i_{2n} = 0 + A_1e^{-t} + A_2e^{-4t}$$

From (16.8), $w(0^-) = 0$; therefore since the energy cannot change instantaneously in the absence of infinite forcing functions, $w(0^+) = 0$. From (16.10), noting that $M < \sqrt{L_1L_2}$, we see that $i_1(0^+) = i_2(0^+) = 0$. To obtain a second initial condition for finding A_1 and A_2, we apply KVL around the primary and secondary at $t = 0^+$, which yields

$$12 = \frac{3}{2}\frac{di_1(0^+)}{dt} - \frac{1}{\sqrt{2}}\frac{di_2(0^+)}{dt}$$

$$0 = -\frac{1}{\sqrt{2}}\frac{di_1(0^+)}{dt} + \frac{di_2(0^+)}{dt}$$

Solving, we find $di_2(0^+)/dt = 6\sqrt{2}$ A/s. Evaluating A_1 and A_2, using $i_2(0^+)$ and $di_2(0^+)/dt$, the solution becomes

$$i_2 = 2\sqrt{2}(e^{-t} - e^{-4t})$$

Suppose we now repeat this example for $M = \sqrt{L_1 L_2} = \sqrt{3/2}$. In this case the network function for $t > 0$ is

$$H(s) = \frac{\sqrt{3/2}\,s}{5s + 4}$$

and the complete response is

$$i_2 = Ae^{-0.8t}$$

We know that $i_1(0^-) = i_2(0^-) = 0$; however, if we take $i_2(0^+) = i_2(0^-)$ as before, then the unreasonable solution $i_2 = 0$ results. Also, applying KVL gives

$$12 = 2i_1 + \frac{3}{2}\frac{di_1}{dt} - \sqrt{\frac{3}{2}}\frac{di_2}{dt}$$

$$0 = -\sqrt{\frac{3}{2}}\frac{di_1}{dt} + 2i_2 + \frac{di_2}{dt}$$

Multiplying the latter equation by $\sqrt{3/2}$ and adding it to the former, we have

$$12 = 2i_1 + \sqrt{6}\,i_2 \tag{16.11}$$

This result contradicts $i_1(0^+) = i_2(0^+) = 0$.

Recalling (16.10), we see that the energy is zero in the unity-coupled case when

$$i_1(0^+) = \frac{M}{L_1}i_2(0^+) = \sqrt{\frac{2}{3}}i_2(0^+)$$

Combining the last result and (16.11) gives $i_2(0^+) = 2.94$ A, and therefore

$$i_2 = 2.94e^{-0.8t}$$

Thus the current in a unity-coupled transformer can change instantaneously as a result of the application of a finite forcing function.

As another example, let us find the steady-state response v_2 in Fig. 16.12. Applying KVL for a complex frequency s, we have

$$V_1 = \left(s + 1 + \frac{1}{s}\right)I_1 - \left(s + \frac{1}{s}\right)I_2$$

$$0 = -\left(s + \frac{1}{s}\right)I_1 + \left(2s + 1 + \frac{1}{s}\right)I_2$$

for which

$$H(s) = \frac{V_2}{V_1} = \frac{I_2}{V_1} = \frac{s^2 + 1}{s^3 + 3s^2 + 2s + 2}$$

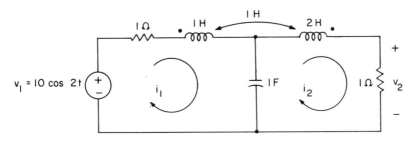

FIGURE 16.12 *Circuit containing a linear transformer*

Substituting $s = j2$ rad/s and $\mathbf{V}_1 = 10$ V, we find $\mathbf{V}_2 = 2.79\underline{/-21.8°}$ V. Thus

$$v_2 = 2.79 \cos (2t - 21.8°) \text{ V}$$

As a final example, let us find the network function $\mathbf{V}_2/\mathbf{V}_1$ for the phasor circuit of Fig. 16.13. Let us first examine the voltage $\mathbf{V}_A$ which appears across winding A in loop 1. We see that the voltage due to $\mathbf{I}_1$ is the self-inductance term $(2s\mathbf{I}_1)$ and the mutual term of winding C $(2s\mathbf{I}_1)$. The voltage due to $\mathbf{I}_2$ arises from the mutual terms of winding B $(-s\mathbf{I}_2)$ and winding C $(-2s\mathbf{I}_2)$. Thus

$$\mathbf{V}_A = (2s + 2s)\mathbf{I}_1 - (s + 2s)\mathbf{I}_2$$

Similarly, we find

$$\mathbf{V}_B = -(s + 3s)\mathbf{I}_1 + (3s + 3s)\mathbf{I}_2$$
$$\mathbf{V}_C = (4s + 2s)\mathbf{I}_1 - (4s + 3s)\mathbf{I}_2$$

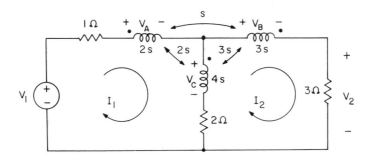

FIGURE 16.13 *Circuit containing a three-winding transformer*

Applying KVL in loops 1 and 2 gives

$$\mathbf{V}_1 = 3\mathbf{I}_1 + \mathbf{V}_A + \mathbf{V}_C - 2\mathbf{I}_2$$
$$0 = -2\mathbf{I}_1 - \mathbf{V}_C + \mathbf{V}_B + 5\mathbf{I}_2$$

or

$$\mathbf{V}_1 = (10s + 3)\mathbf{I}_1 - (10s + 2)\mathbf{I}_2$$
$$0 = -(10s + 2)\mathbf{I}_1 + (13s + 5)\mathbf{I}_2$$

Solving these equations for the ratio I_2/V_1, we find the network function to be

$$H(s) = \frac{V_2}{V_1} = \frac{3I_2}{V_1} = \frac{3(10s + 2)}{30s^2 + 49s + 11}$$

EXERCISES

16.3.1   Determine i_1 for $t > 0$ in the network of Fig. 16.11, given $M = 1/\sqrt{2}$ H.
Assume the circuit is in steady state at $t = 0^-$.   *Ans.* $6 - 4e^{-t} - 2e^{-4t}$ A

16.3.2   Determine the impedance seen from the terminals of the voltage source v_1 in
Fig. 16.12.   *Ans.* $1.48\underline{/37.75}$ Ω

16.3.3   Determine the network function V_2/V_1 for Fig. 16.13 if the polarity dot of coil
B is placed on the other terminal.   *Ans.* $6(s + 1)/(6s^2 + 45s + 11)$

16.4   REFLECTED IMPEDANCE

In this section we shall develop several important impedance relationships for the ac
steady-state case. Let us begin by considering the phasor circuit of Fig. 16.14 having
a practical source V_g with an impedance Z_2 connected in the secondary.

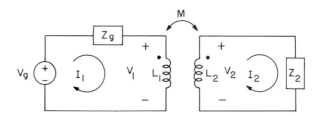

FIGURE 16.14 *Circuit for deriving impedance relationships*

Applying KVL at the primary terminals of the transformer we find

$$\begin{aligned}
V_1 &= j\omega L_1 I_1 - j\omega M I_2 \\
0 &= -j\omega M I_1 + (Z_2 + j\omega L_2)I_2
\end{aligned} \tag{16.12}$$

Eliminating I_2 from these equations we have

$$V_1 = \left[j\omega L_1 - \frac{j\omega M(j\omega M)}{Z_2 + j\omega L_2} \right] I_1$$

so that the input impedance seen at the primary terminals of the transformer is

$$\mathbf{Z}_1 = \frac{\mathbf{V}_1}{\mathbf{I}_1} = j\omega L_1 + \frac{\omega^2 M^2}{\mathbf{Z}_2 + j\omega L_2} \tag{16.13}$$

The first part, $j\omega L_1$, depends entirely on the reactance of the primary. The second part is due to the mutual coupling, and it is called the *reflected impedance*, given by

$$\mathbf{Z}_R = \frac{\omega^2 M^2}{\mathbf{Z}_2 + j\omega L_2}$$

It may be thought of as the impedance inserted into, or *reflected* into, the primary by the secondary.

The input impedance as seen by the source $\mathbf{V}_g$ is evidently

$$\mathbf{Z}_{in} = \mathbf{Z}_g + \mathbf{Z}_1$$

Also the secondary-to-primary current ratio $\mathbf{I}_2/\mathbf{I}_1$ may be found from the second equation of (16.12), and the voltage ratio $\mathbf{V}_2/\mathbf{V}_1$ may be found from

$$\frac{\mathbf{V}_2}{\mathbf{V}_1} = \mathbf{Z}_2\left(\frac{\mathbf{I}_2}{\mathbf{I}_1}\right)\left(\frac{\mathbf{I}_1}{\mathbf{V}_1}\right)$$

using (16.13). The result is

$$\frac{\mathbf{I}_2}{\mathbf{I}_1} = \frac{j\omega M}{\mathbf{Z}_2 + j\omega L_2}$$

$$\frac{\mathbf{V}_2}{\mathbf{V}_1} = \frac{j\omega M \mathbf{Z}_2}{j\omega L_1(\mathbf{Z}_2 + j\omega L_2) + \omega^2 M^2} \tag{16.14}$$

It is interesting to note that $\mathbf{Z}_R$ is independent of the dot locations on the transformer. If either dot in Fig. 16.14 is placed on the opposite terminal, the sign of the mutual term in each equation of (16.11) changes, which is equivalent to replacing M by $-M$. Since $\mathbf{Z}_R$ varies as M^2, its sign is unchanged. A second important property is illustrated by rationalizing $\mathbf{Z}_R$, which gives

$$\mathbf{Z}_R = \frac{\omega^2 M^2}{R_2^2 + (X^2 + \omega L_2)^2}[R_2 - j(X_2 + \omega L_2)]$$

where we have used the relation $\mathbf{Z}_2 = R_2 + jX_2$ for the load impedance. It is seen that the sign of the imaginary part of $\mathbf{Z}_R$ is minus. Therefore the reflected reactance is opposite that of the net reactance $X_2 + \omega L_2$ of the secondary. In particular, if X_2 is a capacitive reactance whose magnitude is less than ωL_2 or if it is an inductive reactance, then the reflected reactance is capacitive. Otherwise, the reflected reactance is either inductive or it is zero. In the latter case, X_2 must be a capacitive reactance $-1/\omega C$ with a resonant frequency $f_0 = \omega_0/2\pi = \frac{1}{2\pi\sqrt{L_2 C}}$. In this case, for $\omega = \omega_0$,

$$\mathbf{Z}_R = \frac{\omega_0^2 M^2}{R_2}$$

and the reflected impedance is purely real.

Inspecting (16.13) and (16.14), we see that if the polarity dot on either winding of Fig. 16.14 occurs on the opposite terminal, then the current and voltage ratios require a sign change, whereas the impedance relations are unaffected.

EXERCISES

16.4.1 Given: In Fig. 16.14, $\mathbf{V}_g = 100\underline{/0°}$ V, $\mathbf{Z}_g = 10\,\Omega$, $L_1 = L_2 = 0.5$ H, $M = 0.1$ H, and $\omega = 100$ rad/s. If $\mathbf{Z}_2 = 10 - (j1000/\omega)\,\Omega$, find (a) $\mathbf{Z}_{in}$, (b) $\mathbf{I}_1$, (c) $\mathbf{I}_2$, (d) $\mathbf{V}_1$, and (e) $\mathbf{V}_2$.

Ans. (a) $48.81\underline{/77.47°}\,\Omega$, (b) $2.05\underline{/-77.47°}$ A, (c) $0.50\underline{/-63.43°}$ A, (d) $97.62\underline{/11.83°}$ V, (e) $7.07\underline{/-108.43°}$ V

16.4.2 Repeat Ex. 16.4.1 if the polarity dot is on the lower terminal of the secondary.

Ans. (a) $48.81\underline{/77.47°}\,\Omega$, (b) $2.05\underline{/-77.47°}$ A, (c) $0.50\underline{/116.57°}$ A, (d) $97.62\underline{/11.83°}$ V, (e) $7.07\underline{/71.57°}$ V

16.4.3 Determine the resonant frequency in Ex. 16.4.1 for which the reflected impedance is real.   *Ans.* 7.12 Hz

16.5 THE IDEAL TRANSFORMER

An *ideal transformer* is a lossless unity-coupled transformer in which the self-inductances of the primary and secondary are infinite but their ratio is finite. Physical transformers which approximate this ideal case are the previously mentioned iron-core transformers. The primary and secondary coils are wound on a laminated iron-core structure such that nearly all of the flux links all of the turns of both coils. The reactances of the primary and secondary self-inductances are very large compared to moderate load impedances, and the coupling coefficient is nearly unity over the frequency range for which the device is designed. The ideal transformer is thus an approximate model for well-constructed iron-core transformers.

An important parameter that is necessary in describing the characteristics of an ideal transformer is the *turns ratio n*, defined by

$$n = \frac{N_2}{N_1} \tag{16.15}$$

where N_1 and N_2 are the number of turns on the primary and secondary, respectively. The flux produced in a winding of a transformer due to a current in the winding is proportional to the product of the current and the number of turns on the winding. Thus, in the primary and secondary windings,

$$\phi_{11} = \alpha N_1 i_1$$
$$\phi_{22} = \alpha N_2 i_2$$

where α is a constant of proportionality which depends on the physical properties of the transformer. Substituting these relations into (16.2) and (16.5), we see that

$$L_1 = \alpha N_1^2, \qquad L_2 = \alpha N_2^2$$

Therefore

$$\frac{L_2}{L_1} = \left(\frac{N_2}{N_1}\right)^2 = n^2 \tag{16.16}$$

In the case of unity coupling, $M = \sqrt{L_1 L_2}$, the second of (16.14) becomes

$$\frac{\mathbf{V}_2}{\mathbf{V}_1} = \frac{j\omega \mathbf{Z}_2 \sqrt{L_1 L_2}}{j\omega L_1 (\mathbf{Z}_2 + j\omega L_2) + \omega^2 L_1 L_2}$$
$$= \sqrt{\frac{L_2}{L_1}}$$
$$= n$$

For the ideal transformer, in addition to unity coupling, the inductances L_1 and L_2 tend to infinity in such a way that the ratio of (16.16) is the constant n^2. From the first equation of (16.14) we see that in this case we have

$$\frac{\mathbf{I}_2}{\mathbf{I}_1} = \lim_{L_1, L_2 \to \infty} \frac{j\omega \sqrt{L_1 L_2}}{\mathbf{Z}^2 + j\omega L_2}$$
$$= \lim_{L_1, L_2 \to \infty} \frac{j\omega \sqrt{L_1 / L_2}}{j\omega + (\mathbf{Z}_2 / L_2)}$$
$$= \lim_{L_1, L_2 \to \infty} \frac{j\omega(1/n)}{j\omega + (\mathbf{Z}_2 / L_2)}$$
$$= \frac{1}{n}$$

Thus the primary and secondary voltages and currents of an ideal transformer are related simply to the turns ratio by

$$\frac{\mathbf{V}_2}{\mathbf{V}_1} = n$$
$$\frac{\mathbf{I}_2}{\mathbf{I}_1} = \frac{1}{n} \tag{16.17}$$

Replacing n by N_2/N_1, we have

$$\frac{\mathbf{V}_2}{\mathbf{V}_1} = \frac{N_2}{N_1}$$

$$N_1\mathbf{I}_1 = N_2\mathbf{I}_2$$

Therefore the voltages are in the same ratio as the turns, and the *ampere turns* (NI) are the same for both primary and secondary.

The symbol for an ideal transformer is shown in Fig. 16.15, connected to a secon-

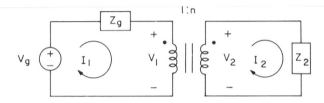

FIGURE 16.15 *Circuit containing an ideal transformer*

dary load $\mathbf{Z}_2$ and a source $\mathbf{V}_g$ with impedance $\mathbf{Z}_g$. The vertical lines are used to symbolize the iron core, and $1:n$ denotes the turns ratio. If one or the other, but not both, of the polarity dots is placed on the opposite terminal, then the ratios of (16.17) become negative.

The primary impedance $\mathbf{Z}_1$ of (16.13), in the case of the ideal transformer, is given by

$$\mathbf{Z}_1 = \frac{\mathbf{V}_1}{\mathbf{I}_1} = \frac{\mathbf{V}_2/n}{n\mathbf{I}_2} = \frac{\mathbf{V}_2/\mathbf{I}_2}{n^2}$$

or

$$\mathbf{Z}_1 = \frac{\mathbf{Z}_2}{n^2}, \qquad \frac{\mathbf{Z}_2}{\mathbf{Z}_1} = n^2 \tag{16.18}$$

where $\mathbf{Z}_2$ is the load impedance of Fig. 16.15. Thus the input impedance viewed from the terminals of the voltage source is

$$\mathbf{Z}_{\text{in}} = \mathbf{Z}_g + \mathbf{Z}_1 = \mathbf{Z}_g + \frac{\mathbf{Z}_2}{n^2} \tag{16.19}$$

The lossless property of an ideal transformer is easily demonstrated using (16.17). Taking the complex conjugate of the current ratio, we have, since n is real,

$$\mathbf{I}_1^* = n\mathbf{I}_2^*$$

from which

$$\mathbf{I}_1^* = \frac{\mathbf{V}_2\mathbf{I}_2^*}{\mathbf{V}_1}$$

or

$$\frac{\mathbf{V}_1\mathbf{I}_1^*}{2} = \frac{\mathbf{V}_2\mathbf{I}_2^*}{2}$$

Thus the complex power applied to the primary is delivered to the load $\mathbf{Z}_2$, and hence the transformer absorbs zero power.

In analyzing networks containing ideal transformers, it is often convenient to remove the transformer before performing the analysis. Let us consider, for example, replacing the transformer and load impedance $\mathbf{Z}_2$ of Fig. 16.15. Clearly, the input impedance seen by the generator $\mathbf{V}_g$ is $\mathbf{Z}_{\text{in}}$ given by (16.19), so that an equivalent circuit insofar as $\mathbf{V}_g$ is concerned is shown in Fig. 16.16. The voltages and currents can now be easily determined from the single-loop circuit.

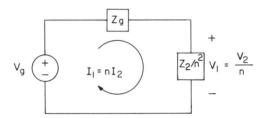

FIGURE 16.16 *Equivalent circuit for Fig. 16.15 obtained by replacing the secondary*

Let us next consider replacing the primary circuit and the transformer of Fig. 16.15 by its Thevenin equivalent. By (16.17) we have

$$\mathbf{I}_1 = n\mathbf{I}_2, \qquad \mathbf{V}_2 = n\mathbf{V}_1$$

so that for $\mathbf{V}_{\text{oc}}$, $\mathbf{I}_2 = 0$ and thus $\mathbf{I}_1 = 0$. Therefore

$$\mathbf{V}_{\text{oc}} = \mathbf{V}_2 = n\mathbf{V}_1 = n\mathbf{V}_g$$

For $\mathbf{I}_{\text{sc}}$, we have $\mathbf{V}_2 = 0$ and thus $\mathbf{V}_1 = 0$. Therefore

$$\mathbf{I}_{\text{sc}} = \mathbf{I}_2 = \frac{\mathbf{I}_1}{n} = \frac{\mathbf{V}_g}{n\mathbf{Z}_g}$$

and

$$\mathbf{Z}_{\text{th}} = \frac{\mathbf{V}_{\text{oc}}}{\mathbf{I}_{\text{sc}}} = n^2\mathbf{Z}_g$$

The resulting equivalent circuit is shown in Fig. 16.17.

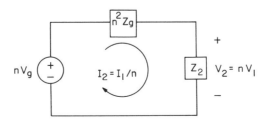

FIGURE 16.17 *Thevenin equivalent circuit of Fig. 16.15*

Inspecting the results for the Thevenin equivalent circuit, we see that each primary voltage is multiplied by n, each primary current is multiplied by $1/n$, and each primary impedance is multiplied by n^2 when we replace the primary circuit and the transformer. It may be shown that this statement holds in general whether the result is the Thevenin circuit or not. On the other hand, if we replace the secondary circuit and transformer by an equivalent circuit, as in Fig. 16.16, we simply multiply each secondary voltage, current, and impedance by $1/n$, n, and $1/n^2$, respectively. If either dot on the transformer is reversed, we simply replace n by $-n$.

In applying the above-described procedure, the student is cautioned that the technique is valid if the transformer divides the circuit into two parts. When external connections exist between the windings, the method in general cannot be used. The equivalent circuits to be discussed in the next section are often useful for networks of this type.

As an example, let us find V_2 in the circuit of Fig. 16.18(a), which contains an ideal transformer and a voltage-controlled current source. In Fig. 16.18(b) an equiva-

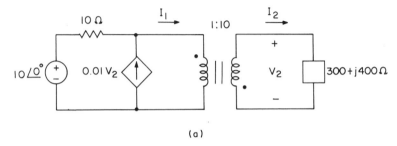

(a)

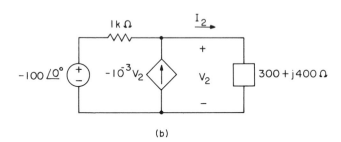

(b)

FIGURE 16.18 (a) Example circuit; (b) equivalent circuit

lent circuit for the primary circuit and the transformer is shown. It should be noted that a minus sign has been used on primary voltages and currents to account for the polarity dots shown on the transformer. The nodal equation for V_2 gives

$$\frac{V_2 + 100/0°}{10^3} + 10^{-3}V_2 + \frac{V_2}{300 + j400} = 0$$

from which

$$V_2 = 27.95/206.6° \text{ V}$$

So far we have considered only the ideal transformer in the ac steady-state case. In the general case, we see from (16.6) that

$$v_1 = L_1 \frac{di_1}{dt} + M\left(\frac{v_2}{L_2} - \frac{M}{L_2}\frac{di_1}{dt}\right)$$

$$= \left(L_1 - \frac{M^2}{L_2}\right)\frac{di_1}{dt} + \frac{M}{L_2}v_2$$

Thus since $M^2 = L_1 L_2$ and $\sqrt{L_2/L_1} = n$, we have

$$v_1 = \frac{v_2}{n}$$

Next, let us rearrange the first equation of (16.6) in the form

$$\frac{v_1}{L_1} = \frac{di_1}{dt} + \frac{M}{L_1}\frac{di_2}{dt}$$

$$= \frac{di_1}{dt} + n\frac{di_2}{dt}$$

Taking the limit as L_1 becomes infinite, we have

$$\frac{di_1}{dt} = -n\frac{di_2}{dt}$$

Integrating both sides, we have

$$i_1 = -ni_2 + C_1$$

where C_1 is a constant of integration. Since dc currents produce no time-varying magnetic flux, they do not contribute to the induced voltages or currents in the ideal transformer. Therefore if we neglect the constant C_1, then

$$i_1 = -ni_2$$

where the minus sign arises due to the direction assigned i_2 in Fig. 16.3(a). Thus the same current and voltage relationships are valid in the time domain as were found in the frequency domain if we neglect any dc currents.

EXERCISES

16.5.1   In Fig. 16.15, $\mathbf{V}_g = 100\underline{/0°}$ V, $\mathbf{Z}_g = 100\ \Omega$, and $\mathbf{Z}_2 = 90\ \text{k}\Omega$. Find n such that $\mathbf{Z}_1 = \mathbf{Z}_g$, and then find the power delivered to $\mathbf{Z}_2$.   *Ans.* 30, 12.5 W

16.5.2 Using Norton's theorem, show that the primary circuit and transformer of Fig. 16.15 is equivalent to a constant current source of V_g/nZ_g A in parallel with an impedance of n^2Z_g Ω.

16.5.3 Find V_2 in Fig. 16.18 (a) by first replacing the secondary circuit and the transformer by an equivalent circuit.

16.6 EQUIVALENT CIRCUITS

Equivalent circuits for linear transformers are easily developed by considering the equations for primary and secondary currents and voltages. In Fig. 16.19(a) we have

$$v_1 = L_1\frac{di_1}{dt} + M\frac{di_2}{dt}$$

$$v_2 = M\frac{di_1}{dt} + L_2\frac{di_2}{dt}$$

It is evident that the circuit of Fig. 16.19(b) satisfies these equations. The dependent voltage sources, however, are controlled by the time derivatives of the primary and secondary currents. In the frequency domain, these sources can be considered as current-controlled voltage sources.

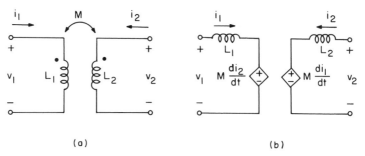

(a)   (b)

FIGURE 16.19 *Linear transformer; (b) equivalent circuit*

Let us now rearrange the equations in the form

$$v_1 = (L_1 - M)\frac{di_1}{dt} + M\left(\frac{di_1}{dt} + \frac{di_2}{dt}\right)$$

$$v_2 = M\left(\frac{di_1}{dt} + \frac{di_2}{dt}\right) + (L_2 - M)\frac{di_2}{dt}$$

These equations are satisfied by the T network of Fig. 16.20(b). Since this circuit is a three-terminal network, it is equivalent to the transformer connection of Fig. 16.20(a). If either polarity dot is changed to another terminal, we must replace M by $-M$ in the equivalent circuits.

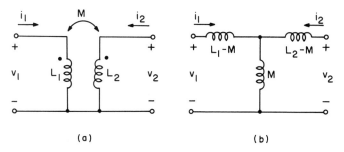

(a)   (b)

FIGURE 16.20 *(a) Linear transformer with common terminals; (b) equivalent T network*

In the case of an ideal transformer as shown in Fig. 16.21(a), the currents and voltages are given by

$$i_2 = \frac{-i_1}{n}, \qquad v_2 = nv_1$$

Clearly, the circuits of Figs. 16.21(b) and (c) satisfy these relations. If either of the dots are reversed, of course, we must replace n by $-n$ in the equivalent circuits.

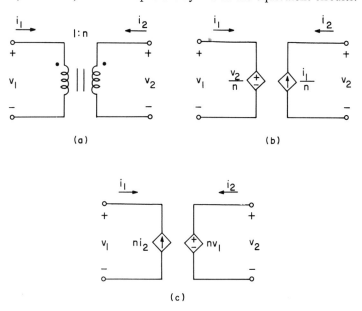

(a)   (b)

(c)

FIGURE 16.21 *(a) Ideal transformer; (b) and (c) equivalent circuits*

EXERCISE

16.6.1 Employing the T equivalent circuit for the linear transformer in Fig. 16.12, determine the phasor current I_1.   *Ans.* $6.77 / -37.75°$ A

PROBLEMS

16.1 In Fig. 16.2 (a), $N_2 = 1000$ turns and $\phi_{21} = 50$ μWb when $i_1 = 1$ A. Determine v_2 if $i_1 = 10 \cos 100t$ A.

16.2 Determine the polarity markings for the coupled circuits.

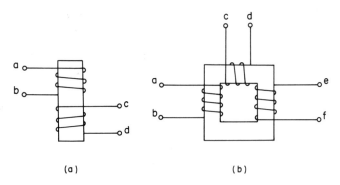

(a)   (b)

PROBLEM 16.2

16.3 In Fig. (a) of Prob. 16.2 it is found that the inductance measured between terminals a and d is 0.2 H when terminals b and c are connected. When terminals b and d are connected, the measured inductance between terminals a and c is 0.6 H. Determine the mutual inductance M.

16.4 Find v if $i = 2u(t)$ A.

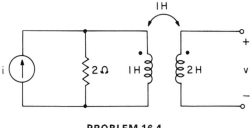

PROBLEM 16.4

16.5 Repeat Prob. 16.4 if $i(t) = 2e^{-t}u(t)$ A.

16.6 Find the ac steady-state value of v.

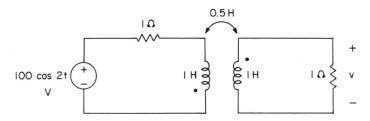

PROBLEM 16.6

16.7 Repeat Prob. 16.6 if a 1-F capacitor is connected in parallel with the 1-Ω secondary resistor.

16.8 Given that $M_{12} = M_{21}$ in (16.7), show that the integral

$$w(t) = \int_{-\infty}^{t} (p_1 + p_2)\, dt$$

yields (16.8).

16.9 Substituting (16.9) into (16.8), show that a transformer satisfies the passivity condition $w(t) \geq 0$.

16.10 Repeat Prob. 16.4 if a 3-Ω resistor is connected to the secondary terminals. (Hint: Replace the current source and 2-Ω resistor by their Thevenin equivalent.)

16.11 Repeat Prob. 16.4 if the mutual inductance is $\sqrt{2}$ H and a 3-Ω resistor is connected to the secondary terminals.

16.12 Find the steady-state value of i_2.

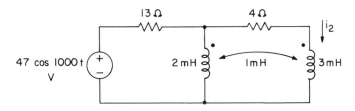

PROBLEM 16.12

16.13 Find $i(t)$ for $t > 0$, given $i(0) = 0$ and $v(0) = 4$ V.

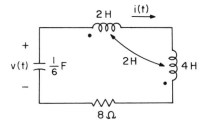

PROBLEM 16.13

16.14 Find $I_1(s)$.

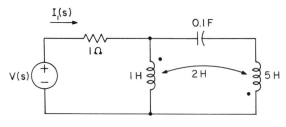

PROBLEM 16.14

16.15 Find the network function, $H(s) = V_2(s)/V_1(s)$.

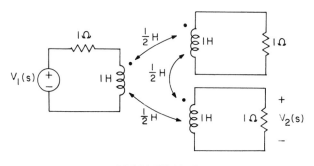

PROBLEM 16.15

16.16 Find the steady-state value of i_1.

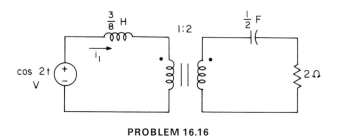

PROBLEM 16.16

16.17 Find the steady-state value of v.

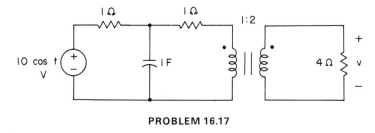

PROBLEM 16.17

16.18   Find the power delivered to the 4-Ω resistor.

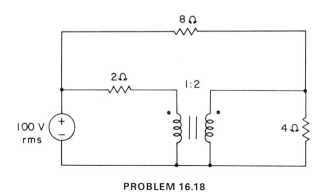

PROBLEM 16.18

16.19   In Fig. 16.20 (a), determine the z-parameters for the transformer for complex frequency s.

16.20   Repeat Prob. 16.19 for the y-parameters.

16.21   Find v using the T equivalent circuit for the transformer.

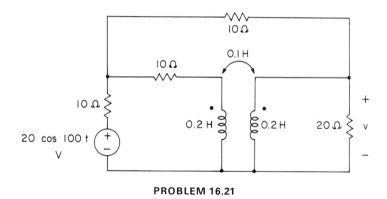

PROBLEM 16.21

17

FOURIER METHODS

Thus far, with few exceptions, we have considered excitations which were exponentials, sinusoids, or damped sinusoids, and the reader may have been impressed by the relative ease with which we can handle such functions. In particular, the concept of impedance allowed us to treat circuits with these excitations as if they were resistive circuits and find both their forced and natural responses.

However, as we all know, there are many functions that are important in engineering which are not exponentials or sinusoids. A few examples are square waves, sawtooth waves, and triangular pulses. Indeed, a function may be represented by a set of data points and have no analytical representation given at all. In this chapter we shall consider these additional functions and show how we may represent them in terms of our familiar sinusoidal functions. The techniques we shall use were first given in 1822 by the great French mathematician and physicist Jean-Baptiste Joseph Fourier, who lived from 1768 to 1830.

17.1 THE TRIGONOMETRIC FOURIER SERIES

Let us begin by considering a function $f(t)$ which is periodic of period T; that is,

$$f(t) = f(t + T)$$

As Fourier showed, if $f(t)$ satisfies a set of rather general conditions, it may be represented by the infinite series of sinusoids

$$f(t) = \frac{a_0}{2} + a_1 \cos \omega_0 t + a_2 \cos 2\omega_0 t + \ldots$$

$$+ b_1 \sin \omega_0 t + b_2 \sin 2\omega_0 t + \ldots$$

or, more compactly,

$$f(t) = \frac{a_0}{2} + \sum_{n=1}^{\infty} (a_n \cos n\omega_0 t + b_n \sin n\omega_0 t) \tag{17.1}$$

where $\omega_0 = 2\pi/T$. This series is called the *trigonometric Fourier series*, or simply the *Fourier series*, of $f(t)$. The a's and b's are called the *Fourier coefficients* and depend, of course, on $f(t)$.

The coefficients may be determined rather easily by the use of Table 12.1, given earlier in Chapter 12. Let us begin by obtaining a_0, which may be done by integrating both sides of (17.1) over a full period; that is,

$$\int_0^T f(t)\, dt = \int_0^T \frac{a_0}{2}\, dt$$

$$+ \sum_{n=1}^{\infty} \int_0^T (a_n \cos n\omega_0 t + b_n \sin n\omega_0 t)\, dt$$

Since $T = 2\pi/\omega_0$, every term in the summation is zero by entry 2 in Table 12.1 ($\alpha = 0$), and therefore we have

$$a_0 = \frac{2}{T} \int_0^T f(t)\, dt \tag{17.2}$$

Next, let us multiply (17.1) through by $\cos m\omega_0 t$, where m is an integer, and integrate. This yields

$$\int_0^T f(t) \cos m\omega_0 t\, dt = \int_0^T \frac{a_0}{2} \cos m\omega_0 t\, dt$$

$$+ \sum_{n=1}^{\infty} a_n \int_0^T \cos m\omega_0 t \cos n\omega_0 t\, dt$$

$$+ \sum_{n=1}^{\infty} b_n \int_0^T \cos m\omega_0 t \sin n\omega_0 t\, dt$$

By entries 2, 4, and 5 of Table 12.1 (for $\alpha = \beta = 0$), every term in the right member is zero except the term where $n = m$ in the first summation. This term is given by

$$a_m \int_0^T \cos^2 m\omega_0 t\, dt = \frac{\pi}{\omega_0} a_m = \frac{T}{2} a_m$$

so that

$$a_m = \frac{2}{T} \int_0^T f(t) \cos m\omega_0 t\, dt, \qquad m = 1, 2, 3, \ldots \tag{17.3}$$

Finally, multiplying (17.1) by $\sin m\omega_0 t$, integrating, and applying Table 12.1, we have

$$b_m = \frac{2}{T} \int_0^T f(t) \sin m\omega_0 t\, dt, \qquad m = 1, 2, 3, \ldots \tag{17.4}$$

We note that (17.2) is the special case, $m = 0$, of (17.3) (which is why we used $a_0/2$ instead of a_0 for the constant term). Also, as the reader may easily show, we may integrate over any interval of length T, such as t_0 to $t_0 + T$, for arbitrary t_0, and the results will be the same. Therefore we may summarize by giving the Fourier coefficients in the form

$$a_n = \frac{2}{T} \int_{t_0}^{t_0+T} f(t) \cos n\omega_0 t \, dt, \qquad n = 0, 1, 2, \ldots$$

$$b_n = \frac{2}{T} \int_{t_0}^{t_0+T} f(t) \sin n\omega_0 t \, dt, \qquad n = 1, 2, 3, \ldots$$

$$(17.5)$$

We have replaced the dummy subscript m by n to correspond to the notation of (17.1).

The term $a_n \cos n\omega_0 t + b_n \sin n\omega_0 t$ in (17.1) is sometimes called the nth *harmonic*, by analogy with music theory. The case $n = 1$ is the first harmonic, or *fundamental*, with fundamental frequency ω_0. The case $n = 2$ is the second harmonic with frequency $2\omega_0$, and so forth. The term $a_0/2$ is the constant, or dc, component.

The conditions that (17.1) is the Fourier series representing $f(t)$, where the Fourier coefficients are given by (17.5), are, as we have said, quite general and hold for almost any function we are likely to encounter in engineering. For the reader's information they are as follows:

In any finite interval $f(t)$ has at most a finite number of finite discontinuities, a finite number of maxima and minima, and

$$\int_0^T |f(t)| \, dt < \infty \qquad (17.6)$$

As an example, suppose $f(t)$ is the sawtooth wave of Fig. 17.1, given by

$$f(t) = t, \qquad -\pi < t < \pi$$

$$f(t + 2\pi) = f(t)$$

$$(17.7)$$

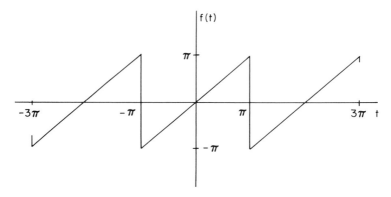

FIGURE 17.1 *Sawtooth wave*

Since $T = 2\pi$, we have $\omega_0 = 2\pi/T = 1$. If we choose $t_0 = -\pi$, then the first equation of (17.5) for $n = 0$ yields

$$a_0 = \frac{1}{\pi} \int_{-\pi}^{\pi} t \, dt = 0$$

For $n = 1, 2, 3, \ldots$, we have

$$a_n = \frac{1}{\pi} \int_{-\pi}^{\pi} t \cos nt \, dt$$

$$= \frac{1}{n^2 \pi} (\cos nt + nt \sin nt) \Big|_{-\pi}^{\pi}$$

$$= 0$$

and

$$b_n = \frac{1}{\pi} \int_{-\pi}^{\pi} t \sin nt \, dt$$

$$= \frac{1}{n^2 \pi} (\sin nt - nt \cos nt) \Big|_{-\pi}^{\pi}$$

$$= -\frac{2 \cos n\pi}{n}$$

$$= \frac{2(-1)^{n+1}}{n}$$

The case $n = 0$ had to be considered separately because of the appearance of n^2 in the denominator in the general case.

From our results the Fourier series for (17.7) is

$$f(t) = 2 \left(\frac{\sin t}{1} - \frac{\sin 2t}{2} + \frac{\sin 3t}{3} - \cdots \right) \tag{17.8}$$

The fundamental and the second, third, and fifth harmonics are sketched for one period in Fig. 17.2. If a sufficient number of terms in (17.8) are taken, the series can be made to approximate $f(t)$ very nearly. For example, the first 10 harmonics are added and the result sketched in Fig. 17.3.

As another example, suppose we have

$$f(t) = 0, \qquad -2 < t < -1$$
$$= 6, \qquad -1 < t < 1$$
$$= 0, \qquad 1 < t < 2$$
$$f(t + 4) = f(t)$$

Evidently, $T = 4$ and $\omega_0 = 2\pi/T = \pi/2$. If we take $t_0 = 0$ in (17.5), we must break each integral into three parts since on the interval from 0 to 4 $f(t)$ has 0, 6, and 0

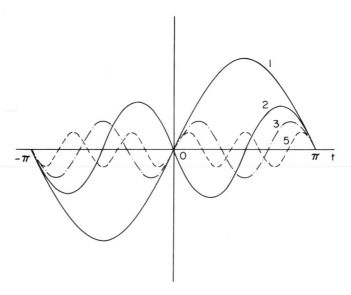

FIGURE 17.2 *Four harmonics of (17.8)*

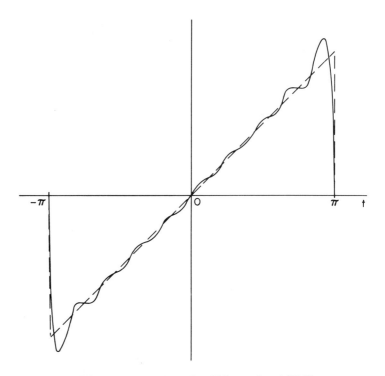

FIGURE 17.3 *Sum of the first 10 harmonics of (17.8)*

values. If $t_0 = -1$, we only have to divide the integral into two parts since $f(t) = 6$ on -1 to 1 and $f(t) = 0$ on 1 to 3. Therefore let us choose this value of t_0 and obtain

$$a_0 = \tfrac{2}{4} \int_{-1}^{1} 6 \, dt + \tfrac{2}{4} \int_{1}^{3} 0 \, dt = 6$$

Also we have

$$a_n = \frac{2}{4} \int_{-1}^{1} 6 \cos \frac{n\pi t}{2} \, dt + \frac{2}{4} \int_{1}^{3} 0 \cos \frac{n\pi t}{2} \, dt$$

$$= \frac{12}{n\pi} \sin \frac{n\pi}{2}$$

and, finally,

$$b_n = \frac{2}{4} \int_{-1}^{1} 6 \sin \frac{n\pi t}{2} \, dt + \frac{2}{4} \int_{1}^{3} 0 \sin \frac{n\pi t}{2} \, dt$$

$$= 0$$

Thus the Fourier series is

$$f(t) = 3 + \frac{12}{\pi} \left(\cos \frac{\pi t}{2} - \frac{1}{3} \cos \frac{3\pi t}{2} + \frac{1}{5} \cos \frac{5\pi t}{2} - \cdots \right)$$

Since there are no even harmonics, we may put the result in the more compact form

$$f(t) = 3 + \frac{12}{\pi} \sum_{n=1}^{\infty} \frac{(-1)^{n+1} \cos \{[(2n-1)\pi t]/2\}}{2n-1}$$

EXERCISES

17.1.1 Find the Fourier series representation of the rectangular wave

$$f(t) = 4, \qquad 0 < t < 1$$
$$= -4, \qquad 1 < t < 2$$
$$f(t+2) = f(t)$$

$$Ans. \ \frac{16}{\pi} \sum_{n=1}^{\infty} \frac{\sin (2n-1)\pi t}{2n-1}$$

17.1.2 Find the Fourier coefficients for

$$f(t) = 3, \qquad 0 < t < 1$$
$$= -1, \qquad 1 < t < 4$$
$$f(t+4) = f(t)$$

$$Ans. \ a_0 = 0, \ a_n = \frac{4}{n\pi} \sin \frac{n\pi}{2}, \ b_n = \frac{4}{n\pi} \left(1 - \cos \frac{n\pi}{2} \right), \ n = 1, 2, 3, \ldots$$

17.1.3 Find the Fourier series for

$$f(t) = 2, \qquad 0 < t < \frac{\pi}{2}$$

$$= 0, \qquad \frac{\pi}{2} < t < \pi$$

$$f(t + \pi) = f(t)$$

$$Ans.\ 1 + \frac{4}{\pi} \sum_{n=1}^{\infty} \frac{\sin 2(2n - 1)t}{2n - 1}$$

17.1.4 Find the Fourier series for

$$f(t) = 0, \qquad -1 < t < -\tfrac{1}{2}$$

$$= 4, \qquad -\tfrac{1}{2} < t < \tfrac{1}{2}$$

$$= 0, \qquad \tfrac{1}{2} < t < 1$$

$$f(t + 2) = f(t)$$

$$Ans.\ 2 + \frac{8}{\pi} \sum_{n=1}^{\infty} \frac{(-1)^{n+1} \cos (2n - 1)\pi t}{2n - 1}$$

17.2 SYMMETRY PROPERTIES

If a function has symmetry about the vertical axis or the origin, then the computation of the Fourier coefficients may be greatly facilitated. A function $f(t)$ which is symmetrical about the vertical axis is said to be an *even* function and has the property

$$f(t) = f(-t) \tag{17.9}$$

for all t. That is, we may replace t by $-t$ without changing the function. Examples of even functions are t^2 and $\cos t$, and a typical even function is shown in Fig. 17.4(a). Evidently we could fold the figure along the vertical axis, and the two portions of the graph would coincide.

A function $f(t)$ which is symmetrical about the origin is said to be an *odd* function and has the property

$$f(t) = -f(-t) \tag{17.10}$$

In other words, replacing t by $-t$ changes only the sign of the function. A typical odd function is shown in Fig. 17.4(b), and other examples are t and $\sin t$. Evidently we can fold the right half of the figure of an odd function and then rotate it about the t-axis so that it will coincide with the left half.

Now let us see how symmetry properties can help us in determining the Fourier coefficients. Evidently, in Fig. 17.4(a) we may see by inspection that for $f(t)$ to be an even function we have

$$\int_{-a}^{a} f(t)\, dt = 2 \int_{0}^{a} f(t)\, dt \tag{17.11}$$

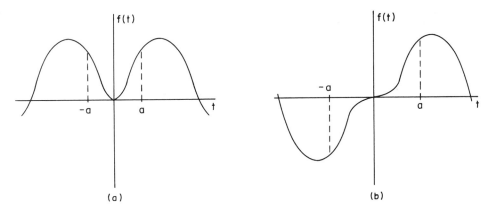

FIGURE 17.4 *(a) Even and (b) odd function*

This is true because the area from $-a$ to 0 is identical to that from 0 to a. This result may also be established analytically by writing

$$\int_{-a}^{a} f(t)\, dt = \int_{-a}^{0} f(t)\, dt + \int_{0}^{a} f(t)\, dt$$

$$= -\int_{a}^{0} f(-\tau)\, d\tau + \int_{0}^{a} f(t)\, dt$$

$$= \int_{0}^{a} f(\tau)\, dt + \int_{0}^{a} f(t)\, dt$$

$$= 2\int_{0}^{a} f(t)\, dt$$

In the case of $f(t)$ an odd function, it is clear from Fig. 17.4(b) that

$$\int_{-a}^{a} f(t)\, dt = 0 \tag{17.12}$$

This is true because the area from $-a$ to 0 is precisely the negative of that from 0 to a. This result may also be obtained analytically, as was done for even functions.

In the case of the Fourier coefficients we need to integrate the functions

$$g(t) = f(t) \cos n\omega_0 t \tag{17.13}$$

and

$$h(t) = f(t) \sin n\omega_0 t \tag{17.14}$$

If $f(t)$ is even, then

$$g(-t) = f(-t) \cos(-n\omega_0 t)$$

$$= f(t) \cos n\omega_0 t$$

$$= g(t)$$

and

$$h(-t) = f(-t) \sin (-n\omega_0 t)$$
$$= -f(t) \sin n\omega_0 t$$
$$= -h(t)$$

Thus $g(t)$ is even and $h(t)$ is odd. Therefore taking $t_0 = -T/2$ in (17.5), we have, for $f(t)$ even,

$$a_n = \frac{4}{T} \int_0^{T/2} f(t) \cos n\omega_0 t \, dt, \qquad n = 0, 1, 2, \dots$$

$$b_n = 0, \qquad\qquad\qquad\qquad n = 1, 2, 3, \dots \tag{17.15}$$

By a similar procedure we may show that if $f(t)$ is odd, then

$$a_n = 0, \qquad\qquad\qquad\qquad n = 0, 1, 2, \dots$$

$$b_n = \frac{4}{T} \int_0^{T/2} f(t) \sin n\omega_0 t \, dt \qquad n = 1, 2, 3, \dots \tag{17.16}$$

In either case one entire set of coefficients is zero, and the other set is obtained by taking twice the integral over half the period, as described by (17.15) and (17.16).

In summary, an even function has no sine terms, and an odd function has no constant or cosine terms in its Fourier series. As examples, the functions given earlier in (17.7) and Ex. 17.1.1 are odd functions, and their Fourier series have only sine terms. The function of Ex. 17.1.4 is even so that its series contains only cosine terms (including the dc case, $n = 0$). The function of Ex. 17.1.2 is neither even nor odd, and consequently both types of terms are present.

To illustrate the procedure, let us find the Fourier coefficients of

$$f(t) = \quad 4. \qquad 0 < t < 1$$
$$= -4, \qquad 1 < t < 2$$
$$f(t + 2) = f(t)$$

which was given earlier in Ex. 17.1.1. Evidently $T = 2$, and thus $\omega_0 = \pi$. The function is odd, and therefore by (17.16) we have

$$a_n = 0, \qquad n = 0, 1, 2, \dots$$

$$b_n = \tfrac{4}{2} \int_0^1 4 \sin n\pi t \, dt$$

$$= \frac{8}{n\pi}[1 - (-1)^n]$$

Therefore $b_n = 0$ for n even and $b_n = 16/n\pi$ for n odd, as indicated in Ex. 17.1.1.

It often happens that the function $f(t)$ is not periodic, and we shall consider an alternative method later for this case. However, as is often the case, we may be interested only in $f(t)$ on some finite interval $(0, T)$, in which case we can consider it as periodic of period T, and find its Fourier series as before. Of course, what we have is not the Fourier series of $f(t)$ but of its *periodic extension*. We may, however, use the result on the interval of interest and not be concerned with its periodic behavior elsewhere. Alternatively, we may consider $(0, T)$ to be *half* the period and expand $f(t)$ as either an even or an odd function using (17.15) or (17.16). The result, called a *half-range* sine or cosine series, is easier to obtain and is equally valid on the interval of interest.

EXERCISES

17.2.1   Find the Fourier series for the function of (17.7) using symmetry properties.

17.2.2   Find the Fourier series for the *half-wave rectified* sinusoid

$$f(t) = 0, \qquad -\frac{\pi}{2} < t < -\frac{\pi}{4}$$

$$= 4 \cos 2t, \qquad -\frac{\pi}{4} < t < \frac{\pi}{4}$$

$$= 0, \qquad \frac{\pi}{4} < t < \frac{\pi}{2}$$

$$f(t + \pi) = f(t)$$

This function, shown in the accompanying figure, is a sinusoid with its negative amplitudes suppressed. Thus it is the output of a circuit that changes ac to pulsating dc.

$$\text{Ans. } a_0 = \frac{8}{\pi}; a_1 = 2; a_n = \frac{8}{\pi(1 - n^2)} \cos \frac{n\pi}{2}, n = 2, 3, 4, \ldots;$$

$$b_n = 0, n = 1, 2, 3, \ldots$$

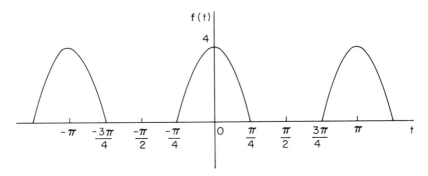

EXERCISE 17.2.2

17.2.3 Find the Fourier series for the *full-wave rectified* sinusoid

$$f(t) = |4 \sin 2t|$$

This function, shown in the accompanying figure, is a sinusoid with its negative amplitudes changed in sign.

$$Ans. \ \frac{16}{\pi}\left(\frac{1}{2} + \sum_{n=1}^{\infty} \frac{1}{1 - 4n^2} \cos 4nt\right)$$

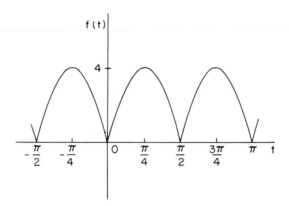

EXERCISE 17.2.3

17.3 THE EXPONENTIAL FOURIER SERIES

Replacing the sinusoidal terms in the trigonometric Fourier series by their exponential equivalents,

$$\cos n\omega_0 t = \frac{1}{2}(e^{jn\omega_0 t} + e^{-jn\omega_0 t})$$

and

$$\sin n\omega_0 t = \frac{1}{j2}(e^{jn\omega_0 t} - e^{-jn\omega_0 t})$$

(see Appendix D), and collecting terms, we have

$$f(t) = \frac{a_0}{2} + \sum_{n=1}^{\infty}\left[\left(\frac{a_n - jb_n}{2}\right)e^{jn\omega_0 t} + \left(\frac{a_n + jb_n}{2}\right)e^{-jn\omega_0 t}\right] \qquad (17.17)$$

If we define a new coefficient c_n by

$$c_n = \frac{a_n - jb_n}{2} \qquad (17.18)$$

and then substitute for a_n and b_n from (17.5), with $t_0 = -T/2$, we have

$$c_n = \frac{1}{T} \int_{-T/2}^{T/2} f(t)(\cos n\omega_0 t - j \sin n\omega_0 t)\, dt$$

or, by Euler's formula,

$$c_n = \frac{1}{T} \int_{-T/2}^{T/2} f(t)e^{-jn\omega_0 t}\, dt \qquad (17.19)$$

We observe also that c_n^* (the conjugate of c_n) is given by

$$c_n^* = \frac{a_n + jb_n}{2}$$

$$= \frac{1}{T} \int_{-T/2}^{T/2} f(t)(\cos n\omega_0 t + j \sin n\omega_0 t)\, dt$$

which is evidently c_{-n} (c_n with n replaced by $-n$). That is,

$$c_{-n} = \frac{a_n + jb_n}{2} \qquad (17.20)$$

Finally, let us observe that

$$\frac{a_0}{2} = \frac{1}{T} \int_{-T/2}^{T/2} f(t)\, dt$$

which by (17.19) is

$$\frac{a_0}{2} = c_0 \qquad (17.21)$$

Summing up, (17.18), (17.20), and (17.21) enable us to write (17.17) in the form

$$f(t) = c_0 + \sum_{n=1}^{\infty} c_n e^{jn\omega_0 t} + \sum_{n=1}^{\infty} c_{-n} e^{-jn\omega_0 t}$$

$$= \sum_{n=0}^{\infty} c_n e^{jn\omega_0 t} + \sum_{n=-1}^{-\infty} c_n e^{jn\omega_0 t}$$

We have combined c_0 with the first summation and replaced the dummy summation index n by $-n$ in the second summation. The result is more compactly written as

$$f(t) = \sum_{n=-\infty}^{\infty} c_n e^{jn\omega_0 t} \qquad (17.22)$$

where c_n is given by (17.19). This version of the Fourier series is called the *exponential Fourier series* and is generally easier to obtain because only one set of coefficients needs to be evaluated.

As an example, let us find the exponential series for the rectangular wave of Ex. 17.1.1, given by

$$f(t) = \quad 4, \qquad 0 < t < 1$$
$$= -4, \qquad 1 < t < 2$$

with $T = 2$. We have $\omega_0 = 2\pi/T = \pi$, and thus by (17.19)

$$c_n = \tfrac{1}{2} \int_{-1}^{1} f(t)e^{-jn\pi t}\, dt$$

For $n \neq 0$ this is

$$c_n = \tfrac{1}{2} \int_{-1}^{0} (-4)e^{-jn\pi t}\, dt + \tfrac{1}{2} \int_{0}^{1} 4e^{-jn\pi t}\, dt$$

$$= \frac{4}{jn\pi}[1 - (-1)^n]$$

Also we have

$$c_0 = \tfrac{1}{2} \int_{-1}^{1} f(t)\, dt$$

$$= \tfrac{1}{2} \int_{-1}^{0} 4\, dt - \tfrac{1}{2} \int_{0}^{1} 4 \cdot dt$$

$$= 0$$

Since $c_n = 0$ for n even and $c_n = 8/jn\pi$ for n odd, we may write the exponential series in the form

$$f(t) = \frac{8}{j\pi} \sum_{n=-\infty}^{\infty} \frac{1}{2n - 1} e^{j(2n-1)\pi t}$$

The reader may verify that this result is equivalent to that obtained in Ex. 17.1.1.

EXERCISES

17.3.1 Find the exponential Fourier series for

$$f(t) = 1, \qquad -1 < t < 1$$
$$= 0, \qquad 1 < |t| < 2$$
$$f(t + 4) = f(t)$$

$$Ans.\ \frac{1}{2} + \frac{1}{\pi} \sum_{\substack{n=-\infty \\ n \neq 0}}^{\infty} \frac{\sin (n\pi/2)}{n} e^{jn\pi t/2}$$

17.3.2 The function

$$Sa\ (x) = \frac{\sin x}{x}$$

called the *sampling function*, occurs frequently in communication theory. Show that if we define

$$c_0 = \lim_{n \to 0} c_n$$

then the result of Ex. 17.3.1 may be given by

$$f(t) = \frac{1}{2} \sum_{n=-\infty}^{\infty} \text{Sa}\left(\frac{n\pi}{2}\right) e^{jn\pi t/2}$$

17.3.3 Generalize the results of Ex. 17.3.2 by finding the exponential series of a train of pulses of width $\delta < T$, described by

$$f(t) = 1, \qquad -\frac{\delta}{2} < t < \frac{\delta}{2}$$

$$= 0, \qquad \frac{\delta}{2} < |t| < \frac{T}{2}$$

$$f(t + T) = f(t)$$

$$Ans. \ \frac{\delta}{T} \sum_{n=-\infty}^{\infty} \text{Sa}\left(\frac{n\pi\delta}{T}\right) e^{j2n\pi t/T}$$

17.4 RESPONSE TO PERIODIC EXCITATIONS

We are now in a position to find the forced response of a circuit to any periodic excitation that has a Fourier series. The excitation is expressed as a Fourier series, and the response to each term of the series may be found by the phasor method. Then the total response is, by superposition, the sum of the individual responses.

To illustrate, suppose we have an RL series circuit, with $R = 6\,\Omega$ and $L = 2$ H, excited by a source

$$v(t) = \quad 4 \text{ V}, \qquad 0 < t < 1 \text{ s}$$

$$= -4 \text{ V}, \qquad 1 < t < 2 \text{ s}$$

$$v(t + 2) = v(t)$$

This excitation is a square wave, and its Fourier series was found earlier in Ex. 17.1.1 to be

$$v(t) = \frac{16}{\pi} \sum_{n=1}^{\infty} \frac{\sin (2n - 1)\pi t}{2n - 1} \text{ V}$$

Therefore we may write

$$v(t) = \sum_{n=1}^{\infty} v_n$$

where

$$v_n = \frac{16}{(2n - 1)\pi} \sin (2n - 1)\pi t$$

$$= \frac{16}{(2n - 1)\pi} \cos [(2n - 1)\pi t - 90°]$$

is the $(2n - 1)$th harmonic of $v(t)$, with frequency $\omega_n = (2n - 1)\pi$ rad/s.

The phasor voltage is

$$\mathbf{V}_n = \frac{16}{(2n - 1)\pi} \, \underline{/-90°}$$

and the impedance seen at the terminals of the source at ω_n is

$$\mathbf{Z}(j\omega_n) = 6 + j2(2n - 1)\pi$$

$$= 2\sqrt{9 + \pi^2(2n - 1)^2} \, \Big/ \tan^{-1} \frac{(2n - 1)\pi}{3}$$

The phasor current is, therefore,

$$\mathbf{I}_n = |\mathbf{I}_n| \underline{/-90° - \theta_n}$$

where

$$|\mathbf{I}_n| = \frac{8}{(2n - 1)\pi\sqrt{9 + \pi^2(2n - 1)^2}} \tag{17.23}$$

and

$$\theta_n = \tan^{-1} \frac{(2n - 1)\pi}{3} \tag{17.24}$$

The time-domain current is given by

$$i_n(t) = |\mathbf{I}_n| \cos (\omega_n t - 90° - \theta_n)$$
$$= |\mathbf{I}_n| \sin (\omega_n t - \theta_n)$$

Therefore, by superposition, the forced component of the current is

$$i_f(t) = \sum_{n=1}^{\infty} |\mathbf{I}_n| \sin (\omega_n t - \theta_n) \tag{17.25}$$

We may obtain the natural response, and thus the complete response, as we have done in the past. A difficulty with the procedure is the necessity for finding the arbitrary constant in the natural response by summing an infinite series. This, of course, also presents difficulties in finding $i(t)$ for a given t. Very often, however, a reasonable number of terms can be used, with the remaining terms ignored, to approximate the series.

The input $v(t)$ in the example we have just considered contains not just one frequency, as was the case for sinusoidal excitations, but an infinite number of frequencies. This is, of course, also true of the output $i(t)$. Since the Fourier series displays the amplitude, as well as the phase, corresponding to each frequency, it is a simple matter to tell which frequencies are playing an important role in the shape of the output and which are not. This, of course, is one of the important applications of Fourier series and will be considered in more detail in the next section.

EXERCISE

17.4.1   Obtain (17.25) using the exponential Fourier series. [*Suggestion*: Recall that the series for $v(t)$ was obtained in Sec. 17.3.]

17.5 FREQUENCY SPECTRA

We may write the nth harmonic in the trigonometric Fourier series in the form of a single sinusoid,

$$a_n \cos n\omega_0 t + b_n \sin n\omega_0 t = A_n \cos (n\omega_0 t + \phi_n) \qquad (17.26)$$

with amplitude

$$A_n = \sqrt{a_n^2 + b_n^2} \qquad (17.27)$$

and phase

$$\phi_n = -\tan^{-1} \frac{b_n}{a_n} \qquad (17.28)$$

In terms of the exponential series, the coefficients are

$$c_n = \frac{a_n - jb_n}{2}$$

or, in polar form,

$$c_n = \frac{\sqrt{a_n^2 + b_n^2}}{2} \left| \underline{-\tan^{-1} \frac{b_n}{a_n}} \right.$$

Therefore we have

$$c_n = \frac{A_n}{2} \left| \underline{\phi_n} \right. \qquad (17.29)$$

   An output function $f(t)$, given in the time domain, may be represented in the frequency domain by A_n and ϕ_n, $n = 0, 1, 2, \ldots$. For example, if we wish to know if the nth harmonic is dominant or not, we need only look at A_n. This information is readily obtained from a plot of A_n versus frequency. Since the frequencies are discrete values such as 0, ω_0, $2\omega_0$, etc., the plot will be a set of lines whose lengths are proportional to A_n. Such a plot is called a *discrete amplitude spectrum* or a *line spectrum* and is analogous to the amplitude response considered in Chapter 15 for the continuous case. A similar line plot of ϕ_n is a *discrete phase spectrum* and is analogous to the phase response in the continuous case. The dc component, of course, is given by $A_0 = a_0/2$ and $\phi_0 = 0$, or if $a_0 < 0$, $A_0 = -a_0/2$ and $\phi_0 = 180°$.

As an example, the rectangular pulse of Ex. 17.3.1 had Fourier coefficients given by

$$c_0 = \tfrac{1}{2}$$

$$c_n = \frac{1}{n\pi} \sin \frac{n\pi}{2}, \qquad n = \pm 1, \pm 2, \pm 3, \ldots$$

Also, from Ex. 17.3.2, we may write this result in terms of the sampling function as

$$A_n = 2|c_n| = \left| \frac{\sin (n\pi/2)}{n\pi/2} \right|$$

$$= \left| \mathrm{Sa}\left(\frac{n\pi}{2}\right) \right|$$

Evidently, $A_n = 0$, n even, and $|2/n\pi|$, n odd, while ϕ_n takes on only the value $0°$ for all n. The amplitude spectrum is shown in Fig. 17.5 by the vertical solid lines. The dotted curve is the absolute value of the sampling function, which is also the *envelope* of the spectrum.

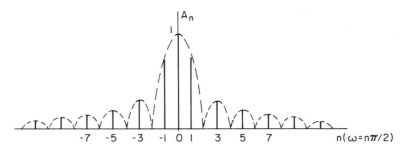

FIGURE 17.5 *Amplitude spectrum*

EXERCISES

17.5.1 Find the amplitude and phase spectra A_n and ϕ_n of the sawtooth wave of (17.7).

Ans. $A_n = |2/n|$, $n = \pm 1, \pm 2, \pm 3, \ldots$
$\phi_n = -90°$, $n = 1, 3, 5, \ldots, -2, -4, -6, \ldots$
$= 90°$, $n = 2, 4, 6, \ldots, -1, -3, -5, \ldots$

17.5.2 Find the amplitude spectrum of

$$f(t) = 1, \qquad -a < t < a$$
$$= 0, \qquad a < |t| < T/2$$
$$f(t + T) = f(t)$$

where $T/2 > a$. Note that this function is a generalization of that of Ex. 17.3.1 and that its spectrum envelope reaches zero at $1/a, 2/a, 3/a, \ldots$, which are independent of T. That is, the "width" of the envelope is independent of the pulse period.   *Ans.* $A_n = (4a/T) \, | \, \text{Sa} \, (2\pi an/T) \, |$

17.6 THE FOURIER TRANSFORM

If a function $f(t)$ is not periodic and is defined on an infinite interval, we cannot represent it by a Fourier series. It may be possible, however, to consider the function to be periodic with an *infinite* period and extend our previous results to include this case. In this section we shall consider this case in a nonrigorous way, but the results may be obtained rigorously if $f(t)$ satisfies the conditions of Sec. 17.1 with (17.6) replaced by

$$\int_{-\infty}^{\infty} |f(t)| \, dt < \infty \tag{17.30}$$

Let us begin with the exponential series for a function $f_T(t)$ defined to be $f(t)$ for $-T/2 < t < T/2$. The result is

$$f_T(t) = \sum_{n=-\infty}^{\infty} c_n e^{j2\pi nt/T} \tag{17.31}$$

where

$$c_n = \frac{1}{T} \int_{-T/2}^{T/2} f_T(x) e^{-j2\pi nx/T} \, dx \tag{17.32}$$

We have replaced ω_0 by $2\pi/T$ and are using the dummy variable x instead of t in the coefficient expression. Our intention is to let $T \to \infty$, in which case $f_T(t) \to f(t)$.

Since the limiting process requires that $\omega_0 = 2\pi/T \to 0$, for emphasis we replace $2\pi/T$ by $\Delta\omega$. Therefore substituting (17.32) into (17.31), we have

$$f_T(t) = \sum_{n=-\infty}^{\infty} \left[\frac{\Delta\omega}{2\pi} \int_{-T/2}^{T/2} f_T(x) e^{-jxn\Delta\omega} \, dx \right] e^{jtn\Delta\omega}$$

$$= \sum_{n=-\infty}^{\infty} \left[\frac{1}{2\pi} \int_{-T/2}^{T/2} f_T(x) e^{-j(x-t)n\Delta\omega} \, dx \right] \Delta\omega \tag{17.33}$$

If we define the function

$$g(\omega, t) = \frac{1}{2\pi} \int_{-T/2}^{T/2} f_T(x) e^{-j\omega(x-t)} \, dx \tag{17.34}$$

then clearly the limit of (17.33) is given by

$$f(t) = \lim_{\substack{T \to \infty \\ (\Delta\omega \to 0)}} \sum_{n=-\infty}^{\infty} g(n\Delta\omega, t)\Delta\omega \tag{17.35}$$

By the fundamental theorem of integral calculus the last result appears to be

$$f(t) = \int_{-\infty}^{\infty} g(\omega, t) \, d\omega \tag{17.36}$$

But in the limit, $f_T \to f$ and $T \to \infty$ in (17.34) so that what appears to be $g(\omega, t)$ in (17.36) is really its limit, which by (17.34) is

$$\lim_{T \to \infty} g(\omega, t) = \frac{1}{2\pi} \int_{-\infty}^{\infty} f(x) e^{-j\omega(x-t)} \, dx$$

Therefore (17.36) is actually

$$f(t) = \frac{1}{2\pi} \int_{-\infty}^{\infty} \left[\int_{-\infty}^{\infty} f(x) e^{-j\omega(x-t)} \, dx \right] d\omega \tag{17.37}$$

As we said, this is a nonrigorous development, but the results may be obtained rigorously.

Let us rewrite (17.37) in the form

$$f(t) = \frac{1}{2\pi} \int_{-\infty}^{\infty} \left[\int_{-\infty}^{\infty} f(x) e^{-j\omega x} \, dx \right] e^{j\omega t} \, d\omega \tag{17.38}$$

Now, let us define the expression in brackets to be the function

$$\mathbf{F}(j\omega) = \int_{-\infty}^{\infty} f(t) e^{-j\omega t} \, dt \tag{17.39}$$

where we have changed the dummy variable from x to t. Then (17.38) becomes

$$f(t) = \frac{1}{2\pi} \int_{-\infty}^{\infty} \mathbf{F}(j\omega) e^{j\omega t} \, d\omega \tag{17.40}$$

The function $\mathbf{F}(j\omega)$ is called the *Fourier transform* of $f(t)$, and $f(t)$ is called the *inverse Fourier transform* of $\mathbf{F}(j\omega)$. These facts are often stated symbolically as

$$\mathbf{F}(j\omega) = \mathfrak{F}[f(t)]$$
$$f(t) = \mathfrak{F}^{-1}[\mathbf{F}(j\omega)] \tag{17.41}$$

Also, (17.39) and (17.40) are collectively called the *Fourier transform pair*, the symbolism for which is

$$f(t) \longleftrightarrow \mathbf{F}(j\omega) \tag{17.42}$$

The expression in (17.37), called the *Fourier integral*, is the analogy for a non-periodic $f(t)$ to the Fourier series for a periodic $f(t)$. Equation (17.40) is, of course, another form of (17.37). The Fourier transform corresponds, in this case, to the

Fourier coefficients in the previous cases. Another description for these two analogies is to say that the Fourier transform is a *continuous* representation (ω being a continuous variable), whereas the Fourier series is a *discrete* representation ($n\omega_0$, for n an integer, being a discrete variable).

As an example, let us find the transform of

$$f(t) = e^{-at}u(t)$$

where $a > 0$. By definition we have

$$\mathscr{F}[e^{-at}u(t)] = \int_{-\infty}^{\infty} e^{-at}u(t)e^{-j\omega t}\, dt$$

$$= \int_{0}^{\infty} e^{-(a+j\omega)t}\, dt$$

or

$$\mathscr{F}[e^{-at}u(t)] = \frac{1}{-(a+j\omega)}e^{-(a+j\omega)t}\Big|_0^\infty$$

The upper limit is given by

$$\lim_{t\to\infty} e^{-at}(\cos \omega t - j \sin \omega t) = 0$$

since the expression in parentheses is bounded while the exponential goes to zero. Thus we have

$$\mathscr{F}[e^{-at}u(t)] = \frac{1}{a+j\omega} \tag{17.43}$$

or

$$e^{-at}u(t) \longleftrightarrow \frac{1}{a+j\omega}$$

As another example, let us find the transform of the single rectangular pulse

$$f(t) = A, \qquad -\frac{\delta}{2} < t < \frac{\delta}{2}$$

$$= 0, \qquad |t| > \frac{\delta}{2}$$

which is shown in Fig. 17.6. By definition we have

$$\mathbf{F}(j\omega) = \int_{-\infty}^{\infty} f(t)e^{-j\omega t}\, dt$$

$$= \int_{-\delta/2}^{\delta/2} Ae^{-j\omega t}\, dt$$

$$= \frac{2A}{\omega}\left(\frac{e^{j\omega\delta/2} - e^{-j\omega\delta/2}}{j2}\right)$$

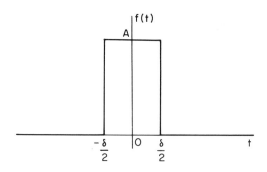

FIGURE 17.6 *Finite pulse of width δ*

or

$$\mathbf{F}(j\omega) = \frac{2A}{\omega} \sin \frac{\omega\delta}{2}$$

Alternatively, we may write

$$\mathbf{F}(j\omega) = A\delta \, \text{Sa}\left(\frac{\omega\delta}{2}\right)$$

where Sa $(\omega\delta/2)$ is the sampling function defined in Ex. 17.3.2.

EXERCISES

17.6.1   Find $\mathfrak{F}[e^{-a|t|}]$, where $a > 0$.   *Ans.* $2a/(\omega^2 + a^2)$

17.6.2   Find $\mathbf{F}(j\omega)$ if

$$
\begin{aligned}
f(t) &= t + 1, & -1 < t < 0 \\
&= -t + 1, & 0 < t < 1 \\
&= 0, & \text{elsewhere}
\end{aligned}
$$

Ans. $[2(1 - \cos \omega)]/\omega^2 = \text{Sa}^2 \, (\omega/2)$

17.6.3   Find $f(t)$ if

$$
\begin{aligned}
\mathbf{F}(j\omega) &= 1, & -1 < \omega < 1 \\
&= 0, & |\omega| > 1
\end{aligned}
$$

Ans. $(1/\pi) \, \text{Sa} \, (t)$

17.7 FOURIER TRANSFORM OPERATIONS

Since the operation of obtaining the Fourier transform is that of performing an integration, it is a *linear* operation. That is, the transform of the combination

$$v(t) = k_1 f_1(t) + k_2 f_2(t)$$

is the combination of the transforms

$$V(j\omega) = k_1 F_1(j\omega) + k_2 F_2(j\omega)$$

where V, F_1, and F_2 are, respectively, the Fourier transforms of v, f_1, and f_2 and k_1 and k_2 are constants. This concept of linearity enables us to find readily transforms of relatively complicated functions from transforms of simpler functions. For example, the transform of the function

$$f(t) = 6(e^{-2t} - e^{-3t})u(t) \tag{17.44}$$

is, by linearity and (17.43),

$$F(j\omega) = \frac{6}{2 + j\omega} - \frac{6}{3 + j\omega}$$

$$= \frac{6}{(2 + j\omega)(3 + j\omega)} \tag{17.45}$$

Another operation involving Fourier transforms that we shall find useful is that of time differentiation. Suppose we wish to find the Fourier transform of the derivative of a function $f(t)$. By definition, if

$$f(t) \longleftrightarrow F(j\omega)$$

then

$$f(t) = \frac{1}{2\pi} \int_{-\infty}^{\infty} F(j\omega)e^{j\omega t}\, d\omega$$

from which we obtain

$$\frac{df(t)}{dt} = \frac{1}{2\pi} \int_{-\infty}^{\infty} \frac{d}{dt}[F(j\omega)e^{j\omega t}]\, d\omega$$

$$= \frac{1}{2\pi} \int_{-\infty}^{\infty} [j\omega F(j\omega)]e^{j\omega t}\, d\omega$$

Therefore we have

$$\frac{df(t)}{dt} \longleftrightarrow j\omega F(j\omega) \tag{17.46}$$

That is, the transform of the derivative of f is found by simply multiplying the transform of f by $j\omega$. This result may be extended readily to the general case

$$\frac{d^n f(t)}{dt^n} \longleftrightarrow (j\omega)^n F(j\omega) \tag{17.47}$$

where $n = 0, 1, 2, \ldots$. We are assuming that the derivatives involved exist and that the interchange of operations of differentiation and integration is valid.

From the results represented by (17.47), we may show that Fourier transforms can be used to extend the concept of network functions, considered in Chapter 14, to the case of nonsinusoidal excitations. For example, taking the Fourier transform of both sides of (14.22) and making use of (17.47) and linearity, we have

$$[a_n(j\omega)^n + a_{n-1}(j\omega)^{n-1} + \ldots + a_1 j\omega + a_0]V_o(j\omega)$$
$$= [b_m(j\omega)^m + b_{m-1}(j\omega)^{m-1} + \ldots + b_1 j\omega + b_0]V_i(j\omega)$$

The functions V_o and V_i are the Fourier transforms of the output and input functions, v_o and v_i. From this result we may write

$$\frac{V_o(j\omega)}{V_i(j\omega)} = H(j\omega)$$

$$= \frac{b_m(j\omega)^m + b_{m-1}(j\omega)^{m-1} + \ldots + b_0}{a_n(j\omega)^n + a_{n-1}(j\omega)^{n-1} + \ldots + a_0} \qquad (17.48)$$

where, by (14.24), $H(j\omega)$ is the network function of Chapter 14, evaluated at $s = j\omega$.

This development indicates that the network function, defined previously as the ratio of the output phasor to the input phasor, is precisely the same as the ratio of the output transform to the input transform, and in the latter case the functions involved do not have to be sinusoids. This generalization may be used also to find the time-domain responses. For example, suppose the input $v_i(t)$ is given by

$$v_i(t) = e^{-3t}u(t)$$

and is related to the output $v_o(t)$ by the equation

$$\frac{dv_o}{dt} + 2v_o = 6v_i$$

Transforming the equation, we have

$$(j\omega + 2)V_o(j\omega) = 6V_i(j\omega)$$

or

$$H(j\omega) = \frac{V_o(j\omega)}{V_i(j\omega)} = \frac{6}{j\omega + 2}$$

Since by (17.43) we have

$$V_i(j\omega) = \frac{1}{3 + j\omega}$$

the transform of the output is

$$V_o(j\omega) = H(j\omega)V_i(j\omega)$$

$$= \frac{6}{(2 + j\omega)(3 + j\omega)}$$

By (17.45) this is

$$\mathbf{V}_o(j\omega) = 6\left(\frac{1}{2 + j\omega}\right) - 6\left(\frac{1}{3 + j\omega}\right)$$

so that by linearity, or by (17.44), we have

$$v_o(t) = 6(e^{-2t} - e^{-3t})u(t)$$

Since the network function concept implies that there is only one input, the time-domain responses obtained in this manner are responses of initially *relaxed* circuits. (No initial energy is stored.) This is, of course, a disadvantage of the Fourier transform technique.

We could develop general methods for using Fourier transforms to find time-domain solutions, but there are many common functions which do not have Fourier transforms, and the methods, as we have just noted, are limited to finding responses of initially relaxed circuits. We prefer, therefore, to consider a more general transform method, involving *Laplace transforms*, in the next chapter. As we shall see, Laplace transforms are more versatile than Fourier transforms, in the sense that more functions possess Laplace transforms. In addition, Laplace transforms may be used to obtain the complete response, taking into account initial conditions.

EXERCISES

17.7.1 If $x(t)$ is the input and $y(t)$ is the output, find the network function, using Fourier transforms, where

$$y'' + 4y' + 3y = 4x$$

Ans. $4/(3 - \omega^2 + j4\omega)$

17.7.2 If in Ex. 17.7.1

$$x = e^{-2t}u(t)$$

find $\mathcal{F}[y(t)]$.   *Ans.* $4/[(j\omega + 1)(j\omega + 2)(j\omega + 3)]$

17.7.3 Verify in Ex. 17.7.2 that

$$\mathcal{F}[y(t)] = \frac{2}{j\omega + 1} - \frac{4}{j\omega + 2} + \frac{2}{j\omega + 3}$$

and find $y(t)$ for the case $y(0) = y'(0) = 0$.

Ans. $(2e^{-t} - 4e^{-2t} + 2e^{-3t})u(t)$

PROBLEMS

17.1 Find the Fourier trigonometric series for the functions
  (a) $f(t) = 1, 0 < t < \pi$
       $= 2, \pi < t < 2\pi$
    $f(t + 2\pi) = f(t)$.
  (b) $f(t) = e^t, -\pi < t < \pi$
    $f(t + 2\pi) = f(t)$.
  (c) $f(t) = t^2, -1 < t < 1$
    $f(t + 2) = f(t)$.

17.2 Find the Fourier exponential series for the functions
  (a) $f(t) = e^{-t}, -1 < t < 1$
    $f(t + 2) = f(t)$.
  (b) $f(t) = t,   -1 < t < 1$
       $= 2 - t, 1 < t < 3$.
    $f(t + 4) = f(t)$
  (c) $f(t) = t, 0 < t < 1$
    $f(t + 1) = f(t)$.

17.3 Obtain the exponential series of the full-wave rectified sinusoid of Ex. 17.2.3 from its trigonometric series.

17.4 Obtain the trigonometric series of $f(t)$ given in Ex. 17.3.1 from its exponential series.

$$Ans. \; \frac{1}{2} + \frac{2}{\pi} \sum_{n=1}^{\infty} \frac{(-1)^{n+1}}{(2n-1)} \cos \frac{(2n-1)\pi t}{2}$$

17.5 Find the first three terms of the forced response $i(t)$ if $v_g(t)$ is the function of Prob. 17.4.

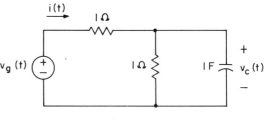

PROBLEM 17.5

17.6 Find the forced response $i(t)$ in Prob. 17.5 if $v_g(t)$ is the function of Prob. 17.1 (c).

17.7 Find the discrete amplitude and phase spectra of the functions of Prob. 17.1.

17.8 If $f(t)$ in (17.22) is the voltage across, or the current through, a 1-Ω resistor, then the instantaneous power is $f^2(t)$, and the average power is

$$P = \frac{1}{T} \int_0^T f^2(t)\, dt$$

Show that the average power is also given by

$$P = \frac{1}{T} \int_0^T f^2(t)\, dt = \sum_{n=-\infty}^{\infty} |c_n|^2$$

where c_n is the exponential Fourier coefficient. This result is known as *Parseval's theorem*. [*Suggestion*: Write $f^2(t)$ in the form

$$f^2(t) = \sum_{n=-\infty}^{\infty} \sum_{m=-\infty}^{\infty} c_n c_m e^{jn\omega_0 t} e^{jm\omega_0 t}$$

and integrate, noting that

$$\int_0^T e^{jn\omega_0 t} e^{jm\omega_0 t}\, dt = 0, \qquad m,\, n = 0,\, \pm1,\, \pm2,\, \ldots$$

unless $m = -n$.]

17.9 From Prob. 17.8 the quantity $|c_n|^2$ is the average power associated with the frequency $n\omega_0$. The plot of $|c_n|^2$ versus frequency is a line spectrum known as the discrete power spectrum of $f(t)$. Plot this function in the cases in Prob. 17.2.

17.10 Find the Fourier transforms of the following functions:
(a) $f(t) = u(t) - u(t - 1)$.
(b) $f(t) = e^{-at} \cos bt\, u(t),\ a > 0$.
(c) $f(t) = e^{-at}[u(t) - u(t - 1)]$.

17.11 Show that
(a) $f(t - \tau) \longleftrightarrow e^{-j\omega\tau} F(j\omega)$.
(b) $F(jt) \longleftrightarrow 2\pi f(-\omega)$.
(c) $f(-t) \longleftrightarrow F(-j\omega)$.
[*Suggestion*: Substitute appropriately into the relations for $F(j\omega)$ or $f(t)$.]

17.12 Find the Fourier transforms of
(a) $f(t) = \dfrac{1}{a^2 + t^2}$.
(b) $f(t) = \dfrac{1}{a + jt},\ a > 0$.
[*Suggestion*: Use the results of Prob. 17.11 (b), Ex. 17.6.1, and (17.43).]

17.13 If the transform of Prob. 17.10 (a) is a network function, find the amplitude and phase responses.

17.14 If the output is $i(t)$ and the input is $v_g(t)$ in the circuit of Prob. 17.5, find the network function.

17.15 If in the circuit of Prob. 17.5 the output is $v_C(t)$ and the input is $v_g(t)$, find (a) the network function $H(j\omega)$ and (b) $v_C(t)$ if $v_C(0) = 0$ and $v_g(t) = e^{-t}u(t)$ V. [*Suggestion*: Note that

$$\frac{1}{(x+1)(x+2)} = \frac{1}{x+1} - \frac{1}{x+2}\bigg]$$

17.16 Find the amplitude and phase response of a network having input $x(t)$ and output $y(t)$ related by

$$y'' + 5y' + 4y = 2x'$$

17.17 If $f(t)$ is the voltage across, or the current through, a 1-Ω resistor, then $f^2(t)$ is the instantaneous power delivered to the resistor. Show that if $f(t)$ is given by (17.40), then the total energy delivered to the resistor is

$$W = \int_{-\infty}^{\infty} f^2(t)\,dt = \frac{1}{2\pi} \int_{-\infty}^{\infty} |F(j\omega)|^2 d\omega$$

where $F(j\omega) = \mathfrak{F}[f(t)]$. This is the continuous version of Parseval's theorem (see Prob. 17.8). *Suggestion*: Write

$$f^2(t) = f(t)\left[\frac{1}{2\pi} \int_{-\infty}^{\infty} F(j\omega)e^{j\omega t}\,d\omega\right]$$

and integrate, changing the order of integration.)

17.18 Since in Prob. 17.17 $|F(j\omega)|^2$ is multiplied by frequency to yield energy, it is sometimes called the *energy density*. The energy in an output $f(t)$ in the band of frequencies $\omega = a$ to $\omega = b$ is then

$$W_{ab} = \frac{1}{2\pi} \int_{a}^{b} |F(j\omega)|^2 d\omega$$

Find the total energy ($a = -\infty$, $b = \infty$) in the output of an ideal low-pass filter having input $e^{-t}u(t)$ and amplitude response

$$|H(j\omega)| = 1, \qquad -1 < \omega < 1$$
$$= 0, \qquad \text{elsewhere}$$

17.19 If the output transform is

$$F(j\omega) = \frac{1}{1 + j\omega}$$

find the energy in the output in $-1 < \omega < 1$ and in $-\infty < \omega < \infty$.

17.20 Find the network function if the output is $y(t)$, the input is $x(t)$, and they are related by other variables q_1, q_2, and q_3 by

$$\frac{dq_1}{dt} = -q_1 - q_3 + x$$

$$\frac{dq_2}{dt} = -q_2 + q_3$$

$$\frac{dq_3}{dt} = \frac{1}{2}(q_1 - q_2)$$

$$y = q_2$$

18

LAPLACE TRANSFORMS

As we have seen in the previous chapter, the Fourier transform may be used to extend the concept of network functions to cases where the excitation is nonsinusoidal. It may also be used in certain applications to find the forced response to an input function which possesses a Fourier transform. With the addition of the Fourier transform to our arsenal we are able to add networks with certain nonsinusoidal, nonperiodic excitations to our set of solvable circuits. We have thus made considerable progress since we first solved resistive circuits with dc excitations in Chapter 2. Along the way we have developed differential equation techniques for circuits containing storage elements, phasor techniques for sinusoidal excitations, generalized phasors for damped sinusoids, trigonometric and exponential Fourier series methods for periodic excitations, and now Fourier transforms for certain nonperiodic excitations.

In this last chapter it remains for us to consider a most elegant procedure, the *Laplace transform* method, which can be applied to a much wider variety of excitation functions than the Fourier transform method and which gives at once both the natural and forced responses for a given set of initial conditions. In the process we shall define a special and highly useful function, known as the *impulse function*, and use its properties to obtain general time-domain solutions to any linear circuit whose excitation possesses a Laplace transform.

The Laplace transform may be obtained as a generalization of the Fourier transform and is developed in this way by many authors. However, to make this chapter self-contained and independent of the previous chapter, we shall give a completely independent treatment of the Laplace transform. A treatment based on the Fourier transform will be left to the problems to give the reader an optional alternative development.

18.1 DEFINITION

The *Laplace transform*, named for the French astronomer and mathematician Pierre Simon, Marquis de Laplace (1749–1827), is defined by

$$\mathcal{L}[f(t)] = \mathbf{F}(s) = \int_0^\infty f(t)e^{-st}\, dt \tag{18.1}$$

The function $\mathbf{F}(s)$ is the Laplace transform of $f(t)$ and is a function of the generalized frequency, $s = \sigma + j\omega$, considered earlier in Chapter 14. We might note that our definition is sometimes referred to as the *one-sided* or *unilateral* transform, as distinguished from the *two-sided* or *bilateral* transform. The latter is defined by (18.1) with the lower limit 0 replaced by $-\infty$. However, we shall not have need for this generalization since for the circuits we shall consider the function $f(t)$ is of interest only for $t \geq 0$. In general, if $f(t)$ has an infinite discontinuity at $t = 0$, we shall understand that the lower limit is 0^-.

Comparing (18.1) with the definition of the Fourier transform of the previous chapter, we see that

$$\mathcal{F}[e^{-\sigma t}f(t)u(t)] = \mathcal{L}[f(t)]$$

The presence of the factors $e^{-\sigma t}$ and $u(t)$ accounts for the greater versatility of the Laplace transform. For example, the Fourier transform of a function $f(t)$ may not exist because the condition

$$\int_{-\infty}^\infty |f(t)|\, dt < \infty \tag{18.2}$$

fails to hold. It is possible, however, that for the same function $f(t)$ a value of σ may be found so that

$$\int_{-\infty}^\infty |e^{-\sigma t}f(t)u(t)|\, dt = \int_0^\infty e^{-\sigma t}|f(t)|\, dt < \infty \tag{18.3}$$

As a general rule, if $f(t) = 0$ for $t < 0$ and $\mathcal{F}[f(t)]$ exists, then the factor $u(t)$ in (18.3) is redundant, and the factor $e^{-\sigma t}$ is not necessary for the existence of the integral. In this case, the Laplace transform with $s = j\omega$ is the Fourier transform.

As mentioned earlier, the Laplace transform can be used to take into account the initial conditions prevailing in the circuit. To gain some insight into how this is accomplished, let us consider the transform of the derivative $f'(t)$, given by (18.1) as

$$\mathcal{L}[f'(t)] = \int_0^\infty f'(t)e^{-st}\, dt$$

Integrating by parts, we have

$$\mathcal{L}[f'(t)] = f(t)e^{-st}\Big|_0^\infty + s\int_0^\infty f(t)e^{-st}\, dt$$

If $f(t)$ and s are such that the integrated part vanishes at the upper limit, as will be the case for virtually all functions we shall encounter in circuit theory, then in view of (18.1), this result becomes

$$\mathcal{L}\left[\frac{df(t)}{dt}\right] = s\mathbf{F}(s) - f(0) \tag{18.4}$$

[We are considering the case $f(t)$ to be a continuous function on $0 \le t < \infty$.] Thus the initial condition $f(0)$ is automatically built into the transform of the derivative. This was not the case for the Fourier transform, as seen in (17.46).

To illustrate the computation of a Laplace transform, let us consider

$$f(t) = e^{-at}u(t)$$

where $a > 0$. Then by (18.1) the transform is

$$\mathbf{F}(s) = \int_0^\infty e^{-at}e^{-st}\, dt$$

$$= -\frac{1}{s+a}e^{-(s+a)t}\Big|_0^\infty$$

If Re $s = \sigma > -a$, then the value at the upper limit is zero, and we have

$$\mathcal{L}[e^{-at}u(t)] = \frac{1}{s+a} \tag{18.5}$$

In general, we shall define $f(t)$ in (18.1) as the *inverse* Laplace transform of $\mathbf{F}(s)$, with the symbolism

$$f(t) = \mathcal{L}^{-1}[\mathbf{F}(s)] \tag{18.6}$$

As in the Fourier transform, we can express the inverse transform explicitly as a function of the transform. (See Prob. 18.6.) However, for our purposes we prefer to obtain the inverse transform by a simpler procedure, which we shall outline in Sec. 18.3. As an example, (18.5) may be written in the form

$$\mathcal{L}^{-1}\left[\frac{1}{s+a}\right] = e^{-at}u(t) \tag{18.7}$$

As another, simpler, but highly useful example, the transform of the step function $u(t)$ is given by

$$\mathcal{L}[u(t)] = \int_0^\infty e^{-st}\, dt = \frac{1}{s} \tag{18.8}$$

provided Re $s > 0$. An alternative statement is, of course,

$$\mathcal{L}^{-1}\left[\frac{1}{s}\right] = u(t)$$

This function further illustrates the versatility of the Laplace transform. The step function does not have a Fourier transform since for it (18.2) does not hold. (That is, there is no Fourier transform in the *conventional* sense. It is possible by the use of generalized functions to derive a Fourier transform of the step function, expressed in terms of an *impulse* function, to be defined in Sec. 18.4. However, we shall not consider transforms of this type here.)

EXERCISES

18.1.1 Find the Laplace transform of (a) $\sin kt\, u(t)$, (b) $\cos kt\, u(t)$, and (c) $(1 + 3e^{-2t})u(t)$.

$$Ans.\ \text{(a)}\ \frac{k}{s^2 + k^2},\ \text{(b)}\ \frac{s}{s^2 + k^2},\ \text{(c)}\ \frac{1}{s} + \frac{3}{s + 2}$$

18.1.2 Find $\mathcal{L}[tu(t)]$ by using (18.4).   *Ans.* $1/s^2$

18.2 SOME SPECIAL RESULTS

As was the case for the Fourier transform, the Laplace transform is an integral and is therefore a linear operation. Thus we may say that

$$\mathcal{L}[c_1 f_1(t) + c_2 f_2(t)] = c_1 \mathbf{F}_1(s) + c_2 \mathbf{F}_2(s) \tag{18.9}$$

where $\mathbf{F}_1$ and $\mathbf{F}_2$ are the Laplace transforms of f_1 and f_2 and c_1 and c_2 are constants. Thus we may obtain a variety of transforms from two known ones. For example, the function of Ex. 18.1.1(c) may be easily transformed by writing

$$\mathcal{L}[u(t) + 3e^{-2t}u(t)] = \mathcal{L}[u(t)] + 3\mathcal{L}[e^{-2t}u(t)]$$
$$= \frac{1}{s} + \frac{3}{s + 2}$$

As another example, since we have

$$\mathcal{L}[e^{jkt}u(t)] = \frac{1}{s - jk}$$

and

$$\mathcal{L}[e^{-jkt}u(t)] = \frac{1}{s + jk}$$

we may write, using (18.9),

$$\mathcal{L}[\cos kt\, u(t)] = \mathcal{L}\left[\frac{e^{jkt} + e^{-jkt}}{2} u(t)\right]$$

$$= \frac{1}{2}\mathcal{L}[e^{jkt}u(t)] + \frac{1}{2}\mathcal{L}[e^{-jkt}u(t)]$$

$$= \frac{1}{2}\left[\frac{1}{s - jk} + \frac{1}{s + jk}\right]$$

$$= \frac{s}{s^2 + k^2} \tag{18.10}$$

In a similar manner we may show that

$$\mathcal{L}[\sin kt\, u(t)] = \frac{k}{s^2 + k^2} \tag{18.11}$$

Both of these results were previously given in Ex. 18.1.1.
    As our next example, let us consider the transform

$$\mathcal{L}[e^{-at}f(t)] = \int_0^\infty e^{-at}f(t)e^{-st}\, dt$$

$$= \int_0^\infty f(t)e^{-(s+a)t}\, dt$$

By comparing this result with (18.1), we see that

$$\mathcal{L}[e^{-at}f(t)] = F(s + a) \tag{18.12}$$

where F is the transform of f. As a consequence of (18.10), (18.11), and (18.12), we may
write the useful transforms

$$\mathcal{L}[e^{-at} \cos bt\, u(t)] = \frac{s + a}{(s + a)^2 + b^2} \tag{18.13}$$

and

$$\mathcal{L}[e^{-at} \sin bt\, u(t)] = \frac{b}{(s + a)^2 + b^2} \tag{18.14}$$

as well as many others.
    To illustrate these results, suppose it is required to find $f(t)$, given that its transform
is

$$F(s) = \frac{6s}{s^2 + 2s + 5}$$

We may complete the square in the denominator, obtaining

$$F(s) = \frac{6s}{(s + 1)^2 + 2^2}$$

which we may write as

$$F(s) = \frac{6(s+1) - 6}{(s+1)^2 + 2^2}$$

$$= 6\left[\frac{s+1}{(s+1)^2 + 2^2}\right] - 3\left[\frac{2}{(s+1)^2 + 2^2}\right]$$

Finally, by (18.13) and (18.14), we have

$$f(t) = (6e^{-t}\cos 2t - 3e^{-t}\sin 2t)u(t) \qquad (18.15)$$

The justification for this last step is given in Ex. 18.2.1.

As a final example, let us find the transform of the function $f(t - \tau)u(t - \tau)$, which is the function $f(t)$ made zero for $t < 0$ and delayed by τ units, as shown in Figs. 18.1(a) and (b). For this case we have

$$\mathcal{L}[f(t-\tau)u(t-\tau)] = \int_0^\infty f(t-\tau)u(t-\tau)e^{-st}\,dt$$

$$= \int_\tau^\infty f(t-\tau)e^{-st}\,dt$$

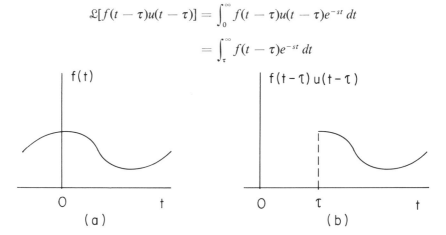

FIGURE 18.1 *(a) Function f(t) and (b) f(t) made zero for t < 0 and translated τ units to the right*

Making the change of variable $t = \tau + x$, we have

$$\mathcal{L}[f(t-\tau)u(t-\tau)] = \int_0^\infty f(x)e^{-s(\tau+x)}\,dx$$

$$= e^{-s\tau}\int_0^\infty f(x)e^{-sx}\,dx$$

or

$$\mathcal{L}[f(t-\tau)u(t-\tau)] = e^{-s\tau}F(s) \qquad (18.16)$$

where, again, F is the transform of f.

To illustrate the use of this result, let us apply it to the problem of finding $y(t)$, the inverse of a product of two transforms, given by

$$y(t) = \mathcal{L}^{-1}[G(s)F(s)]$$

By definition we may write

$$G(s) = \int_0^\infty g(\tau)e^{-s\tau} \, d\tau$$

so that

$$G(s)F(s) = \int_0^\infty g(\tau)[F(s)e^{-s\tau}] \, d\tau$$

By (18.16) this becomes

$$G(s)F(s) = \int_0^\infty g(\tau)\mathcal{L}[f(t - \tau)u(t - \tau)] \, d\tau$$

$$= \int_0^\infty g(\tau)\left[\int_0^\infty f(t - \tau)u(t - \tau)e^{-st} \, dt\right] d\tau$$

Interchanging the order of integration and noting that $u(t - \tau) = 0$ for $\tau > t$, we have

$$G(s)F(s) = \int_0^\infty e^{-st}\left[\int_0^t g(\tau)f(t - \tau) \, d\tau\right] dt$$

But by definition of the transform this is

$$G(s)F(s) = \mathcal{L}\left[\int_0^t g(\tau)f(t - \tau) \, d\tau\right]$$

Thus the function $y(t)$ is given by

$$\mathcal{L}^{-1}[G(s)F(s)] = \int_0^t g(\tau)f(t - \tau) \, d\tau \tag{18.17}$$

This result is known as the *convolution theorem*, and the integral involved is the *convolution* of g and f, symbolized by

$$g(t) * f(t) = \int_0^t g(\tau)f(t - \tau) \, d\tau \tag{18.18}$$

We shall find the convolution theorem very useful later in the chapter.

EXERCISES

18.2.1   Show that the inverse transform is also a linear operation. That is,

$$\mathcal{L}^{-1}[c_1 F_1(s) + c_2 F_2(s)] = c_1\mathcal{L}^{-1}[F_1(s)] + c_2\mathcal{L}^{-1}[F_2(s)]$$

Therefore the last step leading to (18.15) is justified. [*Suggestion*: Invert both sides of (18.9).]

18.2.2 Find the Laplace transform of (a) $te^{-2t}u(t)$, (b) $u(t-3)$, and (c) $f(t) = 1$, $0 < t < 2$, and $f(t) = 0$, elsewhere.

$$\text{Ans. (a) } \frac{1}{(s+2)^2}, \text{ (b) } \frac{e^{-3s}}{s}, \text{ (c) } \frac{1}{s}(1 - e^{-2s})$$

18.2.3 Show that the operation of convolution is commutative; that is,

$$f(t) * g(t) = g(t) * f(t)$$

[*Suggestion*: Make a change of variable, $\tau = t - x$, in (18.18).]

18.2.4 Find $\mathcal{L}^{-1}[1/(s^2 + 1)^2]$. [*Suggestion*: Use the convolution integral, where $\mathbf{F}(s) = \mathbf{G}(s) = 1/(s^2 + 1)$.]   Ans. $\frac{1}{2}(\sin t - t \cos t)u(t)$

18.3 THE INVERSE TRANSFORM

It is evident from the first two sections of this chapter that we may compile a lengthy table of functions and their Laplace transforms by repeated application of (18.1) for various cases of $f(t)$. We could then obtain a wide variety of inverse transforms by matching entries in the table. For example, in Table 18.1 we have collected most of the

TABLE 18.1 *Short table of Laplace transforms*

	$f(t)$	$F(s)$
1.	$\delta(t)$	1
2.	$u(t)$	$\dfrac{1}{s}$
3.	$e^{-at}u(t)$	$\dfrac{1}{s+a}$
4.	$\sin kt\, u(t)$	$\dfrac{k}{s^2 + k^2}$
5.	$\cos kt\, u(t)$	$\dfrac{s}{s^2 + k^2}$
6.	$e^{-at}\sin bt\, u(t)$	$\dfrac{b}{(s+a)^2 + b^2}$
7.	$e^{-at}\cos bt\, u(t)$	$\dfrac{s+a}{(s+a)^2 + b^2}$
8.	$tu(t)$	$\dfrac{1}{s^2}$

common functions $f(t)$ and their transforms $\mathbf{F}(s)$ which we are likely to encounter in circuit theory. We have derived all of these previously except the first entry, a function designated as $\delta(t)$ with transform $\mathbf{F}(s) = 1$, which we shall define and consider in Sec. 18.4.

To illustrate the use of the table, suppose we are required to find the inverse of the transform

$$F(s) = \frac{6}{s+4} + \frac{2}{s^2+9} - \frac{3}{s}$$

Since, as discussed in Ex. 18.2.1, the inverse transform is a linear operation, we have

$$f(t) = 6\mathcal{L}^{-1}\left[\frac{1}{s+4}\right] + \frac{2}{3}\mathcal{L}^{-1}\left[\frac{3}{s^2+3^2}\right] - 3\mathcal{L}^{-1}\left[\frac{1}{s}\right]$$

By entries 3, 4, and 2 of the table we have

$$f(t) = (6e^{-4t} + \tfrac{2}{3}\sin 3t - 3)u(t)$$

As another example, let us invert the transform

$$F(s) = \frac{2(s+10)}{(s+1)(s+4)} \tag{18.19}$$

There is no direct entry in Table 18.1 which we can use to obtain $f(t)$ in this case; however, we may obtain a *partial fraction expansion* of $F(s)$ and apply the table to each term in the expansion. Obtaining a partial fraction expansion is the opposite operation of getting a common denominator. That is, we ask ourselves what simple fractions add together to yield $F(s)$. Since in this case the transform is a proper fraction (the numerator is of lower degree than the denominator), the partial fractions will be proper fractions and must therefore be of the form

$$F(s) = \frac{2(s+10)}{(s+1)(s+4)} = \frac{A}{s+1} + \frac{B}{s+4}$$

The constants A and B are determined so as to make the second and third members identities in s.

The simplest means of determining A and B is to note that

$$(s+1)F(s) = \frac{2(s+10)}{s+4} = A + \frac{B(s+1)}{s+4}$$

and

$$(s+4)F(s) = \frac{2(s+10)}{s+1} = \frac{A(s+4)}{s+1} + B$$

Since these must be identities for all s, let us evaluate the first at $s = -1$, which eliminates B, and the second at $s = -4$, which eliminates A. The results are

$$A = (s+1)F(s)\Big|_{s=-1} = \frac{2(9)}{3} = 6$$

$$B = (s+4)F(s)\Big|_{s=-4} = \frac{2(6)}{-3} = -4$$

Therefore we have

$$F(s) = \frac{6}{s+1} - \frac{4}{s+4}$$

and, by Table 18.1,

$$f(t) = (6e^{-t} - 4e^{-4t})u(t)$$

In the case of a proper fraction with a denominator composed of distinct linear factors, such as (18.19), the partial fraction expansion is a sum of terms which are constants divided by the various linear factors. This is evidently the most general set of terms whose common denominator is the denominator of the transform. If the transform is not a proper fraction, then long division is required to obtain a proper fraction. We shall consider only the case of the numerator not exceeding the denominator in degree, so that at most the long division results in a constant plus a proper fraction. Higher-degree numerators require derivatives of the function $\delta(t)$ in Table 18.1.

As an example of this type, let us invert

$$F(s) = \frac{9s^3}{(s+1)(s^2+2s+10)}$$

We note that the degree of the numerator is the same as that of the denominator, so that long division is required. It is not necessary to actually perform the division since it is evident that it will result in 9 plus a proper fraction. Therefore let us write

$$\begin{aligned}
F(s) &= \frac{9s^3}{(s+1)(s^2+2s+10)} \\
&= 9 + \frac{A}{s+1} + \frac{Bs+C}{s^2+2s+10}
\end{aligned} \qquad (18.20)$$

We know that the linear denominator term has a constant numerator and that the quadratic denominator term has, in the most general proper fraction case, a linear numerator, as indicated. We could factor the quadratic into two linear terms and proceed as in the example of (18.19). However, this will result in complex coefficients, so we choose to leave the quadratic factor intact.

We may find A, as in the previous example, from

$$A = (s+1)F(s)\Big|_{s=-1} = \frac{-9}{1-2+10} = -1$$

To obtain B and C, let us multiply (18.20) by the denominator of $F(s)$, obtaining

$$9s^3 = 9(s+1)(s^2+2s+10) - (s^2+2s+10) + (Bs+C)(s+1)$$

(We have substituted in the value of A.) Since this is an identity in s, the coefficients of like powers must be equal in both members of the equation. Thus the coefficients of

s^2 are

$$0 = 18 + 9 - 1 + B$$

from which $B = -26$. The coefficients of s^0 are

$$0 = 90 - 10 + C$$

yielding $C = -80$. Therefore we have

$$\mathbf{F}(s) = 9 - \frac{1}{s+1} - \frac{26s + 80}{s^2 + 2s + 10}$$

which we rearrange as

$$\mathbf{F}(s) = 9 - \frac{1}{s+1} - 26\left[\frac{s+1}{(s+1)^2 + 3^2}\right] - 18\left[\frac{3}{(s+1)^2 + 3^2}\right]$$

Therefore, by Table 18.1, we have

$$f(t) = 9\delta(t) - e^{-t}(1 + 26\cos 3t + 18\sin 3t)u(t)$$

If the transform has repeated poles, such as

$$\mathbf{F}(s) = \frac{4}{(s+1)^2(s+2)}$$

then the procedure is slightly more complicated. Evidently in this case the partial fraction expansion is

$$\mathbf{F}(s) = \frac{4}{(s+1)^2(s+2)} = \frac{A}{(s+1)^2} + \frac{B}{s+1} + \frac{C}{s+2} \qquad (18.21)$$

since the terms on the right combine to yield the proper common denominator. There is no need to use a linear numerator for the second-degree denominator term, since

$$\frac{K_1 s + K_2}{(s+1)^2} = \frac{K_1(s+1) + K_2 - K_1}{(s+1)^2}$$

$$= \frac{K_1}{s+1} + \frac{K_2 - K_1}{(s+1)^2}$$

We may evaluate A and C from

$$A = (s+1)^2 \mathbf{F}(s)\Big|_{s=-1} = 4$$

$$C = (s+2)\mathbf{F}(s)\Big|_{s=-2} = 4$$

To obtain B, let us clear (18.21) of fractions, obtaining

$$4 = 4(s+2) + B(s+1)(s+2) + 4(s+1)^2$$

Equating coefficients of s^2 yields

$$B + 4 = 0$$

or $B = -4$. The transform is then

$$F(s) = \frac{4}{(s+1)^2} - \frac{4}{s+1} + \frac{4}{s+2}$$

The first term may be inverted by means of entry 8 in Table 18.1 and (18.12) and the other terms by entry 3. The result is

$$f(t) = 4(te^{-t} - e^{-t} + e^{-2t})u(t)$$

As observed earlier, an explicit relation for the inverse transform is given in Prob. 18.6, but for our purposes we may generally use partial fractions and Table 18.1 to obtain the inverse transforms.

EXERCISE

18.3.1   Find the inverse transform of

(a) $\dfrac{2}{(s+1)(s+2)(s+3)}$,

(b) $\dfrac{s}{(s^2+1)(s^2+4)}$, and

(c) $\dfrac{s^4 + 5s^3 + 21s^2 + 47s + 78}{(s^2+9)(s^2+2s+5)}$

Ans. (a) $(e^{-t} - 2e^{-2t} + e^{-3t})u(t)$, (b) $\frac{1}{3}(\cos t - \cos 2t)u(t)$,

(c) $\delta(t) + (\cos 3t + \sin 3t + 2e^{-t}\cos 2t)u(t)$

18.4 THE IMPULSE FUNCTION

The function $\delta(t)$, indicated in Table 1 (entry 1) to have a Laplace transform of 1, is known in the literature as the impulse function. It is not a function in the conventional sense, but it can be defined and its properties established by using generalized, or distribution, function theory. In any case, it is a very useful function in circuit theory, and we shall give a plausibility argument for its existence in this section.

Let us begin by considering the finite pulse of Fig. 18.2, centered about the origin, of width a and height $1/a$, defined by

$$f(t) = \frac{1}{a}, \qquad -\frac{a}{2} < t < \frac{a}{2}$$

$$= 0, \qquad \text{elsewhere}$$

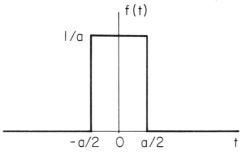

FIGURE 18.2 *Finite pulse*

The area under the pulse is 1 and remains fixed regardless of the value of a. If a is made smaller, then the base of the pulse shrinks and its height increases, maintaining the constant area of 1. In the limit as a tends toward zero, the pulse approaches an infinite pulse occurring over zero time but still associated with an area of 1. Such a function is called an *impulse function*, or a unit impulse function, to indicate its relationship to the unit area, and is symbolized by $\delta(t)$. Graphically it is represented by an arrow erected at $t = 0$, as shown in Fig. 18.3(a). A more general function is $\delta(t - \tau)$, which is an impulse occurring at $t = \tau$, as indicated in Fig. 18.3(b).

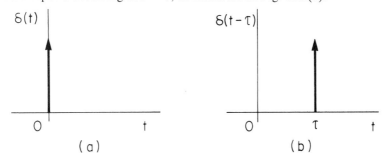

FIGURE 18.3 *Impulses (a) $\delta(t)$ and (b) $\delta(t - \tau)$*

Mathematically, the impulse function is defined by

$$\delta(t) = 0, \qquad t \neq 0$$
$$\int_{-\infty}^{\infty} \delta(t) \, dt = 1 \qquad (18.22)$$

or more generally by

$$\delta(t - \tau) = 0, \qquad t \neq \tau$$
$$\int_{-\infty}^{\infty} \delta(t - \tau) \, dt = 1 \qquad (18.23)$$

Thus the impulse is zero everywhere except at its point of discontinuity, at which it has an area of 1 concentrated. Physically, the impulse describes very well a force of very large magnitude exerted for an extremely short time and in spite of its abstract mathematical definition is quite a useful function.

An important property associated with the impulse function is the *sampling property*, described by

$$\int_a^b f(t)\delta(t - \tau)\, dt = f(\tau) \tag{18.24}$$

where $a < \tau < b$ and $f(t)$ is continuous at $t = \tau$. This property may be made plausible by noting that since $\delta(t - \tau)$ is zero except at $t = \tau$, we may write

$$\int_a^b f(t)\delta(t - \tau)\, dt = \int_{\tau-\epsilon}^{\tau+\epsilon} f(t)\delta(t - \tau)\, dt$$

Thus if ϵ is sufficiently small, since $f(t)$ is continuous at τ, then $f(t)$ is approximately $f(\tau)$ between $\tau - \epsilon$ and $\tau + \epsilon$. Thus we may factor $f(\tau)$ out of the integral, and by the nature of the impulse function we have

$$\int_a^b f(t)\delta(t - \tau)\, dt = f(\tau) \int_{\tau-\epsilon}^{\tau+\epsilon} \delta(t - \tau)\, dt$$
$$= f(\tau)$$

The sampling property is very useful in determining the Laplace transform of the impulse function, given by

$$\mathcal{L}[\delta(t)] = \int_{0^-}^\infty e^{-st}\delta(t)\, dt$$

(Recall that in the event of an infinite discontinuity of the integrand at 0, we are using 0^- as the lower limit.) By the sampling property, with $\tau = 0$, we have

$$\mathcal{L}[\delta(t)] = e^0 \int_{0^-}^\infty \delta(t)\, dt = 1 \tag{18.25}$$

This is entry 1 of Table 18.1.

Another useful result concerning the impulse function is

$$f(t)\delta(t) = f(0)\delta(t)$$

where $f(t)$ is continuous at $t = 0$. This is plausible since $\delta(t) = 0$ for $t \neq 0$. It arises, for example, when we differentiate functions that contain $u(t)$ as a factor, such as in Ex. 18.4.3.

EXERCISES

18.4.1 Evaluate the integral

$$\int_a^3 (t^2 + 3\cos 2t)\delta(t)\, dt$$

where (a) $a = -1$ and (b) $a = 1$.   *Ans.* (a) 3, (b) 0

18.4.2   Show that the pulse of Fig. 18.2 is given by

$$f(t) = \frac{u[t + (a/2)] - u[t - (a/2)]}{a}$$

so that

$$\lim_{a \to 0} f(t) = \frac{du(t)}{dt}$$

This is a plausibility argument for establishing

$$\delta(t) = \frac{du(t)}{dt}$$

18.4.3   Use the results of Ex. 18.4.2 to obtain formally

$$\frac{d}{dt}[f(t)u(t)] = f'(t)u(t) + f(t)\delta(t)$$
$$= f'(t)u(t) + f(0)\delta(t)$$

18.4.4   Use the result of Ex. 18.4.3 and (18.4), which, because of the discontinuity, we write as

$$\mathcal{L}[f'(t)] = s\mathcal{L}[f(t)] - f(0^-)$$

to derive $\mathcal{L}[\delta(t)] = 1$, where $f(t) = e^{-at}u(t)$.

18.5 APPLICATIONS TO DIFFERENTIAL EQUATIONS

The Laplace transform, like the Fourier transform, may be used to solve differential equations. However, the Laplace transform method has the advantage of yielding the complete solution with the initial conditions accounted for automatically, as we shall see in this section.

In Sec. 18.1 we have already established the result of (18.4), which we repeat here as

$$\mathcal{L}[f'(t)] = s\mathbf{F}(s) - f(0) \tag{18.26}$$

If we formally replace f by f', we have

$$\mathcal{L}[f''(t)] = s\mathcal{L}[f'(t)] - f'(0)$$

or, by (18.26),

$$\mathcal{L}[f''(t)] = s^2\mathbf{F}(s) - sf(0) - f'(0) \tag{18.27}$$

We may replace f by f' again in (18.27) to obtain $\mathcal{L}[f'''(t)]$, and so forth, with the general result being

$$\mathcal{L}[f^{(n)}(t)] = s^n\mathbf{F}(s) - s^{n-1}f(0) - s^{n-2}f'(0) - \cdots - f^{(n-1)}(0) \tag{18.28}$$

where $f^{(n)}$ is the nth derivative. The functions $f, f', \ldots, f^{(n-1)}$ are assumed to be continuous on $(0, \infty)$, and $f^{(n)}$ is continuous except possibly for a finite number of finite discontinuities. As in Ex. 18.4.4, if there is an infinite discontinuity present in any of the functions at $t = 0$, then 0 is replaced by 0^-.

Evidently, if we transform both members of a linear differential equation with constant coefficients by (18.28), we see that the result will be an algebraic equation in the Laplace transform. Also, the initial conditions will be accounted for. We may then solve for the transform and invert it to yield the time-domain answer.

As an example, let us find the solution $x(t)$, for $t > 0$, of the system of equations

$$x'' + 4x' + 3x = e^{-2t}$$

$$x(0) = 1, \qquad x'(0) = 2$$

Transforming by means of (18.26) and (18.27), we have

$$s^2 X(s) - s - 2 + 4[sX(s) - 1] + 3X(s) = \frac{1}{s+2}$$

from which

$$X(s) = \frac{s^2 + 8s + 13}{(s+1)(s+2)(s+3)}$$

The partial fraction expansion is

$$X(s) = \frac{3}{s+1} - \frac{1}{s+2} - \frac{1}{s+3}$$

so that the time-domain answer, for $t > 0$, is

$$x(t) = 3e^{-t} - e^{-2t} - e^{-3t}$$

(Since we are interested in $t > 0$, we have not needed to consider the step function involved in the transforms.)

We may also handle certain integrodifferential equations directly without differentiating to remove the integrals. Toward that end we need the transform of $g(t)$, given by

$$g(t) = \int_0^t f(t)\, dt \tag{18.29}$$

Differentiating, we have

$$g'(t) = f(t) \tag{18.30}$$

so that by (18.26), since $g(0) = 0$, we have

$$\mathcal{L}[g'(t)] = s\mathcal{L}[g(t)]$$

or

$$\mathcal{L}[g(t)] = \frac{1}{s}\mathcal{L}[g'(t)]$$

Therefore, by (18.29) and (18.30), we have

$$\mathcal{L}\left[\int_0^t f(t)\,dt\right] = \frac{F(s)}{s} \tag{18.31}$$

To illustrate the use of (18.31), suppose we have an *RLC* series circuit with $R = 2\,\Omega$, $L = 1$ H, and $C = \frac{1}{5}$ F, excited by a unit step function, $v_g(t) = u(t)$ V. Then the current $i(t)$ satisfies the equations

$$\frac{di}{dt} + 2i + 5\int_0^t i\,dt = u(t)$$

$$i(0) = 0$$

Transforming, by (18.26) and (18.31), we obtain

$$s\mathbf{I}(s) + 2\mathbf{I}(s) + \frac{5}{s}\mathbf{I}(s) = \frac{1}{s}$$

or

$$\mathbf{I}(s) = \frac{1}{s^2 + 2s + 5} = \frac{1}{2}\left[\frac{2}{(s+1)^2 + 4}\right]$$

Therefore the step response is

$$i(t) = \tfrac{1}{2}e^{-t}\sin 2t \text{ A}$$

Suppose in the general case that the input v_i and output v_o are related by the differential equation

$$a_n \frac{d^n v_o}{dt^n} + a_{n-1}\frac{d^{n-1}v_o}{dt^{n-1}} + \cdots + a_1 \frac{dv_o}{dt} + a_0 v_o$$

$$= b_m \frac{d^m v_i}{dt^m} + b_{m-1}\frac{d^{m-1}v_i}{dt^{m-1}} + \cdots + b_1 \frac{dv_i}{dt} + b_0 v_i$$

and that the initial conditions are all zero; that is,

$$v_o(0) = \frac{dv_o(0)}{dt} = \cdots = \frac{d^{n-1}v_o(0)}{dt^{n-1}} = v_i(0) = \frac{dv_i(0)}{dt} = \cdots = \frac{d^{m-1}v_i(0)}{dt^{m-1}} = 0$$

Then, transforming, we have

$$(a_n s^n + a_{n-1}s^{n-1} + \cdots + a_1 s + a_0)\mathbf{V}_o(s)$$

$$= (b_m s^m + b_{m-1}s^{m-1} + \cdots + b_1 s + b_0)\mathbf{V}_i(s)$$

or

$$\mathbf{H}(s) = \frac{\mathbf{V}_o(s)}{\mathbf{V}_i(s)} = \frac{b_m s^m + b_{m-1}s^{m-1} + \cdots + b_1 s + b_0}{a_n s^n + a_{n-1}s^{n-1} + \cdots + a_1 s + a_0} \tag{18.32}$$

Comparing this result with (17.48), obtained in the previous chapter, we see that $H(s)$ is the network function, with $j\omega$ generalized to s. Or, going back to (14.24) of Chapter 14, $H(s)$, the ratio of the Laplace transforms of the output to the input, is precisely the network function as it was originally defined. Moreover, since admittances and impedances are network functions, they are precisely the same in the Laplace transform domain as they were in the case of the generalized phasors. With Laplace transforms, however, the situation is more general. We may have any function that is Laplace transformable as the input or output, instead of being restricted to damped sinusoids.

To illustrate with the last example, if the input of the *RLC* series circuit is $v_g(t)$ and the output is $i(t)$, then we have

$$H(s) = \frac{I(s)}{V_g(s)} = \frac{1}{Z(s)}$$

or

$$H(s) = \frac{1}{sL + R + (1/sC)}$$

$$= \frac{s}{s^2 + 2s + 5} \tag{18.33}$$

Then

$$I(s) = H(s)V_g(s) = \frac{H(s)}{s}$$

$$= \frac{1}{s^2 + 2s + 5}$$

as before.

In the general case of (18.32), the output transform is given by

$$V_o(s) = H(s)V_i(s) \tag{18.34}$$

Therefore if $v_i(t) = \delta(t)$, in which case $V_i(s) = 1$, we have

$$V_o(s) = H(s)$$

In this case $v_o(t)$ is called the *impulse response* (response to an impulse) and is denoted by $h(t)$. That is,

$$v_o(t) = h(t) = \mathcal{L}^{-1}[H(s)] \tag{18.35}$$

The impulse response is thus very important since its transform is the network function itself.

Another point should be made in this section. Since, in general, $V_o(s)$ is by (18.34) a product of two transforms, we may write immediately the output $v_o(t)$ by means of the convolution theorem of (18.17). The result is

$$v_o(t) = h(t) * v_i(t) = \int_0^t h(\tau)v_i(t - \tau)\, d\tau \tag{18.36}$$

Therefore if we know the network function and the input, we may find the forced response directly.

As an example, for the *RLC* series circuit considered earlier in this section, the network function given by (18.33) has the inverse transform

$$h(t) = e^{-t}(\cos 2t - \tfrac{1}{2} \sin 2t)u(t)$$

and the input is $u(t)$. Therefore the output $i(t)$ is given by

$$i(t) = h(t) * u(t)$$

$$= \int_0^t e^{-\tau}(\cos 2\tau - \tfrac{1}{2} \sin 2\tau)u(\tau)u(t - \tau)\, d\tau$$

$$= \int_0^t e^{-\tau}(\cos 2\tau - \tfrac{1}{2} \sin 2\tau)\, d\tau$$

$$= \tfrac{1}{2}e^{-t} \sin 2t$$

as before.

The impulse response of this circuit is, of course, $h(t)$. As in the case of the step response, with an impulse input there is no initial stored energy at $t = 0^-$. Thus $h(t)$ is the complete solution.

In the case of a set of simultaneous differential equations, we have only to transform each equation, solve the resulting algebraic equations for the transforms of the unknowns, and find the time-domain solutions by inverting the transforms. For example, suppose we have the system, valid for $t > 0$,

$$2x' + y' - y = 1$$
$$x' + x + y' = 0$$
$$x(0) = 2, \qquad y(0) = -1$$

Transforming, we have

$$2s\mathbf{X}(s) - 4 + s\mathbf{Y}(s) + 1 - \mathbf{Y}(s) = \frac{1}{s}$$

$$s\mathbf{X}(s) - 2 + \mathbf{X}(s) + s\mathbf{Y}(s) + 1 = 0$$

from which

$$\mathbf{X}(s) = \frac{2s + 2}{s^2 + 1}$$

Therefore, for $t > 0$, we have

$$x(t) = 2(\cos t + \sin t)$$

In a similar manner we obtain

$$y(t) = -1 - 4 \sin t$$

EXERCISES

18.5.1 Solve for $x(t)$, for $t > 0$, by using Laplace transforms:
  (a) $x'' + 3x' + 2x = \delta(t)$.
  (b) $x'' + 4x' + 3x = 4e^{-t}$
       $x(0) = x'(0) = 4$.   *Ans.* (a) $e^{-t} - e^{-2t}$, (b) $e^{-t}(2t + 7) - 3e^{-3t}$

18.5.2 Given the network function

$$\mathbf{H}(s) = \frac{2(s + 2)}{(s + 1)(s + 3)}$$

find (a) $h(t)$, (b) the step response, and (c) the forced response to an input e^{-2t}.
              Ans. (a) $(e^{-t} + e^{-3t})u(t)$, (b) $\frac{1}{3}(4 - 3e^{-t} - e^{-3t})u(t)$,
                                                            (c) $(e^{-t} - e^{-3t})u(t)$

18.6 THE TRANSFORMED CIRCUIT

When dealing with phasors we were able to omit the steps of writing the differential equations by going directly to the phasor circuit. The circuit was analyzed by resistive circuit techniques, the phasor output was found, and from this the time-domain solution was obtained. We may perform a similar type of analysis with Laplace transforms by first obtaining a *transformed* circuit. The rest of the procedure is similar to that of the phasor method except that with Laplace transforms we are able to deal with more general functions and obtain the complete solution with the initial conditions satisfied as well.

To see how this may be done, let us consider first a resistance R, with current i_R and voltage v_R, for which

$$v_R = Ri_R$$

Transforming this equation, we have

$$\mathbf{V}_R(s) = R\mathbf{I}_R(s) \tag{18.37}$$

which is represented by the transformed resistor element of Fig. 18.4(a). Next, let us consider an inductor L for which

$$v_L = L\frac{di_L}{dt}$$

Transforming, we have

$$\mathbf{V}_L(s) = sL\mathbf{I}_L(s) - Li_L(0) \tag{18.38}$$

This result is represented by the transformed inductor element of Fig. 18.4(b). The initial condition is taken into account by the included voltage source. In the case of a

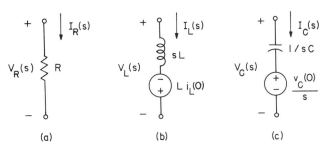

FIGURE 18.4 *Transformed circuit elements*

capacitor C we have

$$v_C = \frac{1}{C} \int_0^t i_C \, dt + v_C(0)$$

Transforming, we obtain

$$\mathbf{V}_C(s) = \frac{1}{sC}\mathbf{I}_C(s) + \frac{1}{s}v_C(0) \qquad (18.39)$$

since for $t > 0$ the transform of a constant K is K/s. The last result is represented by the transformed capacitor of Fig. 18.4(c). The initial condition is provided for by the included voltage source.

If we have dependent sources, we may transform them in the same manner. For example, if we have a controlled voltage source defined by

$$v_1 = Kv_2$$

then in the Laplace transform notation we have

$$\mathbf{V}_1(s) = K\mathbf{V}_2(s)$$

which in the transformed circuit is a source controlled by a transformed variable.

We may write the time-domain equations of a circuit and solve them by Laplace transform methods, or we may replace the circuit by the transformed circuit. In this case the passive elements are replaced by transformed elements such as those of Fig. 18.4, the controlled sources are appropriately transformed, and the independent sources are labeled with the transforms of their time-domain values. Then we may solve the transformed circuit using resistive circuit methods.

As an example, suppose we require $i(t)$, for $t > 0$, in Fig. 18.5(a), given that $i(0) = 4$ A and $v(0) = 8$ V. The transformed network is shown in Fig. 18.5(b), from which we have

$$\mathbf{I}(s) = \frac{[2/(s+3)] + 4 - (8/s)}{3 + s + (2/s)} \qquad (18.40)$$

This may be written

$$\mathbf{I}(s) = -\frac{13}{s+1} + \frac{20}{s+2} - \frac{3}{s+3}$$

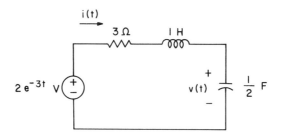

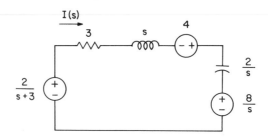

FIGURE 18.5 *(a) Circuit and (b) its transformed counterpart*

so that, for $t > 0$,

$$i(t) = -13e^{-t} + 20e^{-2t} - 3e^{-3t}$$

As a check, the time-domain equations for Fig. 18.5(a) are

$$\frac{di}{dt} + 3i + 2\int_0^t i\, di + 8 = 2e^{-3t}$$

$$i(0) = 4$$

Transforming, we have

$$s\mathbf{I}(s) - 4 + 3\mathbf{I}(s) + \frac{2}{s}\mathbf{I}(s) + \frac{8}{s} = \frac{2}{s+3}$$

from which $\mathbf{I}(s)$ is given by (18.40).

If we are interested in nodal analysis, we need to solve (18.37), (18.38), and (18.39) explicitly for the currents. The results are

$$\mathbf{I}_R(s) = \frac{\mathbf{V}_R(s)}{R}$$

$$\mathbf{I}_L(s) = \frac{1}{sL}\mathbf{V}_L(s) + \frac{1}{s}i_L(0)$$

$$\mathbf{I}_C(s) = sC\mathbf{V}_C(s) - Cv_C(0)$$

As the reader may verify, the transformed elements describing these equations are shown in Fig. 18.6. The labels on the passive elements are impedances, as in Fig. 18.4.

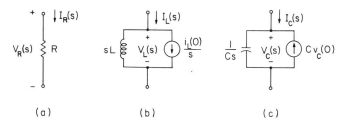

FIGURE 18.6 *Transformed elements useful for nodal analysis*

All the network theorems that apply to resistive and phasor circuits apply to the transformed circuit. An added advantage in the latter case is that initial conditions may be incorporated. As an example, the reader may verify that the elements of Fig. 18.6 are the Norton equivalents of those of Fig. 18.4.

EXERCISES

18.6.1 Solve Prob. 14.19 using the transformed circuit method if $v(0) = 10$ V.

Ans. $10(2 - e^{-t}) + \frac{1}{2}e^{-t} \sin 2t$ V

18.6.2 Find the voltage $v(t)$, for $t > 0$, across the terminals of the RLC parallel circuit, excited by a current source $i_g(t) = e^{-2t}$ A, using the transformed circuit method. The passive elements are $R = 4\,\Omega$, $L = 5$ H, and $C = \frac{1}{20}$ F; the initial inductor current is $i_L(0) = 1$ A; and $v(0) = 2$ V. (The current i_g is directed into the positive terminal, and the current i_L is directed out of the positive terminal of v.)   *Ans.* $-14e^{-t} + 20e^{-2t} - 4e^{-4t}$ V

PROBLEMS

18.1 Find the Laplace transform of the functions
(a) $u(t) - u(t - 1)$.
(b) $e^{-2t}[u(t) - u(t - 1)]$.
(c) $\cosh kt\, u(t)$.
(d) $\sinh kt\, u(t)$.

18.2 Find the inverse Laplace transforms of
(a) $\dfrac{s}{(s + a)(s + b)}$, $b \neq a$.

(b) $\dfrac{s+3}{(s+1)(s+2)}$.

(c) $\dfrac{s+1}{(s+2)^2}$.

(d) $\dfrac{se^{-2s}}{(s+1)(s+2)}$.

18.3 Show that if $f(t+T)=f(t)$, then

$$\mathcal{L}[f(t)] = \frac{\displaystyle\int_0^T e^{-st}f(t)\,dt}{1-e^{-sT}}$$

[*Suggestion*: Write the transform as an infinite series of integrals over $(0, T)$, $(T, 2T)$, . . . , and sum.]

18.4 Use the results of Prob. 18.3 to obtain the transforms of the functions

    (a) $f(t) = 1, \quad 0 < t < 1$
               $= 0, \quad 1 < t < 2$
       $f(t+2) = f(t)$.
    (b) $f(t) = |\sin t|$.

18.5 Note that if $p(t)$ is a pulse of some shape of finite duration, occurring on $0 < t < T$ [$p(t) = 0$ elsewhere], and that if $f(t)$ is defined by

$$f(t) = p(t), \qquad 0 < t < T$$
$$f(t+T) = f(t)$$

then the result of Prob. 18.3 is

$$\mathcal{L}[f(t)] = \frac{\mathcal{L}[p(t)]}{1-e^{-sT}}$$

Use this fact to find the inverse transform of

$$F(s) = \frac{(1-e^{-s})^2}{s(1-e^{-2s})}$$

18.6 In the Fourier integral (17.37), replace $f(t)$ by $e^{-\sigma t}f(t)$, and show that if $f(t) = 0$, for $t < 0$, then

$$f(t) = \frac{1}{2\pi} \int_{-\infty}^{\infty} e^{st} \left[\int_0^{\infty} f(x)e^{-sx}\,dx \right] d\omega$$

In the second integral, replace ω by $(s-\sigma)/j$ and obtain

$$f(t) = \frac{1}{2\pi j} \int_{\sigma-j\infty}^{\sigma+j\infty} e^{st}F(s)\,ds$$

This is the inverse Laplace transform analogous to (17.40) in the Fourier transform case and is valid if σ is sufficiently large to ensure that (18.3) holds.

18.7 Find $f(t) * g(t)$, where $f(t) = tu(t)$ and $g(t) = e^{-t}u(t)$.

18.8 Solve Prob. 18.7 by finding $F(s)$ and $G(s)$ and using the convolution theorem.

18.9 Using Laplace transforms, solve the following for $t > 0$:

(a) $x'' + x = 0$, $x(0) = -1$, $x'(0) = 1$.

(b) $x'' + 2x' + 2x = 0$, $x(0) = 0$, $x'(0) = 1$.

(c) $x''' - 2x'' + 2x' = 0$, $x(0) = x'(0) = 1$, $x''(0) = 2$.

(d) $x'' + 4x' + 3x = 4 \sin t + 8 \cos t$, $x(0) = 3$, $x'(0) = -1$.

(e) $x'' + 4x' + 3x = 4e^{-3t}$, $x(0) = x'(0) = 0$.

(f) $x'' + 4x' + 3x = 4e^{-t} + 8e^{-3t}$, $x(0) = x'(0) = 0$.

18.10 Show by means of the convolution theorem that

$$\mathcal{L}^{-1}\left[\frac{s}{(s^2 + a)^2}\right] = \frac{t}{2a} \sin at\, u(t)$$

Use this result to solve, for $t > 0$,

$$x'' + 9x = 6 \cos 3t, \qquad x(0) = x'(0) = 0$$

18.11 If $x(t)$ is the input and $y(t)$ is the output, find the step response $r(t)$ and the impulse response $h(t)$ for the following:

(a) $y'' + 6y' + 5y = 20x$.

(b) $y'' + 4y' + 13y = 13x$.

18.12 In terms of the network function $H(s)$, the impulse response $h(t)$ and the step response $r(t)$ are given by

$$h(t) = \mathcal{L}^{-1}[H(s)], \qquad r(t) = \mathcal{L}^{-1}\left[\frac{H(s)}{s}\right]$$

so that, by (18.31),

$$r(t) = \int_{0^-}^{t} h(t)\, dt$$

Therefore

$$h(t) = \frac{dr(t)}{dt}$$

Verify the last result for the examples of Prob. 18.11.

18.13 Solve for x, for $t > 0$, if

$$x'(t) + x(t) + \int_{0}^{t} x(\tau)e^{\tau - t}\, d\tau = 0, \qquad x(0) = 1$$

18.14 If a network has an impulse response

$$h(t) = te^{-2t}u(t)$$

find the forced response to an input

$$f(t) = e^{-2t} \cos t\, u(t)$$

18.15 Solve for x and y, valid for $t > 0$:

$$2x' + 4x + y' + 7y = 0$$
$$x' + x + y' + 3y = \delta(t)$$

18.16 Find the network function and the impulse response if the output is $i(t)$.

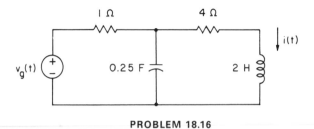

PROBLEM 18.16

18.17 Find the response of a circuit with the network function

$$H(s) = \frac{1}{s+1}$$

and no initial stored energy to the input $f(t)$ of Prob. 18.4(a). *Suggestion:* Find the inverse transform involved by the procedure of Prob. 18.5, but note that if $p(t) \neq 0$, for $t > T$, then the inverse is not periodic but is

$$p(t) + p(t-T) + p(t-2T) + \ldots$$

18.18 Solve Prob. 9.9 using transformed circuits.

18.19 Solve Prob. 9.21 using a transformed circuit.

18.20 Find the network function of Prob. 9.22, and show that the circuit is a band-pass filter. Find also ω_0 and Q.

18.21 With the presence of impulse currents or voltages in a circuit it is possible, theoretically, to instantaneously change inductor currents and capacitor voltages. Demonstrate this by finding $v(t)$ and $i(t)$ for $-\infty < t < \infty$ in the given circuit. The switch is closed at $t = 0$ and $v(0^-) = 0$.

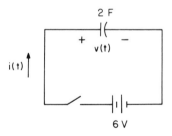

PROBLEM 18.21

18.22 The impulse response of a circuit is

$$h(t) = \sqrt{2}\, e^{-t/\sqrt{2}} \sin \frac{t}{\sqrt{2}}\, u(t)$$

Find the network function and the amplitude response, and thus show that the circuit is a second-order Butterworth low-pass filter with $\omega_c = 1$ rad/s.

A

DETERMINANTS AND CRAMER'S RULE

The solution of simultaneous equations, such as those often encountered in circuit theory, may be obtained relatively easily, in many cases, by the use of *determinants*. We define a determinant as a square array of numbers having a numerical value, such as

$$\Delta = \begin{vmatrix} a_{11} & a_{12} \\ a_{21} & a_{22} \end{vmatrix} \tag{A.1}$$

In this case the determinant is a 2×2 array, with two *rows* and two *columns* and a value Δ defined to be

$$\Delta = a_{11}a_{22} - a_{12}a_{21} \tag{A.2}$$

In the second-order, or 2×2, case of (A.1), the method of obtaining Δ in (A.2) may be thought of as a *diagonal rule*. That is,

$$\Delta = \begin{vmatrix} a_{11} & a_{12} \\ a_{21} & a_{22} \end{vmatrix} = a_{11}a_{22} - a_{12}a_{21} \tag{A.3}$$

or Δ is a difference of the product $a_{11}a_{22}$ of elements down the diagonal to the right and the product $a_{12}a_{21}$ of elements down the diagonal to the left.

As an example, consider

$$\Delta = \begin{vmatrix} 1 & 2 \\ -3 & 4 \end{vmatrix}$$

which is given by

$$\Delta = (1)(4) - (2)(-3) = 10$$

A third-order, or 3×3, determinant, such as

$$\Delta = \begin{vmatrix} a_{11} & a_{12} & a_{13} \\ a_{21} & a_{22} & a_{23} \\ a_{31} & a_{32} & a_{33} \end{vmatrix} \tag{A.4}$$

has three rows and three columns. It may also be evaluated by a diagonal rule, given by

$$\Delta = \begin{vmatrix} a_{11} & a_{12} & a_{13} \\ a_{21} & a_{22} & a_{23} \\ a_{31} & a_{32} & a_{33} \end{vmatrix}$$

$$= (a_{11}a_{22}a_{33} + a_{12}a_{23}a_{31} + a_{13}a_{32}a_{21})$$
$$- (a_{13}a_{22}a_{31} + a_{23}a_{32}a_{11} + a_{33}a_{21}a_{12}) \tag{A.5}$$

Thus the value of the determinant is a difference of products of elements down the diagonals to the right and products of elements down the diagonals to the left.

An example of a third-order determinant and its evaluation is given by

$$\Delta = \begin{vmatrix} 1 & 1 & 1 \\ 2 & -1 & 1 \\ -1 & 1 & 2 \end{vmatrix}$$

$$= [(1)(-1)(2) + (1)(1)(-1) + (1)(1)(2)]$$
$$- [(1)(-1)(-1) + (1)(1)(1) + (2)(2)(1)]$$
$$= -7 \tag{A.6}$$

A general definition of determinants may be given and used to derive a number of evaluation procedures. This is the technique usually given in elementary algebra books. However, for our purposes we shall use the diagonal rules that we have considered for second- and third-order determinants and evaluate higher-order determinants by the method of expansion by *minors*, or *cofactors*.

The *minor* A_{ij} of the element a_{ij} in the *i*th row and *j*th column of a determinant is the determinant left after the *i*th row and the *j*th column are removed. For example, in (A.6) the minor A_{21} of the element $a_{21} = 2$ (second row, first column) is

$$A_{21} = \begin{vmatrix} 1 & 1 \\ 1 & 2 \end{vmatrix} = 2 - 1 = 1$$

The *cofactor* C_{ij} of the element a_{ij} is given by

$$C_{ij} = (-1)^{i+j}A_{ij} \tag{A.7}$$

In other words, the cofactor is the *signed* minor, the minor multiplied by ± 1 with the sign depending on whether the sum of the row number and column number is even or odd.

The value of a determinant is the sum of products of the elements in any row or column and their cofactors. For example, let us *expand* the determinant of (A.4) by cofactors of the first row. The result is

$$\Delta = a_{11}C_{11} + a_{12}C_{12} + a_{13}C_{13}$$

or, by (A.7),

$$\Delta = a_{11}(-1)^{1+1}A_{11} + a_{12}(-1)^{1+2}A_{12} + a_{13}(-1)^{1+3}A_{13}$$
$$= a_{11}A_{11} - a_{12}A_{12} + a_{13}A_{13}$$

Writing out the minors explicitly, we have

$$\Delta = a_{11}\begin{vmatrix} a_{22} & a_{23} \\ a_{32} & a_{33} \end{vmatrix} - a_{12}\begin{vmatrix} a_{21} & a_{23} \\ a_{31} & a_{33} \end{vmatrix} + a_{13}\begin{vmatrix} a_{21} & a_{22} \\ a_{31} & a_{32} \end{vmatrix}$$

By the diagonal rule this may be written

$$\Delta = a_{11}(a_{22}a_{33} - a_{23}a_{32}) - a_{12}(a_{21}a_{33} - a_{23}a_{31}) + a_{13}(a_{21}a_{32} - a_{22}a_{31})$$

which may be simplified to (A.5)

To illustrate expansion by minors, let us evaluate the determinant of (A.6) by applying the technique to the third column. We have

$$\Delta = 1(-1)^{1+3}\begin{vmatrix} 2 & -1 \\ -1 & 1 \end{vmatrix} + 1(-1)^{2+3}\begin{vmatrix} 1 & 1 \\ -1 & 1 \end{vmatrix} + 2(-1)^{3+3}\begin{vmatrix} 1 & 1 \\ 2 & -1 \end{vmatrix}$$
$$= (2 - 1) - (1 + 1) + 2(-1 - 2) = -7$$

A method of obtaining solutions of simultaneous equations by determinants, referred to earlier, is called *Cramer's rule*. We shall illustrate the method by applying it to two systems of equations, a second-order and a third-order system. The examples will be given in such a way that the generalization to higher-order systems should be evident. The second-order system that we shall consider is the set of equations

$$\begin{aligned} x_1 - 2x_2 &= 5 \\ 6x_1 + x_2 &= 4 \end{aligned} \tag{A.8}$$

We define the determinant of the system as the determinant Δ whose first column is the coefficients of x_1, whose second column is the coefficients of x_2, etc. In the case of the system of (A.8) we have

$$\Delta = \begin{vmatrix} 1 & -2 \\ 6 & 1 \end{vmatrix} = 13 \tag{A.9}$$

's rule states that

$$x_1 = \frac{\Delta_1}{\Delta}$$

$$x_2 = \frac{\Delta_2}{\Delta} \tag{A.10}$$

etc.

where Δ_1 is Δ with its first column replaced by the constants in the right members of the system of equations, Δ_2 is Δ with the second column so replaced, etc. In the example of (A.8), the right members of the equations are 5 and 4 so that

$$\Delta_1 = \begin{vmatrix} 5 & -2 \\ 4 & 1 \end{vmatrix} = 13$$

$$\Delta_2 = \begin{vmatrix} 1 & 5 \\ 6 & 4 \end{vmatrix} = -26$$

By Cramer's rule the solution of (A.8) is given by

$$x_1 = \frac{\Delta_1}{\Delta} = \frac{13}{13} = 1$$

$$x_2 = \frac{\Delta_2}{\Delta} = \frac{-26}{13} = -2$$

As an example of a third-order system, let us consider the equations

$$\begin{aligned} x_1 + x_2 + x_3 &= 6 \\ 2x_1 - x_2 + x_3 &= 3 \\ -x_1 + x_2 + 2x_3 &= 7 \end{aligned} \tag{A.11}$$

The solution, by (A.10), is

$$x_1 = \frac{\begin{vmatrix} 6 & 1 & 1 \\ 3 & -1 & 1 \\ 7 & 1 & 2 \end{vmatrix}}{\begin{vmatrix} 1 & 1 & 1 \\ 2 & -1 & 1 \\ -1 & 1 & 2 \end{vmatrix}} = \frac{-7}{-7} = 1$$

$$x_2 = \frac{\begin{vmatrix} 1 & 6 & 1 \\ 2 & 3 & 1 \\ -1 & 7 & 2 \end{vmatrix}}{-7} = \frac{-14}{-7} = 2$$

and

$$x_3 = \frac{\begin{vmatrix} 1 & 1 & 6 \\ 2 & -1 & 3 \\ -1 & 1 & 7 \end{vmatrix}}{-7} = \frac{-21}{-7} = 3$$

APPENDIX

B

GAUSSIAN ELIMINATION

Simultaneous equations, such as (A.8) in Appendix A, may be solved by a method of successively eliminating one unknown at a time, in a certain systematic manner. One such procedure, known as *Gaussian elimination*, will be given here for the special case of (A.11), repeated as

$$x_1 + x_2 + x_3 = 6$$
$$2x_1 - x_2 + x_3 = 3 \tag{B.1}$$
$$-x_1 + x_2 + 2x_3 = 7$$

It will be clear from the procedure how to generalize it to any system.

We may systematically eliminate x_1 from all the equations of (B.1) except the first equation by subtracting twice the first equation from the second and adding the first equation to the third equation. We shall then have the first equation and the two new equations neither of which contains x_1. The result is

$$x_1 + x_2 + x_3 = 6$$
$$-3x_2 - x_3 = -9 \tag{B.2}$$
$$2x_2 + 3x_3 = 13$$

Let us now divide the second equation by -3, resulting in

$$x_1 + x_2 + x_3 = 6$$
$$x_2 + \tfrac{1}{3}x_3 = 3 \tag{B.3}$$
$$2x_2 + 3x_3 = 13$$

We next eliminate x_2 from all the equations except the first two. In the simple case of (B.3), this means eliminating x_2 from the third equation. This may be done by subtracting twice the second equation from the third equation, resulting in

$$x_1 + x_2 + \phantom{\tfrac{1}{3}}x_3 = 6$$
$$x_2 + \tfrac{1}{3}x_3 = 3 \tag{B.4}$$
$$\tfrac{7}{3}x_3 = 7$$

We now divide the third equation by $\tfrac{7}{3}$, yielding

$$x_1 + x_2 + \phantom{\tfrac{1}{3}}x_3 = 6$$
$$x_2 + \tfrac{1}{3}x_3 = 3 \tag{B.5}$$
$$x_3 = 3$$

In a higher-order system we would repeat the procedure. That is, eliminate x_3 from all the equations except the first three. Eventually we reach the point, as in (B.5), where the last unknown is found (x_3 in this case). We then *back-substitute* the last unknown into the next-to-last equation to find another unknown. These two answers are then back-substituted into the next equation to get another unknown, and so forth.

To illustrate the back-substitution process, from the second equation of (B.5) we have

$$x_2 = 3 - \tfrac{1}{3}(x_3)$$
$$= 3 - \tfrac{1}{3}(3)$$
$$= 2 \tag{B.6}$$

The known values of x_2 and x_3 are then substituted into the first equation of (B.5), yielding

$$x_1 = 6 - x_2 - x_3$$
$$= 6 - 2 - 3$$
$$= 1 \tag{B.7}$$

In this example, this last step completes the process, since all the unknowns have been found.

The steps in the Gaussian procedure may be put in compact form by leaving out the symbols for the unknowns, the addition operation signs, and the equal signs. That is, (B.1) may be represented by

$$\begin{bmatrix} 1 & 1 & 1 & 6 \\ 2 & -1 & 1 & 3 \\ -1 & 1 & 2 & 7 \end{bmatrix} \tag{B.8}$$

Such an entity is sometimes called a *matrix*. It has rows and columns like a determinant, but it is not necessarily a square array (the example given is 3×4) and has no number associated with it. The matrix merely represents the equations. That is, its first row means one x_1 plus one x_2 plus one x_3 equals six, etc.

We may eliminate unknowns in (B.1) by manipulating rows in (B.8). Leaving the first row intact, subtracting twice the first row from the second, and adding the first row to the third, we have the new matrix

$$\begin{bmatrix} 1 & 1 & 1 & 6 \\ 0 & -3 & -1 & -9 \\ 0 & 2 & 3 & 13 \end{bmatrix} \tag{B.9}$$

This represents the system of (B.2).

Dividing the second row of (B.9) by -3 yields

$$\begin{bmatrix} 1 & 1 & 1 & 6 \\ 0 & 1 & \frac{1}{3} & 3 \\ 0 & 2 & 3 & 13 \end{bmatrix} \tag{B.10}$$

This represents (B.3).

Subtracting twice the second row of (B.10) from the third row gives

$$\begin{bmatrix} 1 & 1 & 1 & 6 \\ 0 & 1 & \frac{1}{3} & 3 \\ 0 & 0 & \frac{7}{3} & 7 \end{bmatrix} \tag{B.11}$$

which represents (B.4). Dividing its last row by $\frac{7}{3}$ gives

$$\begin{bmatrix} 1 & 1 & 1 & 6 \\ 0 & 1 & \frac{1}{3} & 3 \\ 0 & 0 & 1 & 3 \end{bmatrix} \tag{B.12}$$

The last result is equivalent to (B.5) and may be used, as in (B.6) and (B.7), to obtain the solution.

Alternatively, we may continue the elimination process as follows. In (B.12), subtract from the second row $\frac{1}{3}$ times the last row, and subtract the last row from the first row. This results in

$$\begin{bmatrix} 1 & 1 & 0 & 3 \\ 0 & 1 & 0 & 2 \\ 0 & 0 & 1 & 3 \end{bmatrix}$$

Finally, subtract the second row from the first, obtaining

$$\begin{bmatrix} 1 & 0 & 0 & 1 \\ 0 & 1 & 0 & 2 \\ 0 & 0 & 1 & 3 \end{bmatrix} \tag{B.13}$$

The last result represents $x_1 = 1$, $x_2 = 2$, and $x_3 = 3$, which is the solution.

This procedure of continuing the elimination process to the form (B.13) is known as the Gauss-Jordan method.

APPENDIX

C

COMPLEX NUMBERS

From our earliest training in arithmetic we have dealt with *real* numbers, such as 3, -5, $\frac{4}{7}$, π, etc., which may be used to measure distances in one direction or another from a fixed point. A number such as x that satisfies

$$x^2 = -4 \tag{C.1}$$

is not a real number and is customarily, and unfortunately, called an *imaginary* number. To deal with imaginary numbers, an *imaginary unit*, denoted by j, is defined by

$$j = \sqrt{-1} \tag{C.2}$$

Thus we have $j^2 = -1$, $j^3 = -j$, $j^4 = 1$, etc. (We might note that mathematicians use the symbol i for the imaginary unit, but in electrical engineering this might be confused with current.) An imaginary number is defined as the product of j with a real number, such as $x = j2$. In this case $x^2 = (j2)^2 = -4$, and thus x is the solution of (C.1).

A *complex* number is the sum of a real number and an imaginary number, such as

$$A = a + jb \tag{C.3}$$

where a and b are real. The complex number A has a *real part*, a, and an *imaginary part*, b, which are sometimes expressed as

$$a = \text{Re } A$$

$$b = \text{Im } A$$

It is important to note that both parts are real, in spite of their names.

The complex number $a + jb$ may be represented on a rectangular coordinate plane, or a *complex plane*, by interpreting it as a point (a, b). That is, the horizontal coordinate is a and the vertical coordinate is b, as shown in Fig. C.1, for the case $4 + j3$. Because of this analogy with points plotted on a rectangular coordinate system, (C.3) is sometimes called the *rectangular form* of the complex number A.

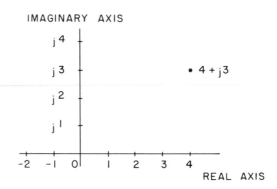

FIGURE C.1 *Graphical representation of a complex number*

The complex number $A = a + jb$ may also be uniquely located in the complex plane by specifying its distance r along a straight line from the origin and the angle θ which this line makes with the real axis, as shown in Fig. C.2. From the right triangle

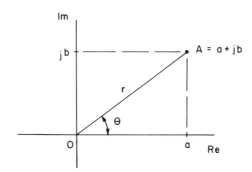

FIGURE C.2 *Two forms of a complex number*

thus formed, we see that

$$r = \sqrt{a^2 + b^2}$$
$$\theta = \tan^{-1} \frac{b}{a} \qquad\qquad \text{(C.4)}$$

and that

$$a = r \cos \theta$$
$$b = r \sin \theta \qquad\qquad \text{(C.5)}$$

We denote this representation of the complex number by

$$A = r\underline{/\theta} \tag{C.6}$$

which is called the *polar form*. The number r is called the *amplitude*, or *magnitude*, and is sometimes denoted by

$$r = |A|$$

The number θ is the *angle* or *argument* and is often denoted by

$$s = \text{ang } A = \text{arg } A$$

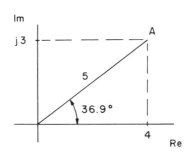

FIGURE C.3 *Two forms of the complex number A*

One may easily convert from rectangular to polar form, or vice versa, by means of (C.4) and (C.5). For example, the number A, shown in Fig. C.3, is given by

$$A = 4 + j3 = 5\underline{/36.9^\circ}$$

since by (C.4)

$$r = \sqrt{4^2 + 3^2} = 5$$
$$\theta = \tan^{-1} \tfrac{3}{4} = 36.9^\circ$$

The *conjugate* of the complex number $A = a + jb$ is defined to be

$$A^* = a - jb \tag{C.7}$$

That is, j is replaced by $-j$. Since we have

$$|A^*| = \sqrt{a^2 + (-b)^2} = \sqrt{a^2 + b^2} = |A|$$

and

$$\text{arg } A^* = \tan^{-1}\left(\frac{-b}{a}\right) = -\tan^{-1}\frac{b}{a} = -\text{arg } A$$

we may write, in polar form,

$$(r\underline{/\theta})^* = r\underline{/-\theta} \tag{C.8}$$

We may note from the definition that if A^* is the conjugate of A, then A is the conjugate of A^*. That is, $(A^*)^* = A$.

The operations addition, subtraction, multiplication, and division apply to complex numbers exactly as they do to real numbers. In the case of addition and subtraction, we may write, in general,

$$(a + jb) + (c + jd) = (a + c) + j(b + d) \qquad \text{(C.9)}$$

and

$$(a + jb) - (c + jd) = (a - c) + j(b - d) \qquad \text{(C.10)}$$

That is, to add (or subtract) two complex numbers, we simply add (or subtract) their real parts and their imaginary parts.

As an example, let $A = 3 + j4$ and $B = 4 - j1$. Then

$$A + B = (3 + 4) + j(4 - 1) = 7 + j3$$

This may also be done graphically, as shown in Fig. C.4(a), where the numbers A and B are represented as vectors from the origin. The result is equivalent to completing the parallelogram or to connecting the vectors A and B in head-to-tail manner, as shown in Fig. C.4(b), as the reader may check by comparing the numbers. For this reason, complex number addition is sometimes called vector addition.

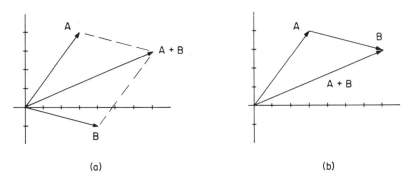

(a)   (b)

FIGURE C.4 *Two methods of graphical addition*

In the case of multiplication of numbers A and B given by

$$A = a + jb = r_1 \cos \theta_1 + jr_1 \sin \theta_1$$
$$B = C + jd = r_2 \cos \theta_2 + jr_2 \sin \theta_2 \qquad \text{(C.11)}$$

we have

$$AB = (a + jb)(c + jd) = ac + jad + jbc + j^2bd$$
$$= (ac - bd) + j(ad + bc) \qquad \text{(C.12)}$$

Alternatively we have, from (C.11),

$$AB = (r_1 \cos \theta_1 + jr_1 \sin \theta_1)(r_2 \cos \theta_2 + jr_2 \sin \theta_2)$$
$$= r_1 r_2 [(\cos \theta_1 \cos \theta_2 - \sin \theta_1 \sin \theta_2) + j(\sin \theta_1 \cos \theta_2 + \cos \theta_1 \sin \theta_2)]$$
$$= r_1 r_2 [\cos (\theta_1 + \theta_2) + j \sin (\theta_1 + \theta_2)]$$

Therefore in polar form we have

$$(r_1 \underline{/\theta_1})(r_2 \underline{/\theta_2}) = r_1 r_2 \underline{/\theta_1 + \theta_2} \qquad (C.13)$$

and hence we may multiply two numbers by multiplying their magnitudes and adding their angles.

From this result we see that

$$AA^* = (r \underline{/\theta})(r \underline{/-\theta}) = r^2 \underline{/0} = |A|^2 \underline{/0}$$

Since $|A|^2 \underline{/0}$ is the real number $|A|^2$, we have

$$|A|^2 = AA^* \qquad (C.14)$$

Division of a complex number by another, such as

$$N = \frac{A}{B} = \frac{a + jb}{c + jd}$$

results in an irrational denominator, since $j = \sqrt{-1}$. We may rationalize the denominator and display the real and imaginary parts of N by writing

$$N = \frac{AB^*}{BB^*} = \frac{a + jb}{c + jd} \cdot \frac{c - jd}{c - jd}$$

which is

$$N = \frac{(ac + bd) + j(bc - ad)}{c^2 + d^2} \qquad (C.15)$$

We may also show by the method used to obtain (C.13) that

$$\frac{r_1 \underline{/\theta_1}}{r_2 \underline{/\theta_2}} = \frac{r_1}{r_2} \underline{/\theta_1 - \theta_2} \qquad (C.16)$$

As examples, let $A = 4 + j3 = 5\underline{/36.9°}$ and $B = 5 + j12 = 13\underline{/67.4°}$. Then we have

$$AB = (5)(13)\underline{/36.9° + 67.4°} = 65\underline{/104.3°}$$

and

$$\frac{A}{B} = \tfrac{5}{13}\underline{/36.9° - 67.4°} = 0.385\underline{/-30.5°}$$

Evidently, it is easier to add and subtract complex numbers in rectangular form and to multiply and divide them in polar form.

If two complex numbers are equal, then their real parts must be equal and their imaginary parts must be equal. That is, if

$$a + jb = c + jd$$

then

$$a - c = j(d - b)$$

which requires

$$a = c, \qquad b = d$$

Otherwise we would have a real number equal to an imaginary number, which, of course, is impossible. As an example, if

$$1 + x + j(8 - 2x) = 3 + jy$$

then

$$1 + x = 3$$
$$8 - 2x = y$$

or $x = 2$, $y = 4$.

D

EULER'S FORMULA

To derive Euler's formula, an important result, let us begin with the quantity

$$g = \cos \theta + j \sin \theta \qquad (D.1)$$

where θ is real and $j = \sqrt{-1}$. Differentiating, we have

$$\frac{dg}{d\theta} = j(\cos \theta + j \sin \theta) = jg$$

as may be seen from (D.1). The variables in the last equation may be separated, yielding

$$\frac{dg}{g} = j \, d\theta$$

Integrating, we have

$$\ln g = j\theta + K \qquad (D.2)$$

where K is a constant of integration. From (D.1) we see that $g = 1$ when $\theta = 0$, which must also hold in (D.2). That is,

$$\ln 1 = 0 = 0 + K$$

or $K = 0$. Therefore we have

$$\ln g = j\theta$$

or

$$g = e^{j\theta} \qquad (D.3)$$

Comparing (D.1) and (D.3), we see that

$$e^{j\theta} = \cos\theta + j\sin\theta \tag{D.4}$$

which is known as *Euler's formula*. An alternative form found by replacing θ by $-\theta$ is

$$e^{-j\theta} = \cos(-\theta) + j\sin(-\theta)$$

which is, equivalently,

$$e^{-j\theta} = \cos\theta - j\sin\theta \tag{D.5}$$

Evidently, (D.4) and (D.5) are conjugates.

Euler's formula provides us with a means of obtaining alternative forms of $\cos\theta$ and $\sin\theta$. For example, adding (D.4) and (D.5) and dividing the result by 2, we have

$$\cos\theta = \frac{e^{j\theta} + e^{-j\theta}}{2} \tag{D.6}$$

Similarly, subtracting (D.5) from (D.4) and dividing the result by $2j$, we have

$$\sin\theta = \frac{e^{j\theta} - e^{-j\theta}}{2j} \tag{D.7}$$

We may use Euler's formula also to clarify the polar representation

$$A = r\underline{/\theta} \tag{D.8}$$

of the complex number

$$A = a + jb \tag{D.9}$$

This was considered in Appendix C, where by (C.5) we had

$$\begin{aligned} a &= r\cos\theta \\ b &= r\sin\theta \end{aligned} \tag{D.10}$$

By Euler's formula we may write

$$\begin{aligned} re^{j\theta} &= r(\cos\theta + j\sin\theta) \\ &= r\cos\theta + jr\sin\theta \end{aligned}$$

which, by (D.10), is

$$re^{j\theta} = a + jb \tag{D.11}$$

Therefore, comparing (D.11) with (D.8) and (D.9), we have

$$A = r\underline{/\theta} = re^{j\theta} \tag{D.12}$$

This result enables us to easily obtain the multiplication and division rules, given in Appendix C in (C.13) and (C.16). Clearly, if

$$A = r_1\underline{/\theta_1} = r_1 e^{j\theta_1}$$

and

$$B = r_2\underline{/\theta_2} = r_2 e^{j\theta_2}$$

then

$$AB = (r_1\underline{/\theta_1})(r_2\underline{/\theta_2})$$
$$= (r_1 e^{j\theta_1})(r_2 e^{j\theta_2})$$
$$= r_1 r_2 e^{j(\theta_1 + \theta_2)}$$
$$= r_1 r_2\underline{/\theta_1 + \theta_2}$$

Similarly, we may obtain

$$\frac{A}{B} = \frac{r_1}{r_2}\underline{/\theta_1 - \theta_2}$$

Euler's formula is illustrated graphically in Fig. D.1. A unit vector is rotating around a circle in the direction shown, with an angular velocity of ω rad/s. Therefore in t seconds it has moved through an angle ωt as shown, and thus the vector may be specified by $1\underline{/\omega t}$ or $e^{j\omega t}$. Its real part is the projection on the horizontal axis, given by cos ωt, and its imaginary part is the projection on the vertical axis, given by sin ωt. That is,

$$e^{j\omega t} = \cos \omega t + j \sin \omega t$$

which is Euler's formula. The projections trace out the cosine and sine waves, as shown, as the vector rotates.

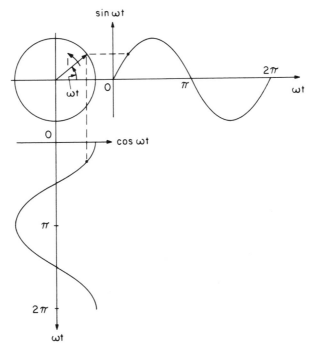

FIGURE D.1 *Graphical illustration of Euler's formula*

INDEX

INDEX